HUMAN MACHINE INTERACTION IN THE DIGITAL ERA

About the Conference

The Human Machine Interaction in the Digital Era (ICHMIDE) 2023 conference aims to address the main issues of concern in the design issues with a particular emphasis on the design and development of interfaces for autonomous robots. Its main objective is to provide an international forum for the dissemination and exchange of up-to-date scientific information on research related to integrated human/machine systems at multiple scales, and includes areas such as human/machine interaction, engineering mathematical models, assistive technologies, system modelling, design, testing and validation. The organization of ICHMS is based on the following Track types: Smart Applications for Digital Era, Computational Mathematical and Electronics, Intelligent Systems in Security and Communication Technologies, Technological Interventions using AI and Machine Learning, Applied Science, and IoT Techniques for Industries.

PROCEEDINGS OF THE INTERNATIONAL CONFERENCE ON
HUMAN MACHINE INTERACTION IN THE DIGITAL ERA (ICHMIDE 2023)

HUMAN MACHINE INTERACTION IN THE DIGITAL ERA

Towards Conversational Artificial Intelligence

Edited by

Prof. J. Dhilipan

Vice Principal(Admin) & Professor and Head
Department of Computer Science and Applications(MCA)
Faculty of Science and Humanities
SRM Institute of Science and Technology
Ramapuram, Chennai

Prof. V. Saravanan

Vice-Principal (Academics) & Professor and Head
Department of Computer Science and Applications(BSc CS)
Faculty of Science and Humanities
SRM Institute of Science and Technology
Ramapuram, Chennai

Prof. R. Agusthiyar

Professor and Head
Department of Computer Science and Applications(BCA)
Faculty of Science and Humanities
SRM Institute of Science and Technology
Ramapuram, Chennai

CRC Press
Taylor & Francis Group
Boca Raton London New York

CRC Press is an imprint of the
Taylor & Francis Group, an **informa** business

First edition published 2024
by CRC Press
4 Park Square, Milton Park, Abingdon, Oxon, OX14 4RN

and by CRC Press
2385 NW Executive Center Drive, Suite 320, Boca Raton FL 33431

British Library Cataloguing-in-Publication Data
A catalogue record for this book is available from the British Library

ISBN: 978-1-032-54998-9 (hbk)
ISBN: 978-1-003-42846-6 (ebk)

DOI: 10.1201/9781003428466

Typeset in Times LT Std
by Aditiinfosystems

Printed and bound in India

SRM Group of Educational Institutions

Dr. R. Shivakumar, M.D.,
Chairman: SRM Group of Institutions
Ramapuram & Trichy

Dr. R. Shivakumar, M.D.,

CHAIRMAN'S MESSAGE

"Education helps us get exposure to new ideas and concepts that we can use to appreciate and improve the world around us and the world within us."

I am delighted to know that the Department of Computer Science and Applications, Faculty of Science & Humanities, SRMIST Ramapuram, is organizing an International Conference on "Human Machine Interaction in the Digital Era - ICHMIDE-2023 in association with AIMST University, Malaysia & Computer Society of India, Chennai Chapter on February 24th, 2023.

Human Machine Interaction (HMI) as a field has made great strides towards understanding and improving our interaction with computer-based technologies.

Human-Computer Interaction is an important part of systems design. User satisfaction requirement is leading to better designs which incorporates intelligent, adaptive, multimodal, natural methods. This conference on a very pertinent topic is the need of the hour.

I am confident that this edited book volume would be a great input for scholars and academicians around the world to unite and exchange their ideas related to study the impact of digitization.

Best Wishes!

R. Shivakumar

CHAIRMAN

SRM Ramapuram & Trichy

Contents

Human Machine Interaction in the Digital Era – Prof. J. Dhilipan et al. (eds)
© 2024 Taylor & Francis Group, London, ISBN 978-1-032-54998-9

List of Figures

Human Machine Interaction in the Digital Era – Prof. J. Dhilipan et al. (eds)
© 2024 Taylor & Francis Group, London, ISBN 978-1-032-54998-9

List of Tables

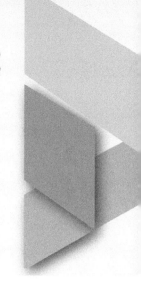

Forward

Human Machine Interaction (HMI) is the study of how people interact and communicate with mechanical systems. That is currently connected to PCs, cutting-edge systems, and devices for the Web of Things (IoT), and has long since stopped being restricted to just traditional industrial machinery. A growing number of devices are linked together and automatically fulfill tasks. Working these tools, frameworks, and gadgets should be instinctual, and clients shouldn't have any unwarranted expectations.

Customers may either provide a spoken command or directly touch the screen of their phone to operate the devices. Alternately, the frameworks naturally separate what each person needs: When a car passes over the inductive circle in the street's surface, traffic lights automatically change variety. Different technological advancements exist more to support our physical organs than to operate devices. Glasses that display augmented reality are one example of it. Advanced employees are also available, such as chatbots that respond appropriately to customer requests and are constantly learning.

This book demonstrates how an interdisciplinary team of HCI and AI experts can foster ground-breaking applications, such as medical care, smart homes, and brilliant medical care, and can also direct Human PC Connection (HCI), a multidisciplinary field that focuses on the design of PC innovation and the communication between clients and PCs in various spaces.

Key features

- Explores a broad range of case studies from across healthcare, industry, and education
- Investigates multiple strategies for designing and developing intelligent user interfaces to solve real-world problems
- Outlines research challenges and future research directions of HCI

Information assurance faces challenging and intriguing issues. To stay up with the always evolving risks, people are working on them with zeal, persistence, and commitment. They are creating new techniques for analysis and offering new solutions. It is essential to arm security practitioners, both professionals and students, with the most up-to-date knowledge on the cutting edge of information assurance in this new era of global interconnection and interdependence. An excellent first step in that approach is this book.

Maj. Dr. M. VENKATRAMANAN
DEAN-Faculty of Science and Humanities

About the Editors

Prof. J. Dhilipan

Dr. J. Dhilipan is working as a professor and Head in the Department of Computer Science and Applications (MCA), SRM Institute of Science and Technology, Ramapuram, Chennai-89. He received Ph.D degree from Manonmaniam Sundranar University, M. Phil degree from Dravidan University and M. C. A degree from Madurai Kamaraj university. He has 21 years of teaching experience in academic institutions. He is passionate on teaching, imparting, training, motivating and helping of young minds with knowledge and good discipline. He holds 47 publications in reputed journals and 60 conference publications. more than 96 papers in National and International reputed journals with 70 citations and 5 h-index. Serving an Organizing Chair, Program Chair and Program Committees of several national and International technical programs. His main areas of research include Artificial Intelligence, Machine Learning and Data Analytics.

Prof. V. Saravanan

Prof. V Saravanan has completed his PhD in Computer Science in the Department of Computer Science and Engineering, Bharathiar University during 2004. His research area includes Artificial Intelligence, Machine Learning and Software Agents. In his credit he has 90+ publications in international journals indexed with ISI, Scopus etc., He has presented many research papers in National, International conferences and Journals and also guiding many researchers leading to their PhD degree. He has totally 25 years' experience in teaching including 3 years as researcher in Bharathiar University. He has guided 18 PhD scholars in Computer Science. As the credit toward funded projects, he had completed with Majmaah University and King Abdulaziz City for Science and Technology, Kingdom of Saudi Arabia with the total project cost of USD $41500. He is a Senior Member of IEEE, life member of Computer Society of India, Indian Society for Technical Education. He worked as Professor & Head of the Department of Computer Applications in Karunya University, Coimbatore and as Director-Computer Applications of Dr. NGP Institute of Technology, Coimbatore, and Dean-Computer Studies, SNS Institutions, Coimbatore. At present he working as Vice-Principal(Academics), Faculty of Science and Humanities, SRM Institute of Science and Technology, Ramapuram, Chennai, Tamilnadu, India.

Prof. R. Agusthiyar

Prof. R. Agusthiyar is well known academician in the field of Higher Education with 21 years of teaching experience and he started his teaching profession from 2002 to till date and now he is working as a Professor and Head in Department of Computer Applications (BCA), Faculty of Science and Humanities, SRMIST, Ramapuram Campus. He has received his Post Graduate MCA degree from Madurai Kamaraj University, Tamil Nadu, India in 2001, and M. Phil degree from Periyar University, Tamil Nadu, India in 2008 respectively. He has received his Ph.D in Data mining at Anna University in the Year 2017 under the guidance of Prof. Dr. K. Narashiman Anna University. His research interests include Data mining, Artificial Intelligence and Machine Learning. He has published 14 research publications in indexed journals and 25 publications in International and National Conferences. He is guiding 8 research scholars in SRM Institute of Science and Technology. He has two patents in the field of IoT and Machine Learning.

Human Machine Interaction in the Digital Era – Prof. J. Dhilipan et al. (eds)
© 2024 Taylor & Francis Group, London, ISBN 978-1-032-54998-9

A Performance of Comparative Study on Hate Speech Detection using Deep Learning Algorithms

1

S. Sankari[1]
Research Scholar,
Bharath Institute of Higher Education and Research, Chennai

S. Silvia Priscila[2]
Associate Professor,
Bharath Institute of Higher Education and Research, Chennai

Abstract Social media has recently grown to be one of the leading entertainment areas. The increased use of social media platforms by people all over the world has given rise to several fascinating NLP challenges. One of the major problems with NLP is detecting taboos in streaming platform chats and comments. In past works, taboos were identified on different datasets using a range of deep learning algorithms and feature engineering techniques to address this growing problem. Although many research have already been conducted to detect hate material, most of these have been conducted in a single context, no study has been done to compare the comparison to machine learning or deep learning models perform the best on common public data. A publicly released dataset with three different classes will be used in this study to compare the performance of two feature engineering techniques and seven deep learning algorithms. The testing findings showed the high accuracy and less time complexity.

Keywords Hate speech detection, Deep learning, Feature selection, LSTM, Social media

1. Introduction

A social media website is an online community where people may communicate and create profiles. Over the past ten years, social media has grown tremendously in both scope and significance as a communication tool. The main focus of a social network will be user-generated content. Users mostly engage with and view content created by other users. They are urged to publish text, status updates, or images for public access. Due to the open nature of social media, anyone may post whatever they want, espousing any point of view, whether it be enlightening or repulsive or anywhere in between.

Internet users now have more freedom of speech than ever thanks to the proliferation of online social networks, the accessibility of information and communication technologies, and the widespread usage of computers and smart phones. Additionally, social media users frequently have the option to conceal their identities, which facilitates the abuse of the tools offered.

Hate speech has been more prevalent both offline and online in recent years. Social media and other internet platforms have a significant role in the creation and transmission of inflammatory content, which ultimately results towards hate crimes [1]. However, it takes a lot of time and effort to manually discover and delete hate speech-related content. Consequently, depending

[1]sankariapr17@gmail.com, [2]silviaprisila.cbsc.cs@bharathuniv.ac.in

DOI: 10.1201/9781003428466-1

on their own definitions, some people may find certain things to be hateful while others may not. According to [2], Hate speech is defined as offensive and violent material that is intended at particular minority, religious, political, sexual, or other groups.

2. Literature Review

Harish Yenala et al., (2017) did research into inappropriate text from 79,041 unique web search queries using python. When using various algorithms, they found that PKF gave 62.5% accuracy, BDT gave 79.25% accuracy, **BDT- DSSM** gave **94.75%** accuracy, SVM gave 83.22% accuracy, SVM- DSSM gave 92.41% accuracy, CNN gave 71.48% accuracy, LSTM gave 88.62% accuracy, C-Bil STM gave 92.46% accuracy, BLSTM gave 80.18% accuracy.[3]

Omar Sharif , Mohammed Moshiul Hoque et al.,(2020) did research to recognise suspicious Bengali text messages on Facebook, Twitter, YouTube using ML. Google colab platform with python is used to analyse the text messages. The algorithm LR gave 84.00% accuracy, DT gave 79.57% accuracy, RF gave 83.71% accuracy, MNB gave 83.78% accuracy, **SGD 84.57%** accuracy. [4]

Janak Sachdeva, Kushank Kumar Chaudhary et al., (2021) did research to find hate speech on Twitter using deep learning. Word2Vec and Google News Vectors is used to find the hate speeches on Twitter. They used many algorithms; LR gave **84.8%** accuracy, **RFC** gave **84.8%** accuracy, KNN gave 75.5% accuracy, L-SVC gave 83.4% accuracy, CNN gave 83.4% accuracy, LSTM gave 84.0% accuracy. [5]

Bhavesh Pariyani, Krish Shah et al., (2021) proposed a research to find hate speech on Twitter using machine learning.Count Vector organizer in python is used. They used algorithms such as SVM gave 95.25% accuracy, LR gave 96.05% accuracy, **RF** gave **96.29%** accuracy. [6]

Ilia Markov, Walter Daelemans., (2022) proposed a research to find hate speech on Facebook using a python. BERT algorithm is used and got 69% accuracy. [12]

 Jaeheon Kim et al., (2022) proposed a research to recognize and find the usage of toxic conversations in Twitch using Python. **LSTM** is used and, got **81%** accuracy. [13]

3. Data and Variables

3.1 Dataset Collection

The dataset contains hate speech tweets has been used for this study and it has been categorised as three categories. This dataset was created and labelled by Crowd Flower. This dataset contains 14502 tweets. 19% of these tweets are classified as hate speech. Furthermore, 52% of tweets are others, the remaining 29% are offensive. Figure 1.2 depicts the specifics of this distribution.

4. Methodology and Model Specifications

This study describes the proposed methodology that has applied to categorize tweets into three separate types as, "hate speech, offensive, and Others". Figure 1.1 shows the entire overview of the process. The research technique is described in this picture, includes six major main steps. They are data collecting, data pre-processing, feature engineering, data splitting, classification design, and classification evaluation. The following sections explain briefly all phases.

4.1 Text Pre-Processing

The results of categorization are improved by text preparation, according to several studies [7]. Therefore, in this dataset, we applied a variety of pre-processing techniques to get rid of tweets' messy and undesirable properties. As part of the pre-processing, we changed the tweets to lower case. Additionally, we removed all names, URL, hash tags, spaces, punctuation, special characters and stop words from the collected tweets using pattern matching algorithms. Additionally, we stemmed and tokenized tweets that had already been processed.

4.2 Feature Engineering

The categorization criteria cannot be understood by deep learning systems from textual content. To grasp classification criteria, these algorithms require quantitative features. As a result, one of the most important processes in text classification is feature

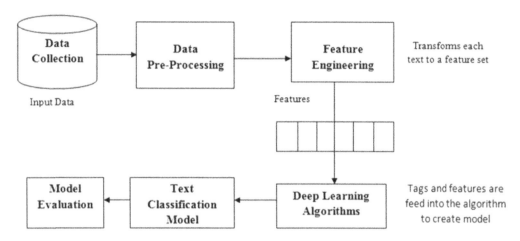

Fig. 1.1 Overall structure of the system

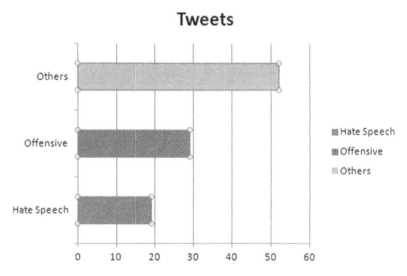

Fig. 1.2 Class-wise distribution

engineering. This phase is used to extract significant attributes from the raw text and numerically represent the data samples. We used two different feature engineering strategies in this study: Word2vec, and Doc2vec.

4.3 Data Splitting

The class-wise distributed training and test sets are displayed in Table 1.1. For training, 80% of the data is used and 20% for testing. The categorization model is evaluated using the test data. The test data is used to evaluate the categorization model and also used to evaluate the categorisation model.

Table 1.1 Dataset details

Class	Total	Training	Test
Hate Speech	2396	1917	479
Offensive	4835	3868	967
Others	7274	5819	1455
Total	14502	11,604	2901

4.4 Deep Learning Models

A master feature should be subjected to the application of several distinct classifiers to see which one yields the best results. We ultimately settled on seven classifiers: NB, SVM, DT, LSTM, CNN, RNN, and RCNN.

4.5 Evaluation of Classifier

In this phase, the created classifier uses the test set to predict the class of unlabeled text. To find Classifier Performance (TP) the following are used: True negatives (TN), true positives (TP), false positives (FP), and false negatives (FN). To evaluate the functionality of the developed classifier, various criteria are used. Some typical text categorization performance measures are addressed briefly below. More information on performance measurements can be found in [8].

Table 1.2 Performance metrics

Performance metrics	
Precision	TP(TP + FP)
Recall	TP(TP + FN)
F-measure	2 × (precision × recall)(precision + recall)
Precision	TP + FP + TN + FN (TP + TN)

5. Empirical Results

Table 1.3 Precision of results

Features	NB	SVM	DT	LSTM	CNN	RNN	RCNN
Word2Vec	0.70	0.67	0.67	0.75	0.65	0.63	0.66
Doc2Vec	0.71	0.66	0.66	0.75	0.70	0.62	0.65

Table 1.4 Recalls of results

Features	NB	SVM	DT	LSTM	CNN	RNN	RCNN
Word2Vec	0.73	0.68	0.69	0.74	0.62	0.64	0.69
Doc2Vec	0.73	0.63	0.68	0.73	0.66	0.64	0.68

Table 1.5 F-Measures of results

Features	NB	SVM	DT	LSTM	CNN	RNN	RCNN
Word2Vec	0.70	0.67	0.67	0.71	0.62	0.61	0.66
Doc2Vec	0.74	0.64	0.69	0.74	0.68	0.65	0.69

Table 1.6 Precision of results

Features	NB	SVM	DT	LSTM	CNN	RNN	RCNN
Word2vec	0.73	0.75	0.69	0.80	0.62	0.67	0.72
Doc2Vec	0.75	0.77	0.68	0.82	0.69	0.71	0.70

Source: Author's compilation

5.1 Discussions

In the study, we tested on a dataset of hate speech with three classes, seven classifiers utilising two different feature engineering techniques were used. The LSTM algorithm in combination with Doc2Vec generated the best outcomes, according to the results of the testing. In order to categorise texts, feature engineering selection is crucial. In this study, we looked at the word2vec and doc2vec feature extraction methods. The experiments' findings showed that doc2vec performed better than the other two methods.

Out Of Vocabulary cannot be handled in Twitter data using Word2vec, which results in poor performance. To understand the complex relationship between words, Word2vec requires additional training data[9]. The Taken dataset contains about 15000 tweets as displayed in Table I, because it is not enough for Word2vec to elicit the difficult word association. Recent studies have revealed that no single deep learning algorithm performs better than most on all types of data. To determine which deep learning algorithm performs the best on the provided dataset, we compared seven different deep learning models to our dataset as a result.

Test findings demonstrated that LSTM and SVM classifiers performed the best because LSTM and SVM employ thresholds to segregate data instead of determining the amount of features boundaries. These results shows SVM is not affected by feature quantity of data [10, 11]. The outcome achieved by the NB and RNN classifiers are slightly lower than the results obtained with the LSTM and SVM classifiers, but way better than the results obtained with CNN, DT, and RCNN.

When compared to the other classes, the category "Hate Speech" has the fewest training sets, as shown in Table 1.1, however this category's misclassification is primarily due to the fact that it is not a class. Many more categories besides hate speech contain word overlapping more frequently.

6. Conclusion

This study used automated text categorization approaches with two feature engineering strategies and seven deep learning algorithms to identify hate speech. The research findings demonstrated that Doc2Vec performed better than word2vec features engineering strategies. Furthermore, LSTM and SVM algorithms outperformed NB, DT, CNN, RNN, and RCNN. DT shows less performance than LSTM and SVM. In Hate Speech detection, this study results are significant since this acts as a benchmark for assessing subsequent investigations into various automatic text categorization methods. This study is also useful for a lot of automatic hate speech detection because of its experimental results from a scientific standpoint. This research contains two significant drawbacks. The First limitation is real-time prediction precision of the data, the suggested deep learning model is inefficient. Finally, it merely separates hate speech messages into three categories and is incapable of determining the severity of the dataset. Prediction of the severity of hate speech using deep learning techniques is the expected next step. The classification performance of the recommended model will also be improved using two techniques .Second, additional information examples will be gathered so that the categorization rules can be learned quickly.

References

1. Abro, S., Shaikh, S., Ali, Z., Khan, S., Mujtaba, G., & Khand, Z. H. (2020). Automatic hate speech detection using machine learning: A comparative study. *International Journal of Advanced Computer Science and Applications*, *11*(8).
2. Khan, S., Kamal, A., Fazil, M., Alshara, M. A., Sejwal, V. K., Alotaibi, R. M., Baig, A. R., & Alqahtani, S. (2022). HCovBi-Caps: Hate Speech Detection Using Convolutional and Bi-Directional Gated Recurrent Unit With Capsule Network. *IEEE Access*, *10*.
3. Yenala, H., Jhanwar, A., Chinnakotla, M. K., & Goyal, J. (2018). Deep learning for detecting inappropriate content in text. *International Journal of Data Science and Analytics*, *6*(4).
4. Sharif, O., Hoque, M. M., Kayes, A. S. M., Nowrozy, R., & Sarker, I. H. (2020). Detecting suspicious texts using machine learning techniques. *Applied Sciences (Switzerland)*, *10*(18).
5. Senarath, Y., & Purohit, H. (2020). Evaluating semantic feature representations to efficiently detect hate intent on social media. *Proceedings - 14th IEEE International Conference on Semantic Computing, ICSC 2020*.
6. Pariyani, B., Shah, K., Shah, M., Vyas, T., & Degadwala, S. (2021). Hate speech detection in twitter using natural language processing. *Proceedings of the 3rd International Conference on Intelligent Communication Technologies and Virtual Mobile Networks, ICICV 2021*.
7. Shaikh, S. and S.M. Doudpotta, Aspects Based Opinion Mining for Teacher and Course Evaluation. Sukkur IBA Journal of Computing and Mathematical Sciences, 2019. 3(1): p. 34–43
8. Seliya, N., T.M. Khoshgoftaar, and J. Van Hulse. A study on the relationships of classifier performance metrics. in 2009 21st IEEE international conference on tools with artificial intelligence. 2009. IEEE.
9. Li, Y. and T. Yang, Word embedding for understanding natural language: a survey, in Guide to Big Data Applications. 2018,Springer.p.83-104.
10. Cavnar, W. B. and J. M. Trenkle. N-gram-based text categorization. in Proceedings of SDAIR-94, 3rd annual symposium on document analysis and information retrieval. 1994. Citeseer.
11. Zhang, M.-L. and Z.-H. Zhou, A k-nearest neighbour based algorithm for multi-label classification. GrC, 2005. 5: p. 718–721.
12. Wang, S., Mazumder, S., Liu, B., Zhou, M., & Chang, Y. (2018). Target-sensitive memory networks for aspect sentiment classification. *ACL 2018 - 56th Annual Meeting of the Association for Computational Linguistics, Proceedings of the Conference (Long Papers)*, *1*.
13. Kim, J., Wohn, D. Y., & Cha, M. (2022). Understanding and identifying the use of emotes in toxic chat on Twitch. *Online Social Networks and Media*, *27*.

Note: All the figures and tables in this chapter were made/compiled by the Author

Human Machine Interaction in the Digital Era – Prof. J. Dhilipan et al. (eds)
© 2024 Taylor & Francis Group, London, ISBN 978-1-032-54998-9

A Study on Technology-Related Distraction in College Student's Academics Using Data Mining Techniques

2

S. Sankari[1]
Research Scholar, Bharath Institute of Higher Education and Research, Chennai

S. Silvia Priscila[2]
Associate Professor, Bharath Institute of Higher Education and Research, Chennai

Abstract Technology has advanced significantly in today's modern era. It is open to everyone, including students. Students now have easier access to the technological world through the internet. Students are becoming more interested in social media. Long-term use of social media can lead to apathy toward studies. When students are distracted by social media and other factors that they often overlook, it can have a negative impact on their academic performance. The aim of this study is to identify the major factors that contribute to distractions to students' studies and also identify the social media usage among students for lowering their grades. The responses of the 465 students who participated from various colleges reveal that social networking sites are very popular, and the majority of the responses prove that they are distracting from studies due to excessive internet usage.

Keywords Social Media, Educational Data Mining, Internet, Student Performance, Mobile Phone Activity, Students Distraction

1. Introduction

Students are more enticed than ever before to engage in media multitasking with the emergence of personal digital gadgets. However, it is unclear whether all digital devices interfere with student learning. [1] With the advancement of the Internet and digital technologies, society is becoming increasingly connected.

Personal page of the user can be set to semi-public or protected, depending on the user's preferences. Instagram, Facebook, YouTube, and TikTok are some of the most popular sites.

Teachers at most colleges are constantly competing for attention in the classroom with cell phones and other technologically - advanced gadgets. While the teacher is instructing, students text, surf the web, and post on social media. As a result, they are only half-present in the classroom for the majority of the time, which automatically leads to academic distraction.

2. Literature Review

2.1 Students Distraction During Lectures

Abraham E., Scott (2020) conducted a survey with 100 participants, to find whether the students get distracted from taking lectures while using laptops. The participants were made to watch a 15 minute seminar and, in between the time, they conducted

[1]sankariapr17@gmail.com, [2]silviaprisila.cbsc.cs@bharathuniv.ac.in

DOI: 10.1201/9781003428466-2

various tests to analyse the distraction using text messages sent during lectures. They found that making the students finish the notes is a better way to reduce distraction [2].

2.2 College Students Distraction of Two Universities

Leida, Ravi, Zhenya (2019) conducted a survey with the students of two universities. They created a scale by which they analysed the distraction in college students and the intensity of it using ratings. There were 425 students who participated in this study. They found the students had impetuous personalities due to mobile usage [3].

2.3 Student Distraction

McKinley Green (2019) conducted a study with the responses of 143 students and 10 students were interviewed. In his research, he used nine activities to find distractions for students by finding which activities are done by many students. The study reveals many distracting activities and the students' attitude trends due to technology usage [4].

Fayyaz Ahmad Faize, Muhammad Nawaz, 2020 The author used 196 students' data to find their attitude toward the online classes during the COVID-19 pandemic. The students' satisfaction levels, challenges in learning, suggestions, modifications for improvement were found during this study. It was found that there was 12.7% distraction during online classes. The students identified some challenges faced during the online classes[5].

Suad A. A. Al-Furaih, Hamed M. Al-Awidi, (2020) surveyed the data of 2084 undergraduate students for analysis. The data was analysed using SPSS to find the t-Test and correlation between the variables. The results found by the analysis show that students are distracted and they are not attentive , the attitude of students changes due to distraction caused by smart phones. The fear of missing out was found affecting the learning ability of students[6].

2.4 Attitude Towards Active Learning

Madison E. Andrews , Matthew Graham, et al, (2019) conducted a inspection on the data collected from 27 instructors and 758 students. The data was collected from STEM instructors and students to find their attitude toward active learning. The data suggests that students show a more open mindset towards active learning, and the instructors were a bit hesitant to adopt active learning [7].

Data and Variables

The students data is gathered from a variety of undergraduate students. The total number of records is 465 contains 251 males and 214 females.

3.1 Data Preparation

The survey research study method was used to collect data, and a specific questionnaire was created and distributed to 465 undergraduate students. The questionnaire was divided into four sections: (1) Device Usage (2) Purpose for using SNS (3) Digital Distraction (4) Reason for using social media.

3.2 Pre-Processing

The students' data was collected from the pre-processing step. Pre-processing has the ability to increase the accuracy of categorisation. The proposed pre-processing system is divided into two stages.

1. Remove any values that are missing.
2. Categorization

4. Methodology

4.1 Feature Selection

Data mining provides a number of feature selection strategies. The fields of pattern recognition, statistics, and data mining have all actively studied feature selection. Attribute selection is another name for feature selection [8]. Several feature selection strategies were applied to the pre-processed data set of 465 samples in order to select the important qualities and eliminate the irrelevant features, with a focus on picking the top attributes.

4.2 Classification

Classification refers to the method of categorising data objects using predefined class labels. One of the ML supervised methods is classification [9].

K-means

K-means refers as simplistic recursive algorithm used for clustering. It calculate the distance mean, produces initial centroid, using distance as the metric and the K classes in the data. With respect to a given data set X with n multidimensional data points and a category K to be classified, the Euclidean distance is selected as the similarity index, and the clustering objectives minimize squares sum of the different kinds.[10]

Decision Tree

A supervised classification method called a decision tree forecasts both the classifier and the regression model. The basic purpose of classification trees is to allocate an object to a specified class based on its attributes. The root of the tree is the absence of any incoming edges, the internal node is the only node with an outgoing edge, and all other nodes are known as leaf nodes because they lack any outgoing edges [11].

4.3 Data Analysis

This survey uses Knime analytic platform to analyze the data collected from 465 college students. Over 3000 organizations utilize Knime, an open source graphical workbench for data mining, visualization, and reporting. The basic Knim open source version is called Knime desktop. Its foundation is the renowned and well-liked Eclipse IDE platform, giving it both a development environment and a platform for data mining.[12]

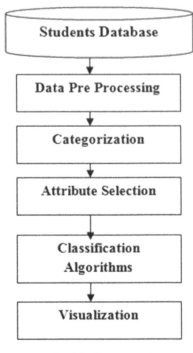

Fig. 2.1 Framework

5. Results

For the question of whether students have social media accounts, the responses were 89.2% (415) of students had social media accounts. This indicates that most of the students have social media accounts. (Fig. 2.2).

The most used devices used by college students. The result was found that most students use mobile phones, the percent of mobile usage is 98% (455). Laptops are used rarely by college students. The data indicates that only 2% (10) use laptops. (Fig. 2.3).

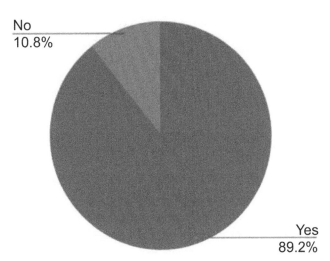

Fig. 2.2 Indicates whether the students have social media accounts or not

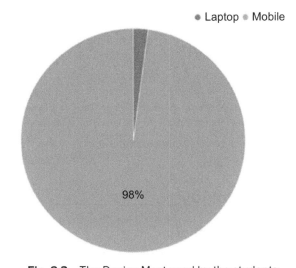

Fig. 2.3 The Device Most used by the students

The social media platforms on which students have an account are found. Instagram is mostly used by students with 54% (163),followed by Whatsapp with 21% (69), followed by Snapchat with 6% (17), then Facebook and other applications . It is found that most of them have an account on Instagram.

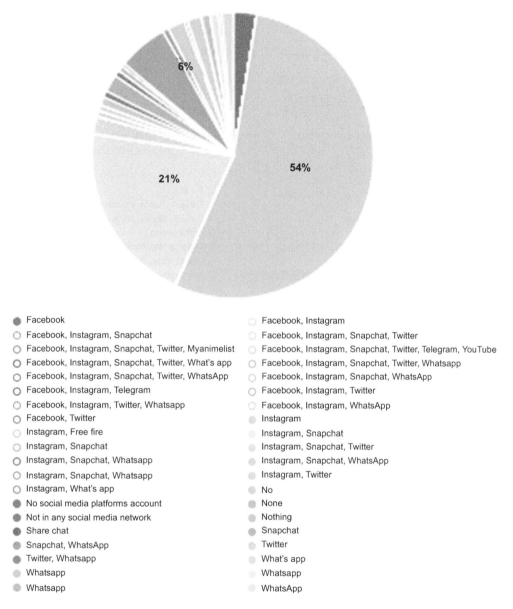

● Facebook	○ Facebook, Instagram
○ Facebook, Instagram, Snapchat	○ Facebook, Instagram, Snapchat, Twitter
○ Facebook, Instagram, Snapchat, Twitter, Myanimelist	○ Facebook, Instagram, Snapchat, Twitter, Telegram, YouTube
○ Facebook, Instagram, Snapchat, Twitter, What's app	○ Facebook, Instagram, Snapchat, Twitter, Whatsapp
○ Facebook, Instagram, Snapchat, Twitter, WhatsApp	○ Facebook, Instagram, Snapchat, WhatsApp
○ Facebook, Instagram, Telegram	○ Facebook, Instagram, Twitter
○ Facebook, Instagram, Twitter, Whatsapp	○ Facebook, Instagram, WhatsApp
○ Facebook, Twitter	● Instagram
○ Instagram, Free fire	Instagram, Snapchat
○ Instagram, Snapchat	Instagram, Snapchat, Twitter
○ Instagram, Snapchat, Whatsapp	Instagram, Snapchat, WhatsApp
○ Instagram, Snapchat, Whatsapp	Instagram, Twitter
○ Instagram, What's app	● No
● No social media platforms account	● None
● Not in any social media network	Nothing
● Share chat	Snapchat
Snapchat, WhatsApp	Twitter
● Twitter, Whatsapp	What's app
Whatsapp	Whatsapp
Whatsapp	WhatsApp

Fig. 2.4 Social media accounts that students have

It was found that 33.3% (155) use gadgets for more than 3 hours a day, followed by 25.2% (117) who use about 1 hour to 2 hours each day, followed by 24.5% (114) who use for 2 hours to 3 hours each day and finally only about 17% (79) use their gadgets for less than a hour. From this we can find that most students are immersed in social media for a prolonged time.

It was found that 75% (349) carry their smart gadgets with them every day. Followed by 25% (116) did not carry their gadgets with them on a daily basis. (Fig. 2.6).

To find the reason for the usage of mobile phones and other gadgets, we find that 31% (62) use it only for entertainment, followed by 23% (46) use it only for chatting, 17% (34) use it for studies and work, then online gaming 10% (19). In this representation, the combinations were removed to find usage for individual reasons. From this data we can see that most usage of social media is entertainment.

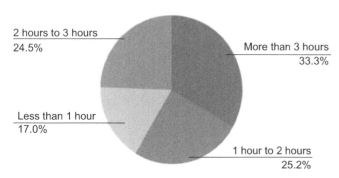

Fig. 2.5 Duration spent by students on devices

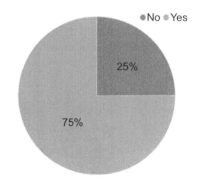

Fig. 2.6 Indicates whether the students carry their smart gadgets with them on a daily basis or not

- Entertainment
- For work, Studies
- For work, Studies, Social media and messengers for chatting
- Multi purpose
- Online gaming, Entertainment
- Online gaming, For work, Studies, Entertainment
- Online gaming, For work, Studies, Social media and messengers for chatting, Entertainment
- Online gaming, Social media and messengers for chatting, Entertainment
- Social media and messengers for chatting, Entertainment

- For all
- For work, Studies, Entertainment
- For work, Studies, Social media and messengers for chatting, Entertainment
- Online gaming
- Online gaming, For work, Studies
- Online gaming, For work, Studies, Social media and messengers for chatting
- Online gaming, Social media and messengers for chatting
- Social media and messengers for chatting
- Studies and relaxing

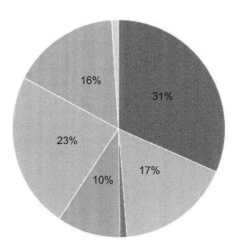

Fig. 2.7 Reason for the usage of Technology

The Overall representation of collected data has been represented using sunburst chart

6. Conclusion

This study has explored why college students are distracted from academics. Even though most students did not find social media to be distracting them from studies, when they found the usage of social media applications, it was evident that most students use social media when they get bored. Then the most used social media platform is Instagram. From this study we found that more students use gadgets for a prolonged time and it could cause damage to their learning ability. This research finds the various social media applications used by students, devices mostly used, the usage of smart gadgets. This research could be used to find and reduce academic distraction for students.

Sunburst Chart

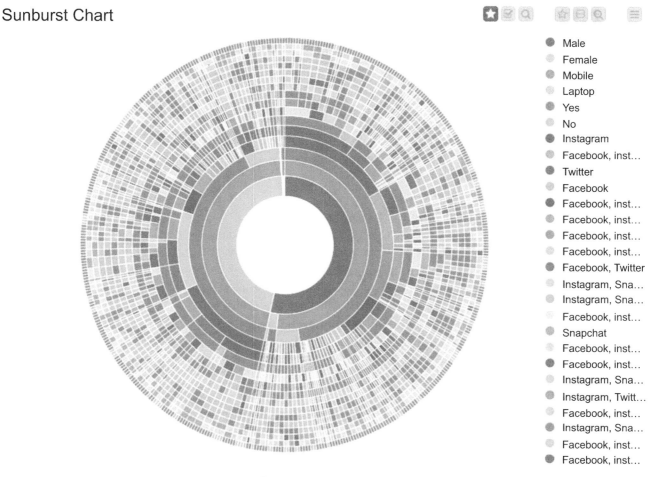

- Male
- Female
- Mobile
- Laptop
- Yes
- No
- Instagram
- Facebook, inst...
- Twitter
- Facebook
- Facebook, inst...
- Facebook, inst...
- Facebook, inst...
- Facebook, inst...
- Facebook, Twitter
- Instagram, Sna...
- Instagram, Sna...
- Facebook, inst...
- Snapchat
- Facebook, inst...
- Facebook, inst...
- Instagram, Sna...
- Instagram, Twitt...
- Facebook, inst...
- Instagram, Sna...
- Facebook, inst...
- Facebook, inst...

Fig. 2.8 Sunburst chart

References

1. Dontre, A. J. (2021). "The influence of technology on academic distraction: A review". In Human Behavior and Emerging Technologies (Vol. 3, Issue 3).
2. Abraham E. Flanigan , Scott Titsworth, "The impact of digital distraction on lecture note taking and student learning", Instructional Science 2020.
3. Leida Chen, Ravi Nath, Zhenya Tang, "Understanding the Determinants of Digital Distraction: An Automatic Thinking Behavior Perspective", Computers in Human Behavior 2019.
4. McKinley Green, "Smartphones, Distraction Narratives, and Flexible Pedagogies: Students' Mobile Technology Practices in Networked Writing Classrooms", 2019.
5. Fayyaz Ahmad Faize, Muhammad Nawaz, "Evaluation and Improvement of students' satisfaction in Online learning during COVID-19", Open Praxis, vol. 12 issue 4, 2020.
6. Shihui Feng, et al, "The Internet and Facebook Usage on Academic Distraction of College Students", 2019.
7. Zahra Vahedi, Lesley Zannella, Stephen C Want, "Students' use of information and communication technologies in the classroom: Uses, restriction, and integration", 2019.
8. http://en.wikipedia.org/wiki/Feature_selection
9. http://www.saedsayad.com/classification.htm
10. Yuan, Chunhui, and Haitao Yang. 2019. "Research on K-Value Selection Method of K-Means Clustering Algorithm" *J* 2, no. 2: 226-235.
11. Quinlan, J. R. "Induction of decision trees." Machine learning 1.1 1986, pp. 81-106
12. Naik, Amrita; Samant, Lilavati (2016). *Correlation Review of Classification Algorithm Using Data Mining Tool: WEKA, Rapidminer, Tanagra, Orange and Knime. Procedia Computer Science, 85(), 662–668.*

Note: All the figures in this chapter were made/compiled by the Author

Human Machine Interaction in the Digital Era – Prof. J. Dhilipan et al. (eds)
© 2024 Taylor & Francis Group, London, ISBN 978-1-032-54998-9

A Study of Various Deep Learning Methods Used for Image Splicing Detection

3

Lisha Varghese[1]

P.hD Research Scholar, Department of Computer Applications,
SRM IST, Ramapuram Campus, Chennai

J. Jebamalar Tamilselvi[2]

Associate Professor, Department of Computer Science,
SRM IST, Ramapuram Campus, Chennai

Rubin Thottupurathu Jose[3]

Associate Professor & Head, Department of Computer Applications,
Amal Jyothi College of Engineering, Kanjirappally

Abstract A lot of fraudulent and manipulated photographs have been created in recent years and disseminated through the media and the Internet due to the accessibility and simplicity of image editing softwares. To judge the veracity of an image and, in some situations, to identify the parts that have been altered (forged), a variety of methods have been put forth. The most prevalent copy-move and splicing attacks are the focus of this paper's examination of some of the most recent methods for detecting image forgeries that specifically rely on DL techniques. Deep learning-powered techniques seem to be the most relevant at the moment, as evidenced by the best overall performances they exhibit on the benchmark datasets that are now available, so this survey is especially pertinent at this time. These approaches' key characteristics were discussed in addition to the datasets that were used to train and test them.

Keywords: Deep learning, Classification algorithm, Image forgery, Principal component analysis, Copy-move and splicing attacks

1. Introduction

Images and videos are consequently frequently shared and used as information sources in a variety of scenarios. In fact, a lot of commonplace information is recorded using smartphones, even by experts. A wide range of digital technologies, including efficient compression techniques, quick networks, and specifically developed user apps, enable widespread sharing of visual content. Web platforms, such as social networks like Instagram and forums like Reddit, which enable the almost instantaneous dissemination of user-generated photographs and video, are among the latter group.

This paper's aim is to provide an overview of a few forgery detection techniques, devoting strong emphasis to deep learning (DL) algorithms that have recently gained prominence.

[1]lv5742@srmist.edu.in, lishavarghese@amaljyothi.ac.in; [2]jebamalj@srmist.edu.in, [3]rubinthottupuram@amaljyothi.ac.in

DOI: 10.1201/9781003428466-3

2. Literature Review

2.1 General Image Forgery Detection System

In order to make a few ideas clearer, we first provide a broad overview of the application in question. The types of forgeries that are most frequently found are briefly described in the section below that.

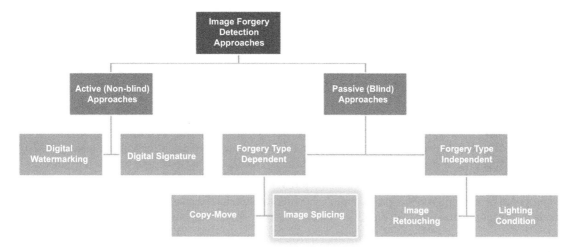

Fig. 3.1 Types of forgeries [1], [2], [3], [4], [6], [7], [8], [9]

To modify digital photos, three methods are frequently utilised.

1. Tampering
2. Splicing
3. Cloning or Copy-Move.

Tampering

Image Tampering is altering an image to produce a new outcome. It is a form of digital art that requires a strong sense of visual inventiveness and an understanding of picture attributes. One tampers images for a variety of reasons, such as for entertainment purposes when using digital tools to create amazing photos or to create fake proof. Passive image tampering detection validates the legitimacy of digital photos without having any prior information of the source images.

Image splicing

It is a photo editing technique that involves digitally fusing multiple independent images into a composite. This is a key stage in the popular digital photomontage technique used for modifying pictures in digital form, that is also known as a paste-up because it was created by joining the photos using tools like Photoshop.

(a) Original image 1   (b) Original image 2   (c) Duplicated image (Forged)

Fig. 3.2 Original image and forged image

Source: Original images taken on Oneplus Mobile Phone

The most popular way to modify a digital image is through image forging. Copy-move techniques basically include inserting picture blocks into an existing image while obscuring crucial details or objects. As a result, the image's validity and uniqueness are both compromised. Forgery is exceedingly challenging to spot because the cloned blocks have the same attributes as the other image blocks because they are from the same image.

Detecting image forgery with splicing

The spliced image is used in various contexts, including news coverage, photography competitions, essential evidence in academic papers, and more, which could have unfavourable effects. It follows that determining the image authenticity and spotting altered portions of an image become essential and difficult research topics as digital images become more susceptible to malevolent alteration than their non-digital equivalents.

Some techniques that have been mentioned in the literature include the following; categorized as feature-based methods, estimation of illumination colour, inconsistent noise levels in the source image, the source image's statistical characteristics, as well as other techniques.

Illumination colour estimation

Among other red flags, lighting discrepancies may be useful for splicing detection in determining the validity of a digital image. This is because it's challenging to create a forged image with the right adjustment of the lighting circumstances.

For the purpose of identifying differences between local picture regions, a physics-based illuminant colour model was devised in the method proposed by C. Riess And E. Angelopoulou. In order to estimate the results and use them in forensic investigation, the authors constructed an illumination map based on distance measurement. Users must get involved with this strategy.

Used fluctuations in the colour of the illuminant in the area surrounding the object to identify region splicing forgeries using local illumination estimate was proposed by Yu Fan, Philippe Carré and Christine FernandezMaloigne. To estimate the illuminant of each horizontal and vertical band, they suggested combining five simple statistics-based techniques.

C. Riess And E. Angelopoulou provided a novel method for identifying faked photos of individuals using the illuminant colour for further research. Using statistical grey edge estimation and a physics-based technique that takes use of the inverse intensity-chromaticity colour space, they calculated the illuminant colour.

2.2 Forgeries Based on Inconsistency in Image Noise Levels

Image noise can be used to improve accuracy in spliced image region detection by forging documents with inconsistent noise levels. It is clear that every digital camera image is susceptible to some form of noise, which may occur during the photon entry process and up until the camera outputs the image.

In their study, Xunyu Pan, Xing Zhang and Siwei Lyu, a blind forgery detection method based on local noise inconsistencies is suggested to find small regions that have been tainted by local noise. In order to achieve the greatest resolution with no overlapping blocks, the technique employs high pass diagonal wavelet coefficients. To identify spliced forgeries, the image was segmented using a straightforward region merging method based on the homogeneity requirement into a number of homogenous subregions. When the authenticated image contains the same amount of noise as the image with the uniform noise level, the approaches do not operate as well.

The writers of this research suggested a practical way as an improvement. The image is first separated into non-overlapping blocks, and clustering is used to make the blocks neat and altered.

An automatic method for spliced forgery detection in raw photos is suggested by Thibaut Julliand, Vincent Nozick And Hugues Talbot in their paper. By examining image discrepancies from quad-tree decomposition, they were able to identify potential sliced images by using the relative consistency of the noise parameters.

A method for detecting region splicing based on the phenomena of projection kurtosis concentration was proposed by Siwei Lyu ,Xunyu Pan And Xing Zhang.

A closed-form solution exists for the optimization problem of noise statistics estimation. These methods are all based on differences in noise on a single scale.

A technique where the image is split into super pixels of many scales and then noise level function is applied to each individual scale was proposed by Chi-Man Pun, Bo Liu and Xiao-Chen Yuan, taking advantage of multi scales as an indicator for identifying spliced image counterfeiting. The Optimal Parameter Combination Searching algorithm is then used to further process the segments that are not restricted by the noise level function in order to identify the spliced sections.

Yu-Feng Hsu And Shih-Fu Chang developed a technique based on determining the camera features that are consistent across different portions of an image. A camera response function (CRF) is computed from each segment of a segmented image using geometric invariants from LPIPs. SVM-based classifier is provided with computed CRF scores and area intensity characteristics.

Zhenhua Qu1, Guoping Qiu2, And Jiwu Huang suggested a machine learning method that uses the human visual system model. There is a strong link between the first few fixation sites determined by edge sharpness and the spliced borders.

The approach proposed by H. R. Chennamma, Lalitha Rangarajan uses line-based calibration to measure the image's lens radial distortion. When there are straight edges in the image, the approach recognises them successfully, but a poor image will result disturbances along with the straight lines, leading to an incorrect assessment of radial distortions.

Kunj Bihari and Vipin Tyagi in their study of a deep learning-based technique to identify picture splicing in the photographs used a method called "Noiseprint". The average classification accuracy for the suggested approach is 97.24%.

This study by Muhammad Hameed Siddiqi,Khurshed Asghar, Umar Draz,4Amjad Ali, Madallah Alruwaili, Yousef Alhwaiti, Saad Alanazi, M. M. Kamruzzaman presented a novel approach for the detection of fraudulent images using the Discrete Wavelet Transform, image splicing, and histograms of robust discriminative local binary patterns. In the end, a support vector machine is used to build an image forgery detection model.

Ali Retha Hasoon Khayeat, Ahmed Abdulhadi Al-Moadhen and Mustafa Ridha Hassoon Khayyat used the semantic segmentation of the decomposed image.

Yuan Rao; Jiangqun Ni proposed a method that extracts dense features from the test pictures using the pre-trained CNN as a patch descriptor. The final discriminative features for SVM classification are obtained by investigating feature fusion techniques.

Souradip Nath & Ruchira Naskar proposed a blind image splicing detection method that uses a fully connected classifier network to distinguish between authentic and spliced images as its backbone and a deep convolutional residual network architecture as its front end.

3. Conclusion

In this review various Image Splicing forgery techniques were studied and summarised. The differences in the image properties or camera characteristics are the primary source in identifying the forged sections of the photographs because the spliced images are created from distinct images. We further categorise the pixel-based and statistically based methods into noise inconsistencies, illumination colour estimation, statistical aspects of the image, and other feature-based approaches. The literature may contain a variety of strategies, but each one has drawbacks. Notwithstanding its shortcomings, the research topic of image forensics holds great promise for enhancing forgery detection thanks to competition between forgery producers and detectors.

References

1. Vidhi P. Raval. (2013). Analysis and Detection of Image Forgery Methodologies. International Journal for Scientific Research & Development| Vol. 1, Issue 9.
2. C. Riess And E. Angelopoulou. (2010). Scene Illumination As An Indicator Of Image Manipulation. Inf. Hiding, Vol. 6387, Pp. 66–80.
3. Yu Fan, Philippe Carré, Christine FernandezMaloigne. (2015). Image Splicing Detection With Local Illumination Estimation. ICIP.
4. C. Riess And E. Angelopoulou. (2013) Exposing Digital Image Forgeries By Illumination Color Classification. IEEE Transactions On Information Forensics And Security, Vol. 8, No. 7.
5. Xunyu Pan, Xing Zhang, Siwei Lyu. (2012). Exposing Image Forgery With Blind Noise Estimation. Thirteenth ACM Multimedia Workshop On Multimedia And Security.
6. Xunyu Pan, Xing Zhang, Siwei Lyu. (2012). Exposing Image Forgery With Blind Noise Estimation. IEEE International Conference On Computational Photography (ICCP).
7. Thibaut Julliand, Vincent Nozick And Hugues Talbot. (2015). Automated Image Splicing Detection From Noise Estimation In Raw Images", 6th International Conference On Imaging For Crime Prevention And Detection (ICDP15).
8. Siwei Lyu, Xunyu Pan And Xing Zhang. (2014). Exposing Region Splicing Forgeries With Blind Local Noise Estimation. Springer, Int J Comput Vis.
9. Chi-Man Pun, Bo Liu, Xiao-Chen Yuan. (2016). Multi-scale noise estimation for image splicing forgery detection. Journal Of Visual Communication Image Representation, Elsevier.
10. Yu-Feng Hsu And Shih-Fu Chang. (2007). Image Splicing Detection Using Camera Response Function Consistency And Automatic Egmentation. IEEE International Conference On Multimedia And Expo.

11. Zhenhua Qu1, Guoping Qiu2, And Jiwu Huang1. (2009). Detect Digital Image Splicing With Visual Cues", LNCS, Springer.
12. H. R. Chennamma, Lalitha Rangarajan. (2010). Image Splicing Detection Using Inherent Lens Radial Distortion. International Journal Of Computer Science.
13. Kunj Bihari,Vipin Tyagi. (2021). A Deep Learning based Method for Image Splicing Detection. Journal of Physics Conference Series.
14. Muhammad Hameed Siddiqi, Khurshed Asghar, Umar Draz,4Amjad Ali,Madallah Alruwaili, Yousef Alhwaiti, Saad Alanazi, M. M. Kamruzzaman. (2021). Image Splicing-Based Forgery Detection Using Discrete Wavelet Transform and Edge Weighted Local Binary Patterns. Security and Communication Networks.
15. Akram Hatem Saber, Mohd Ayyub Khan, Basim Galeb Mejbel. (2020). A Survey on Image Forgery Detection Using Different Forensic Approaches, Volume 5, 3: 361–370. Ali Retha Hasoon Khayeat, Ahmed Abdulhadi Al-Moadhen and Mustafa Ridha Hassoon Khayyat. (2020). Splicing detection in color image based on deep learning of wavelet decomposed image. AIP Conference Proceedings 2290.
16. Siddhi Gaur, Shamik Tiwari. (2017). Image Splicing Forgery Detection. ICRIET Proceedings.
17. Souradip Nath & Ruchira Naskar. (2021). Automated image splicing detection using deep CNN-learned features and ANN-based classifier.
18. Guangjie Liu, Yuewei Dai. (2014). Detecting Image Splicing Using Merged Features in Chroma Space

Human Machine Interaction in the Digital Era – Prof. J. Dhilipan et al. (eds)
© 2024 Taylor & Francis Group, London, ISBN 978-1-032-54998-9

ASO-GRU-NN: Atom Search Optimized Gated Recurrent Units based Neural Network for the Severity Classification of COVID-19 Using Different Feature Extraction Methods

4

J. Kalaivani[1]

Research Scholar, Tamil Nadu, Vels Institute of Science,
Technology and Advanced Studies (VISTAS)

A. S. Arunachalam[2]

Associate Professor, Tamil Nadu, Vels Institute of Science,
Technology and Advanced Studies (VISTAS)

Abstract The recent pandemic COVID-19 had invaded all the countries without any exceptions. The main reason of the virus to become pandemic is its intensity is infection contagiousness. Hence, rapid diagnosis is to be carried out to prevent the infection from spreading. As a rapid diagnosis framework, this research proposes an Artificial Intelligence (AI) based framework for identifying the presence and the severity of COVID-19 with the help of chest CT slices of the subject. This paper proposes a new AI severity classifier called Atom Search Optimized Gated Recurrent Units based Neural Network (ASO-GRU-NN), which is a GRU model with recurrent weights optimized by ASO algorithm. This model is trained for severity classification using different feature extraction techniques on the infection segmented images. The feature extraction phase aims to estimate the statistical features, gray-level co-occurrence matrix (GLCM) features, and region features of the identified infected areas in the lungs. The feature vectors are given as one feature group and as combinations of three to assess the performance of the ASO-GRU-NN model for the identification and the severity classification of COVID-19. The performance of the classification model trained with different features are analyzed and compared with existing models with the help of various performance measures such as accuracy, error rate, sensitivity, specificity, and execution time.

Keywords COVID-19, Feature extraction, Gated recurrent unit, Gray-level co-occurrence matrix, Severity classification

1. Introduction

The advent of COVID-19 has brought a huge health catastrophe to the world in 2019, resulting in a pandemic. The rapid diagnosis of COVID-19 is essential to stay precautious and to stop the spread of infection. Medical images like MRI and CT-Scans are used as sources of information for the diagnosis. With help of automated diagnostic models, it is easier to overcome the difficulties encountered by non-expert radiologists in the accurate diagnosis process [2]. There are several kinds of AI classification models which were reported in the research works dealing with real datasets of COVID-19 comprising of different targets and case studies [5]. Even though, AI methods can be found beneficial in the process of diagnosis and detection of the presence of COVID-19, it is challenging to develop an appropriate AI model which can yield precise results [7 and 8].

The proposed research work focuses on evaluating the different feature extraction methods, and their performance in the prediction and classification of COVID-19 severity. The effect of statistical features, textural features, region-based features,

[1]kalaivanij03@gmail.com, [2]arunachalam1976@gmail.com

DOI: 10.1201/9781003428466-4

and their combinations are evaluated in this study. Furthermore, this work proposes a novel classifier model that includes the concept of Computational Intelligence (CI). The main paradigms of CI are the metaheuristic algorithms, which are fundamentally used for solving the problems of optimization [9]. In this regard, latest metaheuristic approach called Atom Search Optimization (ASO) [3] is employed in the proposed model to optimize the weights of Gated Recurrent Unit based Neural Network (GRU-NN). The proposed model of Atom Search Optimized GRU-NN (ASO-GRU-NN) helps in enhancing the performance of COVID-19 and its severity classification with respect to accuracy, specificity, sensitivity, and error rate.

The major contributions of the proposed research work are listed below.

- A novel CI based AI model of classifier called ASO-GRU-NN is developed for the prediction of COVID-19 and classification of the severity of COVID-19.
- The study evaluates the impact of statistical features, GLCM features, and infected region-based shape features on COVID-19 detection and classification.
- The combined effects of the three features are evaluated for detection and classification of the severity of COVID-19. The performance of the proposed method is validated by comparing it with the existing state-of-the-art models of classifiers.

2. Literature Review

Al-Areqi, F., & Konyar, M. Z. (2022) [1] extracted texture features using gray-level co-occurrence matrix (GLCM), shape features, and first order statistical features using the CT slices. The feature matrix for training the model is developed using one group of feature or a combination of more number of groups. The performance of the feature extraction techniques was evaluated and compared with different ML classifiers such as SVM, Random Forest, k-nearest neighbor and XGBoost.

Dey et al. [6] considered 400 CT images (200 non-COVID-19 and 200 COVID-19 subjects). The authors recommended a system which segments the lesions of COVID-19 and then extract features of those segmented regions. Four classifiers were employed for classification, namely, RF, KNN, SVM and DT. A highest accuracy of 87% was achieved for KNN model.

Qiblawey, Y., et al. (2021) [4] proposed a cascaded model to segment the lungs, detect, localize, and quantify COVID-19 lesions from CT slices. An extensive set of experiments were accomplished with the help of Encoder–Decoder CNNs (ED-CNNs), UNet, and Feature Pyramid Network (FPN), with diverse structures of backbone (encoder) employing the variants of ResNet and DenseNet.

3. Proposed Framework of Severity Classification of COVID-19

The proposed framework of detection of COVID-19 and its severity level classification is given in Fig. 4.1. The input lung CT image is pre-processed for noise removal using Contrast Limited Adaptive Histogram Equalization based non-local means Filter (CLAHEN Filter). The pre-processed image is subjected to segmentation of lungs using Logarithmic Non-Maxima Suppression (LNMS) method. The segmented lungs are passed through the trained Inf-Seg-Net model for determining the infections present in the lungs. The segmented infections are superimposed over the original input CT image to get the infection extracted image. Three types of features are estimated from the segmented output image. The estimated features are given separately and in combination for training the proposed ASO-GRU-NN classification model to detect and classify the severity of COVID-19.

Feature extraction plays a crucial roles which is performed by eliminating the irrelevant and redundant information from the data. The extracted features are used in training the classifier models to classify the incoming data. The performance of a classifier depends directly on the choice of feature estimation techniques employed on the data. Hence, three major kinds of features such as statistical features, shape features, and texture features using gray-level co-occurrence matrix (GLCM) are extracted from the infection segmented output to train the classification model.

3.1 Statistical Features

The statistical features extracted in this study are mean, median, standard deviation, and variance. Mean can be referred as the average of all the pixels in the image estimated using the following equation (1).

$$\text{Mean} = \frac{\sum_{i,j=1}^{N} P_{i,j}}{N} \tag{1}$$

Where, $P_{i,j}$ is the image pixel and N is the total count of pixels in the matrix of the image.

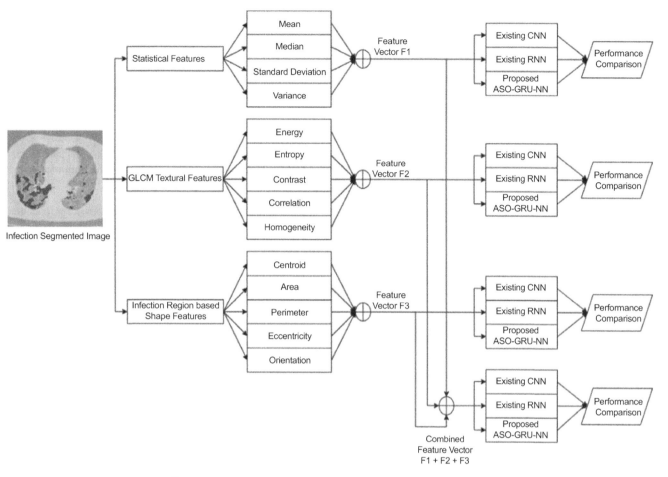

Fig. 4.1 Proposed framework of COVID-19 severity classification

Median represents the mid pixel of the image matrix, which is estimated using the equation (2).

$$\text{Median} = \frac{\left(P_{\frac{N}{2},\frac{N}{2}}\right) + \left(P_{\frac{N+1}{2},\frac{N+1}{2}}\right)}{2} \tag{2}$$

Standard deviation can be defined as the dispersion measure of observation in the dataset relative to the mean value. This can be expressed as given in equation (3).

$$SD = \sqrt{\frac{\sum_{i,j=1}^{N}\left(P_{i,j} - \text{Mean}\right)^2}{N}} \tag{3}$$

Variance can be defined as the numerical value that describes the variability of the observations from the mean value. Variance can be estimated using equation (4).

$$\text{Variance} = SD^2 \tag{4}$$

3.2 Gray Level Co-occurrence Matrix (GLCM) Features

The textural features estimated using GLCM in this research work are energy, entropy, contrast, correlation, and homogeneity. The GLCM function characterizes the textural features of the image by assessing how often the pixel pairs with specified values and in a specific spatial relationship exist in an image, with the help of a GLCM to extract statistical features from this matrix. Every element (i, j) of the GLCM is the total number of times the pixel of intensity i existed in the specific spatial relationship to a pixel of intensity j in the input matrix of image.

Energy estimates the uniformity in the texture (repetitions in pixel pair). Energy can be expressed using equation (5).

$$\text{Energy} = \sum_i \sum_j (g_{i,j})^2 \tag{5}$$

Where, $g_{i,j}$ is the element of the GLCM G.

Entropy calculates the complexities or disorders in the image. The value of entropy is high when the uniformity is not found in the image and when the elements of GLCM are very small. Entropy is inversely and strongly proportional to energy. Entropy can be calculated using the equation (6).

$$\text{Entropy} = -\sum_i \sum_j g_{i,j} \log_2 g_{i,j} \tag{6}$$

Contrast calculates the difference moment of GLCM and the spatial frequency of an image. It is defined as the difference between the lowest and highest values of the pixels in the matrix. Contrast of the image can be estimated using equation (7).

$$\text{Contrast} = \sum_i \sum_j (i - j)^2 \, g_{i,j} \tag{7}$$

Correlation value refers to the measure of image's gray tone linear dependencies. The correlation feature can be estimated using equation (8).

$$\text{Correlation} = \frac{\sum_i \sum_j (ij) g_{i,j} - \text{mean}_g}{SD_g} \tag{8}$$

Where SD_g and mean_g are the standard deviation and mean of g.

Homogeneity is referred as the Inverse Difference Moment. It estimates the homogeneity of image as it considers higher values for lesser gray tone differences in element pair. This can be measured using equation (9).

$$\text{Homogeneity} = \sum_i \sum_j \frac{1}{1 + (i - j)^2} g_{i,j} \tag{9}$$

3.3 Infection Region Features

Shape features are physical dimensional measures that characterize the appearance of the lesions. The shape features extracted from the infected regions are centroid, area, perimeter, eccentricity, and orientation. Centroid is the region's center of mass, returned as a 1-by-Q vector, where Q denotes the dimensionality of the image. Area is the actual number of pixels in the region, returned as a scalar value. Perimeter is the boundary distance of the region returned as a scalar value. Eccentricity defined as the ratio of the distance between the lesions' foci and the length of the main axis. Orientation is the angle swept by the x-axis from the main axis of the ellipse which has the second-moments same as the region, returned as a scalar.

4. Proposed ASO-GRU-NN Classification Model

4.1 ASO Optimizer

ASO is a novel physics-inspired computational metaheuristic optimization approach. ASO was formulated by the inspiration of basic molecular dynamics. This algorithm was formulated for addressing global optimization problems [19]. ASO optimizer is employed in this work to optimize the weights of the GRU-NN model during the training phase. In ASO, the location of the atoms in the search space represent a solution set (neural network weights) measured by means of their masses, with a solution that is better in terms of heavier masses, and vice versa. In order to solve this unconstrained optimization, an atom population with N atoms is considered. The atom population assumed for this application is the N input data from the N neurons in input layer. The location of the ith atom can be expressed as given in equation (10)

$$x_i = [x_i^1, x_i^2, \ldots, x_i^D], \, i = 1, 2, \ldots, N \tag{10}$$

where $x_i^d (d = 1, 2, \ldots, D)$ represents the dth position component of the ith atom in a D-dimensional search space. Each population is evaluated for fitness. The objective function used in this optimization for fitness evaluation is the Mean Square Error (MSE) function. The minimum MSE value of the population is the best fit value F_{best} and is expressed as given in equation (11),

$$F_{\text{best}} = \min F = \min MSE \tag{11}$$

On the other hand, the worst fit value F_{worst} can be expressed as given in equation (12),

$$F_{worst} = \max F = \max MSE \tag{12}$$

Where, $MSE = \dfrac{1}{n}\sum_{i=1}^{n}\left(Y_{pred} - Y_{actual}\right)^{2}$, Y_{pred} is the predicted class, Y_{actual} is the actual class, and n is the total count of data samples. The weight matrix (atom population set) x_{best} for which F_{best} is obtained is the optimal solution of weights for the neural network. The mass of the ith atom which can be assessed at the simplest stage is calculated with the help of the following equation (13) that involve fitness values.

$$M_{i}\left(t\right) = \dfrac{m_{i}\left(t\right)}{\sum_{j=1}^{N} m_{j}\left(t\right)} \tag{13}$$

Where, $m_{i}\left(t\right) = e^{\frac{F_{i}(t) - F_{best}(t)}{F_{worst}(t) - F_{best}(t)}}$, $F_{i}(t)$ is the fitness value of the ith atom in the tth iteration. In this algorithm, every atom interacts with many atoms so as to find better values of fitness as its k neighbors. This interaction process enhances the exploration phase in the initial stage of iterations. For the purpose of enhancing the exploitation in the final phase of iterations, the atoms are made to interact with only a few atoms possess better values of fitness as their k neighbors. Hence, as a function of time, k declines gradually as the iterations increase. This k value can be measured using equation (14),

$$k\left(t\right) = N - \left(N - 2\right) \times \sqrt{\dfrac{t}{T}} \tag{14}$$

Where, t is the current iteration and T is the maximum iteration.

The force of interaction that results from the L-J potential function is defined as the atomic motion's priming power. The force of interaction acted on the ith atom from the jth atom at the tth iteration is expressed as given in equation (15),

$$F_{i,j}\left(t\right) = \dfrac{24\varepsilon\left(t\right)}{\sigma\left(t\right)}\left[2\left(\dfrac{\sigma\left(t\right)}{r_{i,j}\left(t\right)}\right)^{13} - \left(\dfrac{\sigma\left(t\right)}{r_{i,j}\left(t\right)}\right)^{7}\right]\dfrac{r_{i,j}\left(t\right)}{r_{i,j}^{d}\left(t\right)} \tag{15}$$

Where, $r_{i,j}$ is the Euclidean distance between i^{th} atom and j^{th} atom, ε is the potential well's depth, σ denotes the length scale referring to the diameter of the collision.

The total number of components with random weights performed at the ith atom from the rest of the atoms can be measured as the total force, expressed as given in equation (16).

$$F_{i}^{d}\left(t\right) = \sum_{j \in k_{best}} rand_{j} F_{i,j}^{d}\left(t\right) \tag{16}$$

Here, $rand_{j}$ represents a random number generated within the range [0,1].

The constraint of geometry on molecular dynamics plays a significant role in the motion of atoms. It is assumed that every atom in the algorithm shares a covalence bond with the best fit atom. Therefore, every atom is acted on by a constraint force from the best fit atom. The geometric constraint of the ith atom is expressed as given in equation (17),

$$G_{i}^{d}\left(t\right) = \lambda\left(t\right)\left(x_{best}^{d}\left(t\right) - x_{i}^{d}\left(t\right)\right) \tag{17}$$

Where $x_{i}^{d}(t)$ denotes the position of the ith atom in t^{th} iteration and $x_{best}^{d}(t)$ denotes the position of the best atom in t^{th} iteration, and λ represents the Lagrangian multiplier.

The atom's acceleration is then estimated using the interaction force and the geometric constraint as shown in equation (18).

$$a_{i}^{d}\left(t\right) = \dfrac{F_{i}^{d}\left(t\right)}{M_{i}^{d}\left(t\right)} + \dfrac{G_{i}^{d}\left(t\right)}{M_{i}^{d}\left(t\right)} \tag{18}$$

The velocity and the position of the atoms are updated for the next iteration $(t + 1)$ using the equations (19 and 20).

$$v_i^d\left(t+1\right) = rand_i^d\, v_i^d\left(t\right) + a_i^d\left(t\right) \tag{19}$$

$$x_i^d\left(t+1\right) = x_i^d\left(t\right) + v_i^d\left(t+1\right) \tag{20}$$

This procedure is repeated until the maximum search iteration T is reached. The final best solution is given as the weights W to the GRU-NN classifier model.

4.2 GRU-NN Model of COVID-19 Severity Classification

The final best solution is given as the weights W to the GRU-NN classifier model.

GRU is a newer version of RNN. Unlike Long Short Term Memory (LSTM) cells, GRUs do not contain cell states. GRU makes use of its hidden states for transporting information. It comprises of only 2 gates (Update gate and Reset gate). GRUs are faster than LSTM as they have lesser tensor operations. Update Gate is a combination of input and forget gates. Forget gate chooses what data to add in memory and what to ignore. Reset gate resets the past data for the purpose of eliminating gradient explosion. Reset gate estimates quantity of past data to be forgotten. The update gate and reset gate equations are expressed in equations (21) and (22) respectively.

$$r_t = \text{sigma}\left(x_t W_r + h_{t-1} U_r\right) \tag{21}$$

$$u_t = \text{sigma}\left(x_t W_u + h_{t-1} U_u\right) \tag{22}$$

Where, *sigma* is the activation function applied at the gates, x_t is the input data, h_{t-1} is the hidden state data, W_r and U_r are the weights passed for reset gate, and W_u and U_u are the weights passed for update gate.

The GRU is trained using backpropagation algorithm by feeding the extracted features as inputs and ASO solution set as weights. This work considers 80% of dataset for training and 20% of data for testing. The training of the model performed with training features and corresponding labels. The labels considered are: Normal, COVID-19 with mild infections, COVID-19 with moderate infections, COVID-19 with severe infections. The testing phase is carried out with testing inputs to evaluate the performance.

5. Results and Discussion

This section of the research paper presents the classification results and performances of different classifiers with different feature vectors. The simulation of these models was conducted in MATLAB R2020a platform with Windows 10 Pro edition of operating system, Intel Core i3 processor, and 8 GB RAM. The results obtained from the simulation are presented and discussed in the following sub sections.

The experimentation was conducted for different models of classifiers such as, conventional CNN, conventional RNN, and the proposed ASO-GRU-NN. These models were trained with different feature vectors such as F1 (statistical feature vector), F2 (GLCM textural feature vector), F3 (Shape feature vector), and combined F1 + F2 + F3 feature vector. The performance metrics considered for the comparison of the models are accuracy, error rate, sensitivity, specificity, and execution time.

Accuracy of classification is referred as the number of correct predictions to the total sum of predictions. This is expressed as given in equation (23),

$$\text{Accuracy} = \frac{T_P + T_N}{T_P + T_N + F_P + F_N} \tag{23}$$

Where,

T_P is True Positives, T_N is True Negatives, F_P is False Positives, and F_N is False Negatives.

Error rate is referred as the ratio of the total count of incorrect predictions to the total count of observations. This can be expressed in terms of accuracy given in equation (24).

$$\text{Error} = 1 - \text{Accuracy} \tag{24}$$

Sensitivity is defined as the total count of correct COVID-19 class predictions to the sum of correct COVID-19 class class predictions and the normal class predictions. Sensitivity can be calculated using equation (25).

$$\text{Sensitivity} = \frac{TP}{TP + FN} \tag{25}$$

Specificity is defined as the total count of correct negative class predictions to the correct negative class predictions and the incorrect positive class predictions. Specificity can be calculated using equation (26).

$$\text{Specificity} = \frac{TN}{TN + FP}$$

(26)

Total execution time (in s) is the time taken from giving an input image to getting the classification results. The effects of different feature sets were assessed in this research work and the obtained values of performance measures are shown in the following Tables 4.1–4.4.

Table 4.1   Performance with statistical features

Model	Accuracy	Error	Sensitivity	Specificity	Total Execution Time (s)
CNN	0.2	0.8	0	1	66.79
RNN	0.45	0.55	0.67	0.647	63.95
Proposed ASO-GRU-NN	0.5	0.5	1	0.647	61.23

Table 4.2   Performance with GLCM features

Model	Accuracy	Error	Sensitivity	Specificity	Total Execution Time (s)
CNN	0.35	0.65	0.25	1	65.58
RNN	0.45	0.55	0.25	0.81	62.87
Proposed ASO-GRU-NN	0.7	0.3	0.25	1	60.71

Table 4.3   Performance with infection region features

Model	Accuracy	Error	Sensitivity	Specificity	Total Execution Time (s)
CNN	0.5	0.5	0.67	0.857	66.22
RNN	0.8	0.2	0.67	0.928	63.16
Proposed ASO-GRU-NN	0.85	0.15	0.834	0.928	60.34

Table 4.4   Performance with statistical features + GLCM features + infection region features

Model	Accuracy	Error	Sensitivity	Specificity	Total Execution Time (s)
CNN	0.5	0.5	0.6667	1	70.71
RNN	0.85	0.15	0.8333	1	60.57
Proposed ASO-GRU-NN	0.95	0.05	1	1	58.65

Table 4.1 shows the classification results of statistical feature set. The proposed ASO-GRU-NN was found to produce best results showing 0.5 accuracy, 0.5 error rate, 1 sensitivity, and 61.23s total execution time. However, in terms of specificity, the CNN model showed best score of 1.

Table 4.2 presents the classification results of GLCM textural feature set. The proposed ASO-GRU-NN was found to produce best results showing 0.7 accuracy, 0.3 error rate, 0.25 sensitivity, 1 specificity, and 60.71s total execution time.

Table 4.3 presents the classification results of region-based shape feature set. The proposed ASO-GRU-NN was found to produce best results showing 0.85 accuracy, 0.15 error rate, 0.834 sensitivity, 0.928 specificity, and 60.34s total execution time.

Table 4.4 shows the classification results of combined feature vector (F1+F2+F3). The proposed ASO-GRU-NN was found to produce best results showing 0.95 accuracy, 0.05 error rate, 1 sensitivity, 1 specificity, and 58.65s total time taken for execution.

6. Conclusion

This research work aims to assess the effects of three main feature estimation techniques, namely, statistical features, GLCM textural features, and shape features. The effect of combined feature vector was also evaluated in this research. The comparison of the effects of different feature sets on the classification recommends that the combined feature vector has greater potential to achieve the best performance of detection and severity classification of COVID-19. This research also proposes a new model of classifier called ASO-GRU-NN, which uses GRU as the base model and the optimization of the weights are carried out using CI algorithm called ASO. This ASO-GRU-NN model was found to provide the best performance of classification when compared to conventional CNN and conventional RNN models. This is because, the ASO-GRU-NN optimally tunes the weights during the training period, resulting in reduced training losses and increased accuracy. In the future, the ASO can be employed in a CNN to evaluate its classification performance on image feature maps of COVID-19 CT slices.

References

1. Al-Areqi, F., & Konyar, M. Z. (2022). Effectiveness evaluation of different feature extraction methods for classification of covid-19 from computed tomography images: A high accuracy classification study. Biomedical Signal Processing and Control. 76: 103662.
2. Mohammad-Rahimi, H., Nadimi, M., Ghalyanchi-Langeroudi, A., Taheri, M., & Ghafouri-Fard, S. (2021). Application of machine learning in diagnosis of COVID-19 through X-ray and CT images: a scoping review. Frontiers in cardiovascular medicine.8: 638011.
3. Abed, M., Mohammed, K. H., Abdulkareem, G. Z., Begonya, M., Salama, A., Maashi, M. S., ... & Mutlag, L. (2021). A comprehensive investigation of machine learning feature extraction and classification methods for automated diagnosis of COVID-19 based on X-ray images. Computers, Materials, & Continua. 66: 3289–3310.
4. Qiblawey, Y., Tahir, A., Chowdhury, M. E., Khandakar, A., Kiranyaz, S., Rahman, T., ... & Ayari, M. A. (2021). Detection and severity classification of COVID-19 in CT images using deep learning. Diagnostics.11(5): 893.
5. Shi, F., Wang, J., Shi, J., Wu, Z., Wang, Q., Tang, Z., ... & Shen, D. (2020). Review of artificial intelligence techniques in imaging data acquisition, segmentation, and diagnosis for COVID-19. IEEE reviews in biomedical engineering. 14: 4–15.
6. Dey, N., Rajinikanth, V., Fong, S. J., Kaiser, M. S., & Mahmud, M. (2020). Social group optimization–assisted Kapur's entropy and morphological segmentation for automated detection of COVID-19 infection from computed tomography images. Cognitive Computation. 1011–1023.
7. Alsalem, M. A., Zaidan, A. A., Zaidan, B. B., Albahri, O. S., Alamoodi, A. H., Albahri, A. S., ... & Mohammed, K. I. (2019). Multiclass benchmarking framework for automated acute Leukaemia detection and classification based on BWM and group-VIKOR. Journal of medical systems, 43(7), 1–32.
8. Alsalem, M. A., Zaidan, A. A., Zaidan, B. B., Hashim, M., Albahri, O. S., Albahri, A. S., ... & Mohammed, K. I. (2018). Systematic review of an automated multiclass detection and classification system for acute Leukaemia in terms of evaluation and benchmarking, open challenges, issues and methodological aspects. Journal of medical systems. 42: 1–36.
9. Abdel-Basset, M., Abdel-Fatah, L., & Sangaiah, A. K. (2018). Metaheuristic algorithms: A comprehensive review. Computational intelligence for multimedia big data on the cloud with engineering applications. 1: 185–231.

Note: All the figures and tables in this chapter were made by the Authors

Human Machine Interaction in the Digital Era – Prof. J. Dhilipan et al. (eds)
© 2024 Taylor & Francis Group, London, ISBN 978-1-032-54998-9

Deep Learning Based Brain Tumour Detection Using Image Processing

5

Asma Parveen A.[1]

Research Scholar,Department of computer science, Vels institute of Science,
Technology and Advanced Studies, Chennai, India

Kamalakannan T.[2]

HOD, Department of computer science, Vels institute of Science,
Technology and Advanced Studies, Chennai, India

Abstract In order to improve pictorial information for human understanding and to process picture data for transmission, storage, and retrieval of meaningful information, the intriguing and inspiring discipline of Image Processing was developed. Raw data from cameras and sensors, such as those used in everyday life, can be processed using a method called "image processing." In order to do image analysis, it is necessary to analyse the images and divide them into meaningful segments, a process known as image segmentation. Anatomical and physiological structures of the human body can be studied with the aid of medical picture segmentation. Multiple high-resolution photographs of the human brain are created using cutting-edge technology, allowing researchers to track its development and spot a wide range of disorders. Brain cancer is the most common malignant disease to strike children and young people. This study will aid the public in determining whether or not a brain tumour is present in an MRI of the brain. Other researchers' ground-breaking efforts have served as inspiration for this investigation. As a result, their work has been a major inspiration for this study. In addition, I want to get involved in the medical domain by learning about and eventually analysing brain tumours.

Keywords Data transmission, Tumor, Processing, MRI, Image segmentation

1. Introduction

Many scientists in the field of medicine are using cutting-edge methods for disease forecasting. These methods have consistently helped medical professionals with diagnosis, prevention, and therapy preparation. Studying how to diagnose brain tumours from brain MRI scans using Image Processing techniques is one of the most prominent and challenging areas of study now underway. However, people of all ages are among the brain tumor's many victims. Human lives can be saved with early detection or prediction of brain tumours. Brain tumours are among the most lethal disorders a person may have, say doctors and scientists in the field of medicine. Researchers are therefore driven to find ways to detect and anticipate tumours in the area of the brain where they are most likely to arise, based on the earliest symptoms, and to determine the specific type of tumour using MRI scans of the brain. One of the most in-demand areas of medicine, this work investigates the use of picture segmentation techniques to detect brain tumours. However, there is no correlation between age and the prevalence of brain tumours of all sorts. While some of these tumours are surgically treatable benign variants, others are malignant and can eventually cause brain cancer.

[1]kalaivanij03@gmail.com, [2]arunachalam1976@gmail.com

DOI: 10.1201/9781003428466-5

2. Literature Review

Ingenious image processing involves digitising an image and running it through a series of mathematical operations in order to either improve it or extract relevant data from it. There are two main categories of IP: the first deals with enhancing the appearance of images for human viewing, and the second measures various aspects and structures inside those images. Depending on the body part, diagnostic task, viewing preferences, and so on, a digital image can be customised for applications by segmenting or modifying the look of structures within it. Radiologists will benefit from computerised image analysis since it allows them to quickly and accurately locate questionable areas in an image and provide an accurate diagnosis. In this section, we'll talk about the research that has been done on image processing techniques, classification algorithms, and diagnosing brain tumours. Moreover, it provides a comprehensive analysis of the preprocessing methods and picture segmentation algorithms employed by various researchers. Information from the survey was also cited in pieces discussing a variety of brain disorders and tumours. Algorithms for segmenting data, classifying it, and clustering it have all found applications in the field of medicine. Super pixel-based brain tumours segmentation using RF and SVM classifiers was evaluated and found to favour RF in terms of accuracy. Combining C-MRI with DTI, as reported in [1], and feeding the data to support vector machines (SVM) allowed researchers to identify distinct types of tumour tissue [2]. An advantage of the super pixel based method, as suggested by [3], is that it greatly reduces the amount of computing needed for classification in the new feature space.

3. Proposed Methodology

Researchers have tried a wide variety of approaches to analyse MRI brain pictures in search of tumours. Preprocessing and Segmentation are the two main phases of the suggested approach. Locating an MRI brain picture, eliminating noise, and

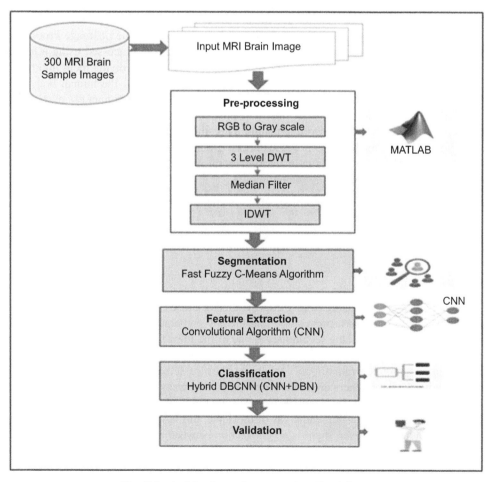

Fig. 5.1 Architecture of proposed methodology

Source: Made by Author

improving image quality all require the use of image processing techniques. In order to precisely locate the tumour area and damaged regions in the photos, we extract them using segmentation methods.

What follows is a breakdown of the suggested procedure's steps.

- Step 1: Train the system using 3064 MRI scans of the brain saved in MATLAB format, dividing them evenly between normal and pathological examples of types 1, 2, 3, and 4.
- Step 2: Upload a fresh.jpg of your MRI brain scan
- Step 3: This image has been grayscaled from its original RGB format.
- Step 4: Median Filter with Three Levels for Preprocessing
- Step 5: This noise has been reduced by using a median filter.
- Step 6: The data was preprocessed with an IME, an inverse median filter. Part 5: Segmenting Images using the Fast Fuzzy C-means Algorithm
- Step 7: "Feature Extraction using a Convolutional Neural Network" (CNN)
- Step 8: Deep Belief Convolutional Neural Network Hybrid Algorithm Classification.

3.1 Description of Dataset

Imaging planes, or directions, are used to characterise brain MRI scans in three different ways: horizontally, vertically, and cross-sectionally. The axial plane, the coronal plane, and the sagittal plane are all terms used in the field of medical imaging. According to neurosurgeons, the sagittal plane's description of the brain's left and right sides tells us nothing about the tumor's location. Images in both the axial and coronal planes can be utilised to pinpoint the tumor's afflicted area for use in treatment planning and clinical diagnosis. Any algorithm requiring an MRI picture as input requires a conversion step before it can be implemented. The images generated by MRI scanners are in the DICOM (Digital Imaging and Communication in Medicine) format, thus a conversion is required before they can be processed visually in a CAD programme. Sample photos were also acquired from each of the four groups of aberrant patients (Types 1, 2, 3, and 4).

Table 5.1 Description of patients data

Attribute	Description	Remarks
Normal Brain Image	1000 Patients	No further analysis
Abnormal Brain Image	2000 Patients	Further Analysis for Types
Type I	32 patients	Very less patients visit hospital in early
Type II	205 patients	Stage Moderate patients visit hospital
Type III	772 patients	More patients visit hospital in Type III
Type IV	991 patients	More patients visit hospital in finalstage

Source: Dataset prepared and collected by the Author

Brain MRI scans were used in this study's preprocessing, segmentation, feature extraction, classification, and validation phases (as shown in Table 5.2). There are a total of six photos, three each from the normal and pathological groups and from patients of varying ages and imaging modalities. Images can be processed with MATLAB 2021a after being saved as.mat files.

Table 5.2 Details of MRI brain image

S. No.	Name	Age	Gender	Category	Imagingplanes
1	Narayanan	52	Male	Normal	T2Axial
2	Radhika	46	Female	Normal	T2Axial
3	Rajasekar	26	Male	Type1	T2AxialFIL
4	AbdulRahman	48	Male	Type2	T2AxialFIL
5	Karthikeyan	9	Male	Type3	T2Axial
6	AchiraPar	9	Female	Type4	T2FlairAxial

Source: Prepared by the Author

Normal brain images from a variety of patients are depicted in Fig. 5.2 below. Pictured in Fig. 5.2 (a) NOR IMG 01 is a normal MRI brain image of a male in the T2 axial plane, and in Figure 2 (b) NOR IMG 02 is a normal MRI brain image of a female in the T2 axial plane.

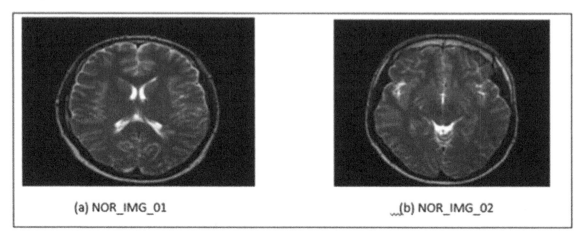

(a) NOR_IMG_01   (b) NOR_IMG_02

Fig. 5.2   Normal brain image

Source: Made by Author

3.2 Steps in the Filtering Process

When an MRI picture is acquired, it is contaminated by many types of noise due to the environment. Results from noisy images tend to be subpar. This means that prior to implementing any technique, noise reduction and image quality enhancement must occur.

- **Gaussian filter:** It is common practise to employ a Gaussian filter to clean up an image by decreasing noise and boosting contrast. Images are preprocessed using the Gaussian filter's closeness and similarity functions to reduce noises before further analysis.
- **Bilateral filter:** is an edge-preserving, noise-reduction, non-linear smoothing filter for pictures. It does this by substituting the average intensity of neighbouring pixels for the intensity of the current pixel. There is room for a Gaussian distribution in this weighting.
- **Median filter:** The goal of the preprocessing procedure is to modify the photos such that they have higher contrast and are more legible. Median has a tendency to retain the crispness of image edges while smoothing out the noise. It is a method that helps get rid of impulse noise.

RGB to Gray Scale

A colour picture, abbreviated as RGB, consists of the primary colours red, green, and blue. Each pixel in an RGB image encapsulates the red, green, and blue components of the image at a given point in M x N x 3 array of colour pixels.

- Step 1: Read RGB color image
- Step 2: Take an RGB image and separate its red, green, and blue components into three separate 2-dimensional matrices.
- Step 3: Generating a new matrix with the same amount of rows and columns as the RGB image, but filled with zeroes.
- Step 4: To create grayscale values, create a weighted sum of the Red, Green, and Blue colour components and assign it to the ijth position in the new matrix.

3.3 Segmentation

Without altering the original image's structures or attributes, segmentation serves to separate the images for processing in order to recover information utilising algorithms or image segmentation approaches. Segmentation is the process of dividing a digital image into multiple non-overlapping parts based on characteristics of the image, such as its pixel size, texture, depth, and grayscale. These are some of the many methods used in segmentation::

- Threshold-based segregation
- Clustering based on features to create subgroups

- Automatic Region Delimitation Using Edge Detection
- Compression-based techniques

3.4 Feature Extraction

There are primarily four phases involved in image processing. After the images have been preprocessed and segmented, the next stage is Feature extraction, which is crucial in obtaining the features. The final level or step in processing an image is classification, which receives its input from the results of both the segmentation and feature extraction processes. After an image has been subjected to various processing steps—including pre-processing, segmentation, feature extraction, and classification—its results must be validated. While many different approaches exist for Feature Extraction, the ones listed here are tried and true methods for extracting and choosing features.

- The Principal Components Method (PCA)
- Matrix of Grey-Level Occurrences (GLCM)

4. Conclusion

The proposed method outperforms previously developed systems that attempted to accomplish the same goal. Previous efforts relied on a simpler 2-level discrete wavelet transform (DWT), but the implementation of the 3-level Median Filter has improved contrast and allowed for the easy suspension of signals within an image. The image became too blurry at levels 4 and 5, and the classification accuracy rate was too low, so we settled on level 3. While both the Chanvese algorithm and CNN were examined for their potential in segmentation, the former yielded less encouraging findings. The fast Fuzzy C-means method improved the results, and it became an innovative tool in the scientific community. Last but not least, a hybrid method called Deep belief convolutional neural network (DBCNN) was created in MATLAB 2021a for classification, bringing together CNN (Convolutional Neural Network) and DBN (Deep belief Neural network) for superior results.

References

1. B. Cheng, C. Bing, T. H. Chu, S. Alzahrani, S. Pichardo and G. B. Pike, "Simultaneous Localized Brain Mild Hyperthermia and Blood-Brain Barrier Opening via Feedback-Controlled Transcranial MR-Guided Focused Ultrasound and Microbubbles," in IEEE Transactions on Biomedical Engineering, vol. 69, no. 6, pp. 1880–1888, June 2022.
2. M. A. Ottom, H. A. Rahman and I. D. Dinov, "Znet: Deep Learning Approach for 2D MRI Brain Tumor Segmentation," in IEEE Journal of Translational Engineering in Health and Medicine, vol. 10, pp. 1–8, 2022, Art no. 1800508.
3. A. Sekhar, S. Biswas, R. Hazra, A. K. Sunaniya, A. Mukherjee and L. Yang, "Brain Tumor Classification Using Fine-Tuned GoogLeNet Features and Machine Learning Algorithms: IoMT Enabled CAD System," in IEEE Journal of Biomedical and Health Informatics, vol. 26, no. 3, pp. 983–991, March 2022.
4. Y. Ding et al., "MVFusFra: A Multi-View Dynamic Fusion Framework for Multimodal Brain Tumor Segmentation," in IEEE Journal of Biomedical and Health Informatics, vol. 26, no. 4, pp. 1570–1581, April 2022.
5. M. Ismail et al., "Radiomic Deformation and Textural Heterogeneity (R-DepTH) Descriptor to Characterize Tumor Field Effect: Application to Survival Prediction in Glioblastoma," in IEEE Transactions on Medical Imaging, vol. 41, no. 7, pp. 1764–1777, July 2022.
6. M. Rahimpour et al., "Cross-Modal Distillation to Improve MRI-Based Brain Tumor Segmentation With Missing MRI Sequences," in IEEE Transactions on Biomedical Engineering, vol. 69, no. 7, pp. 2153–2164, July 2022.
7. A. Bs, A. V. Gk, S. Rao, M. Beniwal and H. J. Pandya, "Electrical Phenotyping of Human Brain Tissues: An Automated System for Tumor Delineation," in IEEE Access, vol. 10, pp. 17908–17919, 2022.
8. Y. Liu, F. Mu, Y. Shi and X. Chen, "SF-Net: A Multi-Task Model for Brain Tumor Segmentation in Multimodal MRI via Image Fusion," in IEEE Signal Processing Letters, vol. 29, pp. 1799–1803, 2022.
9. S. Montaha, S. Azam, A. K. M. R. H. Rafid, M. Z. Hasan, A. Karim and A. Islam, "TimeDistributed-CNN-LSTM: A Hybrid Approach Combining CNN and LSTM to Classify Brain Tumor on 3D MRI Scans Performing Ablation Study," in IEEE Access, vol. 10, pp. 60039–60059, 2022.
10. F. Xing, X. Liu, C. .-C. J. Kuo, G. E. Fakhri and J. Woo, "Brain MR Atlas Construction Using Symmetric Deep Neural Inpainting," in IEEE Journal of Biomedical and Health Informatics, vol. 26, no. 7, pp. 3185–3196, July 2022.
11. A. S. Musallam, A. S. Sherif and M. K. Hussein, "A New Convolutional Neural Network Architecture for Automatic Detection of Brain Tumors in Magnetic Resonance Imaging Images," in IEEE Access, vol. 10, pp. 2775–2782, 2022.
12. A. Vidyarthi, R. Agarwal, D. Gupta, R. Sharma, D. Draheim and P. Tiwari, "Machine Learning Assisted Methodology for Multiclass Classification of Malignant Brain Tumors," in IEEE Access, vol. 10, pp. 50624–50640, 2022.

13. H. Fu et al., "HMRNet: High and Multi-Resolution Network With Bidirectional Feature Calibration for Brain Structure Segmentation in Radiotherapy," in IEEE Journal of Biomedical and Health Informatics, vol. 26, no. 9, pp. 4519–4529, Sept. 2022.

14. M. Rizwan, A. Shabbir, A. R. Javed, M. Shabbir, T. Baker and D. Al-Jumeily Obe, "Brain Tumor and Glioma Grade Classification Using Gaussian Convolutional Neural Network," in IEEE Access, vol. 10, pp. 29731–29740, 2022.

15. R. Mehta et al., "Propagating Uncertainty Across Cascaded Medical Imaging Tasks for Improved Deep Learning Inference," in IEEE Transactions on Medical Imaging, vol. 41, no. 2, pp. 360–373, Feb. 2022.

16. B. Deepa, M. Murugappan, M. G. Sumithra, M. Mahmud and M. S. Al-Rakhami, "Pattern Descriptors Orientation and MAP Firefly Algorithm Based Brain Pathology Classification Using Hybridized Machine Learning Algorithm," in IEEE Access, vol. 10, pp. 3848–3863, 2022.

17. A. Bs, A. B, H. R S, V. V, A. Mahadevan and H. J. Pandya, "Electromechanical Characterization of Human Brain Tissues: A Potential Biomarker for Tumor Delineation," in IEEE Transactions on Biomedical Engineering.doi: 10.1109/TBME.2022.3171287

Human Machine Interaction in the Digital Era – Prof. J. Dhilipan et al. (eds)
© 2024 Taylor & Francis Group, London, ISBN 978-1-032-54998-9

Detection and Notification System in Emergency Vehicles for Unidentified Accident Victims

6

V. Dinesh[1]

Assistant Professor, Easwari Engineering College, Anna University Chennai

R. Krithika[2]

Easwari Engineering College, Anna University Chennai

S. Kirti Sri[3]

Easwari Engineering College, Anna University Chennai

Abstract The intent here is to shorten the time taken to detect the basic information of the unidentified accident victims and notify the patient's guardian and police control room regarding the current situation. This device is installed in emergency vehicles. The implemented design stores a specific amount of data about each individual along with their fingerprints so that the data can be quickly accessed by scanning the unidentified accident victim's fingerprint through the fingerprint scanner. Following that, utilizing the data acquired, the victim's guardian and the police control room are informed regarding the present status via an SMS notification system using a GSM module.

Keywords: Notification, Fingerprint, Emergency vehicle SMS, GSM

1. Introduction

India is the second most populated country, and as a result, several essential aids and emergency medical procedures are disregarded. Every day, several accidents take place. When we come across an unfamiliar person, we always wait for them to become conscious before attempting to identify their blood relatives. This may possibly be more dangerous if the person has already been missing for some time. The goal here is to prevent having to wait while notifying the patient's guardian and the police station. The victim's fingerprint is scanned in order to notify those in need. According to the method described in this system, a GSM module and a fingerprint sensor R307 are added to the emergency vehicle so that the victim can be quickly identified after an accident by having their fingerprint scanned, having access to their data, and SMS notifications are sent to the appropriate parties. The data which is needed to identify the victim is already stored in the memory. The fingerprint is scanned using the optical fingerprint sensor R307. The device contains a built-in SIM within the GSM module that uses mobile telephone technology to contact patients' family members and the police control room. The R307 fingerprint sensor, LCD display, nodemcu, and the GSM module are interfaced through the Arduino Uno microcontroller.

---

[1]kalaivanij03@gmail.com, [2]arunachalam1976@gmail.com

DOI: 10.1201/9781003428466-6

2. Literature Review

2.1 Automatic Messaging System for Vehicle Tracking and Accident Detection

With the use of a GPS module, GSM module, and accelerometer, which are connected to an Arduino Uno, which serves as the controller, a system is created that can alert the concerned parties about an accident via SMS services.

Efficient accident detection and notification system

The Arduino board which is installed in the vehicle has GSM and GPS. The Android app present is the software component. In the proposed system, the gsm module is triggered only after the victim's fingerprint is scanned.

An approach to making way for intelligent ambulance Using IoT

With the idea of a "black box" in a vehicle, it is possible to use GPS and GSM technologies to find out what transpired. Controlling the traffic signals will enable the ambulance department to quickly admit the sufferer to the hospital.

Person's identity verification using Arduino UNO

In most cases, fingerprint scanning involves comparing several aspects of the print pattern, which is necessary when analyzing fingerprints for matching purposes. These consist of tiny points, individual qualities contained within the patterns as an aggregate property of ridges.

Feasibility study on applications of GSM–SMS technology

GSM-SMS technology is used to control, communicate, and transmit data across two systems. A GSM-SMS-based communication architecture provides a package format of short messages which is appropriate for keeping track of a wide range of applications.

3. System Design and Implementation

By using a fingerprint sensor to access the data and a GSM module to automatically send an SMS message to the victim's guardian and police control room, we have built an automated identification and notification system for the needy. This approach is primarily intended to speed up the process of finding the fundamental information on an accident victim who is found inside an emergency vehicle. Checking for adequate network connectivity and a reliable power supply might serve as the first step in starting this procedure. The victim's name, emergency contact number information, and address will be shown on the LCD screen by just scanning the victim's fingerprint in the device, using the R307 fingerprint sensor. An SMS message is then sent immediately to the designated emergency contact using the necessary information. In order to report a road accident, the system is set up in such a manner that an E-Service SMS complaint will automatically be triggered and transmitted to the police control center. This method is done using a GSM module SIM900A, which has an inbuilt SIM, to contact the patient's relatives and the police control room, it offers a wireless data connection to a network using GSM mobile telephone technology. The biggest benefit of employing this system is how easy it is to install in all ambulances. There is no waiting period before treating unidentified accident victims. Mechanisms for exchanging information are a lot more quicker and automated.

4. Block Diagram

From Fig. 6.1 the microcontroller used in this device is an Arduino Uno (ATmega328P). The device is supplied with a +5v of DC power supply. The corner of the Arduino microcontroller has a power supply port. Flash memory, EEPROM, and SRAM are the onboard memories that are available here. The Arduino IDE software can be utilized to execute the required program for our system. Following that, this program can be uploaded through the power jack cable to the Arduino microcontroller. The Arduino uno acts as an interface between the fingerprint sensor (R307) and the GSM module (SIM900A) as it stores the captured fingerprints and the data of the group of people, which is required to match when the fingerprint is scanned. The victim's fingerprint is scanned using the fingerprint sensor R307 to obtain his basic information, which is then utilized to notify the victim's guardian and the police control center. This fingerprint sensor module connects directly to microcontrollers through a TTL UART channel. If the fingerprint of the victim matches the existing data, the Arduino Uno is programmed in such a way that the victim's name and emergency contact number are displayed on the 2x16 LCD display.

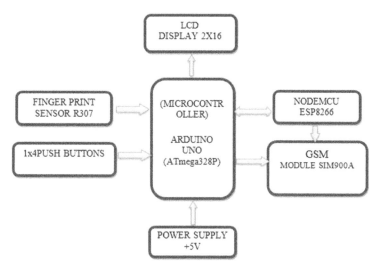

Fig. 6.1 Block diagram of detection and notification system for unidentified accident victims

After the data has been shown on the LCD screen for a short while, the GSM module, which has an integrated SIM, uses the data to send an SMS notice to the victim's legal representative and the police control room. The subscriber identification module (SIM) is inserted into the GSM network to give personal mobility through the worldwide system for mobile telecommunications (GSM). The nodemcu (ESP8266) is an open-source IOT platform that connects items and enables Wi-Fi protocol data transfer, Arduino. The push button here is utilized to perform device enrollment, deletion, and transmission of up and down commands.

5. Result and Discussions

The scanning of a fingerprint is displayed alongside the final output image.

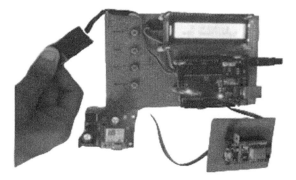

Fig. 6.2 Positioning of the finger in the fingerprint sensor present in the hardware kit

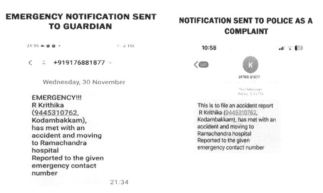

Fig. 6.3 SMS notification sent via GSM module

6. Conclusion and Future Scopes

The "Detection and Notification System for Unidentified Accident Victims" prototype has been constructed. By scanning the victim's fingerprints, this prototype potentially recognizes an unidentified accident victim and contacts the victim's guardian, and the police control center. In this complete system, the utilization of a fingerprint sensor and GSM mobile technology is fundamental. By placing this equipment in the emergency vehicle, the proposal is thus advanced to shorten the response time for an unidentified accident.

In addition to the current system, a few more records can be added by continuously monitoring the patient's vitals using the sensors connected to the victim's body, such as heart rate, blood pressure, and pulse. Then the records will be delivered to the hospital through a live web server after the vital data has been acquired, allowing for a quicker start to the treatment planning process.

7. Acknowledgement

My sincere gratitude goes out to my mentor, Mr. V. Dinesh, for his constant guidance and supervision throughout the procedure. In addition to thanking all the faculty at Easwari engineering college's electronics and communication department.

References

1. R. Rishi, S. Yede, K. Kunal and N. V. Bansode, "Automatic Messaging System for Vehicle Tracking and Accident Detection," 2020 pp. 831–834, doi: 10.1109/ICESC48915.2020.9155836.
2. Abilash V V, Aljo Emerson , Tony Sebastian , Vishnu C, Rashmi P C, 2021, Efficient Accident Detection and Notification System, (2021), Volume 10, Issue 07
3. Venkatesh H, Shrivatsa D Perur, Jagadish M - An Approach to Make Way for Intelligent Ambulance Using IoT, 2015 (International Journal of Electrical and Electronics Research; Vol. 3, Issue 1, pp: (218–223))
4. Chwan-Lu Tseng, Joe-Air Jiang, Ren-Guey Lee, Fu-Ming Lu, Cheng-Shiou Ouyang, Yih-Shaing Chen, Chih-Hsiang Chang,Feasibility study on application of GSM–SMS technology to field data acquisition,Computers and Electronics in Agriculture,Volume 53, Issue1, 2006, ISSN0168-1699.
5. Suruchi Kaushik, Nishant Arora, Rakshit Sachdev, Vaibhav Ahluwalia, Person's Identity Verification using Arduino Uno,2019, (IJERT) Volume 08, Issue 07
6. B.Malarvizhi V. Dinesh M.Janani, R.Gunaseeli, B.Abarna, "IoT based Staple Food Endowment System and Waste Management System for Foster Care," 2020 Sambodhi- UGC Care Journal, ISSN:2249-6661, Volume-43, UGC Care
7. Gaikwad, Sharmila S., Mahek Khanna, Shreyansh Kothari and Ashutosh Kudale. "Systematic Literature Survey on Accident Alert & Detection System." (2021).Vol2889/PAPER_01.
8. B. Malarvizhi, V. Dinesh, M. Janani, R. Gunaseeli, B. Abarna, "An IoT based Staple Food Endowment System and Waste Management System for Foster Care using Arduino and Blockchain," 2020, ICRCSIT-20, ISBN No. 978-93-80831-66-4
9. Hossam M. Sherif, Hossam M. Sherif, Samah A. Senbel, "Real Time Traffic Accident Detection System using Wireless Sensor Network",2014, International Conference of Soft Computing and Pattern Recognition.
10. Sharmila Gaikwad, "Mobile Agents in Heterogeneous Database Environment for Emergency Healthcare System",2008 at ITNG, Las Vegas, Nevada, USA. Published in IEEE Computer society, ISBN: 978-0-7695-3099-4, PP. 1220–1221
11. Roberto G. Aldunate, Oriel A. Herrera, Juan Pablo Cordero,(2013)Context-Awareness and Context-Driven Interaction, Volume 8276ISBN: 978-3-319-03175-0
12. Gadekar, Santosh and Kolpe, Gauri and Rutuja, Gosavi and Fatate, Vaishnavi and Rohit, Rohit and Chate, Shriprasad and Lad, Akshay, Arduino Uno-ATmega328 Microcontroller Based Smart Systems(2021). (ICCIP) SSRN:/http://dx.doi.org/10.2139/ssrn.3920231.
13. Sudhan, R.H., Kumar, M.G., Prakash, A., & Devi, S. .ARDUINOATMEGA-328 MICROCONTROLLER.(2015)DOI:10.17148/ijireeice.2015.3406,CorpusID: 212479591
14. Murru, Ganesh & Kakollu, Chetan & Kenguva, Ashok & surya chandra, Podugu. . Door Unlocking System using Fingerprint Sensor for Home Automation(2020). 6. 2395–566.
15. Nicky Kattukkaran, Arun George, Mithun Haridas T.P, "Intelligent Accident Detection and Alert System for Emergency Medical Assistance ", International Conference on Computer Communication.
16. M. Yuchun, H. Yinghong, Z. Kun and L. Zhuang, "General Application Research on GSM Module," *2011 International Conference on Internet Computing and Information Services*, 2011, pp. 525–528, doi: 10.1109/ICICIS.2011.137.
17. M. Rahnema, "Overview of the GSM system and protocol architecture," in IEEE Communications Magazine, vol. 31, no. 4, pp. 92–100, April 1993, doi: 10.1109/35.210402.22
18. Accident Detection and Alert System T Kalyani, S Monika, B Naresh, Mahendra Vucha, International Journal of Innovative Technology and Exploring Engineering (IJITEE) ISSN: 2278-3075, Volume-8 Issue-4S2 March, 2019

Note: All the figures in this chapter were made by the Author

Human Machine Interaction in the Digital Era – Prof. J. Dhilipan et al. (eds)
© 2024 Taylor & Francis Group, London, ISBN 978-1-032-54998-9

E-Learning Technologies with Interactive Factors and Elements on the Perspective of Knowledge Management

7

C. Rajeshkumar[1]
Research Scholar, Department of MCA,
Bharath Institute of Higher Education and Research (BIHER), Chennai

K. Rajakumari[2]
Associate Professor, Department of CS,
Bharath Institute of Higher Education and Research (BIHER), Chennai

Abstract: E-Learning is used to connect the students and teachers in virtual manner. In this learning concept teaching can be done with interactive manner and mass range of the students can be reached with this learning technique. ELearning through eMail, Internet, World_Wide_Web (WWW), multimedia become more popular as the result developing technologies in digital and delivering higher education on interest of computerized manner. Questions like how confidently college instructors use technology in their instruction. After the pandemic situation E-Learning applications become very familiar to the students and the teachers. It makes the user to make the communication easier and handle the class in easy manner. Major features of E-Learning are a virtual classroom, adaptable designs and website layouts with learners in mind, Reminders, compliments, and additional pertinent push alerts a variety of educational resources, including images, movies, podcasts, etc. principles for holistic micro learning, online content that can be downloaded for future reviewing and different quiz formats. In this research paper, different types of E-Learning applications are compared to know the features and usages of that application in various approach manners.

Keyword: Digital, E-Learning and virtual classroom

1. Introduction

An interactive website that lets users input data and interact to receive desired outcomes on ELearning using webapp. From distant locations, students use internet connection for web browser to active participate on accessing the application. The above 90 percentage of students believing that the online learning consider as superior learning in traditional concept, according to SmallBizTrends. To keep the students engaged the online elearning community using elearning apps on web.

Web programs for e-learning are kept on the web server. Apps on web, are opposed to websites on elearning, let students are participating in interactive tasks including the test, writing an essay, giving a presentation, solving a math_problem, and more digitally connected on the classroom. One such example is Zoom, a video conferencing tool that enables users to host and join remote meetings. By doing this, teachers and students can engage in a live learning environment without entering the actual school building.

[1]crkrajrajesh@gmail.com, [2]rajakumari.mca@bharathuniv.ac.in

DOI: 10.1201/9781003428466-7

The major features of E-Learning are a virtual classroom, adaptable designs and website layouts with learners in mind, Reminders, compliments, and additional pertinent push alerts a variety of educational resources, including images, movies, podcasts, etc. principles for holistic micro learning, online content that can be downloaded for future reviewing and different quiz formats.

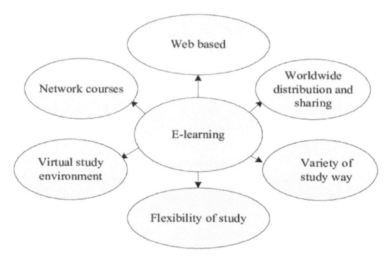

Fig. 7.1 Features of E-Learning

Source: Image is drawn using MS Word tools

The above Fig. 7.1 express the various features of E-Learning. In current situation after this pandemic period online class plays an important role in education and reaches the mass range of the students in urban and rural arrears. Most commonly used technologies among the users are Google Classroom, Kahoot and Zoom online platform.

2. Google Classroom

Google Apps is the newest product utilized in education, and Classroom helps teachers manage their time, organize their classes, and encourage student involvement. Teachers may quickly create and arrange assignments, provide feedback, and engage with their classes using Classroom. Students can use Classroom to organize their Google Drive work, finish and submit it, and interact with their teachers and peers in real time.

Fig. 7.2 User view of Google Classroom

Source: Image is taken from the official Google Class Room

Learning Method: Allows for efficient feedback and online teamwork. The social learning aspect of online learning are increased, enabling teachers to design collaborative digital learning activities and allowing students to benefit from their peers' skills and knowledge. *User Friendly:* Easy to set up, log in, receive, and submit assignments, quick and convenient.

Access: Google Classroom can be accessible with all mode of device such as PC, laptop, Tab and Mobile.

3. Kahoot

The learning process done in the manner of gamming terms which become the host of the students' personal learning experiences: Engage audiences to foster curiosity, and evaluate prior learning using word clouds, brainstorming, and open-ended question types.

Learning Method: It is a platform for game-based learning that may be used to check students' knowledge, revise their understanding, or simply give pupils a break from routine lessons.

User Friendly: Easy to set up, log in, receive, and submit assignments, quick and convenient.

Access: It can be accessible with all mode of device such as PC, laptop, Tab and Mobile.

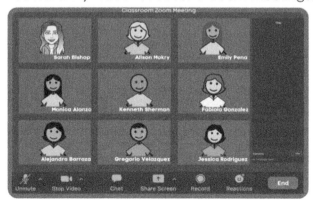

Fig. 7.3   User view of Kahoot

Source: Image is taken from the official Kahoot App

4. Zoom

Zoom is the video communication platform that provides a click-and-connect conferencing option. It can be used for group conversations, one-on-one office hours, and classroom instruction and learning.

Modified by Sarah Bishop

Fig. 7.4   User view of Zoom

Source: Image is taken from the official Zoom App

The framework of zoom is designed with android & Ionic of local applications. Android Studio, Android SDK, Visual Studio, Apple Code, and iOS SDK are among the essential tools. Swift and Objective C are utilized for iOS backends, whereas Kotlin and Java are used for Android apps.

Learning Method: Enables students to benefit from their peers expertise and skills by increasing the social learning component of online learning for the medical industry.

User Friendly: Easy to set up, log in, receive, and submit assignments, quick and convenient.

Access: It can be accessible with all mode of device such as PC, laptop, Tab and Mobile.

Table 7.1 Interactive factors of e-Learning technologies

E-Learning Technologies	Year	Access	Privacy	Prize	User Friendly
Google Classroom	2014	All Device	High	Free for limited user.	High
Kahoot	2013	All Device	Medium	Free for limited user.	Medium
Zoom	2011	All Device	High	Free for limited user.	Medium

Source: Described with the survey analysis of all the internet tools

5. Conclusion

The eLearning technologies play an important role in pandemic period for the teachers and students. This paper resolves the various approach used in learning technology for the user. Zoom technique most commonly used for scheduling the meeting among organization and students. In Kahoot technique, it mostly used among medical field to work on the live process of operations which invoke the gaming technology for the user. Organization of class in Google classroom will be user friendly, in single user login it allows multiple class to create and at a same time work can be assigned for each individual students separately. With this comparison of this three technology Google classroom is very user friendly among the user and easy to learn for the students and easy for the teachers to communicate with students.

Reference

1. H. Takabi, J. B. Joshi, and G.-J. Ahn, "Security and privacy challenges in cloud computing environments," IEEE Security & Privacy, no. 6, pp. 24–31, 2010
2. K. Ren, C. Wang, and Q. Wang, "Security challenges for the public cloud," IEEE Internet Computing, no. 1, pp. 69–73, 2012
3. Wen, X., et al, Comparison of open-source cloud management platforms: "OpenStack and OpenNebula. Fuzzy Systems and Knowledge Discovery (FSKD)," 2012 9th International Conference on. IEEE
4. E.Srimathi and Dr. SP.Chokkalingam, "OpenKey-Generation for Enabling Cloud Storage Security in Open Source Cloud Computing", JARDCS, Vol. 9. Sp-17/2017.
5. Yang Luo, Wu Luo, Tian Puyang, Qingni Shen, Anbang Ruan†, Zhonghai Wu, "OpenStack Security Modules: a Least-Invasive Access Control Framework for the Cloud,"2016 IEEE 9th International Conference on Cloud Computing
6. E.Srimathi and Dr. S.P. Chokkalingam "Securing Open Source Cloud Storage on OPenStack Cloud Computing Platform",IJRTE, Vol-7, April/2019.
7. Dr. SP.Chokkalingam and E.Srimathi "Efficiency on Public Cloud Storage Providers in Cloud Computing", IJRTE,Vol-7,April/2019.
8. Suryadipta Majumdar, Taous Madi, Yushun Wang, Yosr Jarraya, Makan Pourzandi, Lingyu Wang and Mourad Debbabi, "Security Compliance Auditing of Identity and Access
9. Secured Data Communication in Cloud Computing using Channel API with MD5 Hashing, ISSN: 2320-1363, journal of International Journal of Merging Technology and Advanced Research in Computing
10. Management in the Cloud: Application to OpenStack," 2015 IEEE 7th International Conference on Cloud Computing Technology and Science

Human Machine Interaction in the Digital Era – Prof. J. Dhilipan et al. (eds)
© 2024 Taylor & Francis Group, London, ISBN 978-1-032-54998-9

Expert System in Wearables and Web of Things Using Artificial Neural Network

8

P. Vichitra[1]
Research Scholar, Department of Computer Science,
School of Computing Sciences, VISTAS, Pallavaram, Chennai, India

S. Mangayarkarasi[2]
Associate professor, Department of Computer Science,
School of Computing Sciences, VISTAS, Pallavaram, Chennai, India

Abstract: Wearable and intelligent systems provide the impersonation that it's everything regarding the hardware, web, electric eye, interactive router and values, but the real time data will be in understanding. In this proposed work, we traverse both the intelligent retrieval and machine learning which get into a crucial tool for wisdom, perspective on AI and a hands-on experience on how to do artificial intelligence as a part of the work.

Keywords: Artificial intelligence, Web of things, Wearable, Electric eye, Data

1. Introduction

The web of things (WoT) arrived in the year 1990, the first thing was generated is a microwave, which was merged along with the network to authorize it with the faraway jurisdiction.

The web of things plays a vital role in the sector of interaction, arithmetic calculations, services with multiple usage of the realm which includes wholesomeness or fitness, manufacturing, farming and so on. In each sector, several researches have taken place and where creativities and machinery have taken everything to the next level due to the connectivity of the internet.

Wearable is connected to the network which is referred to as BAN (Body Area Network). WoT can be utilized to do tiny jobs in our day to day lives at home as well as in offices.

The Fig. 8.1 shows that, the future technology will be completely an AI, which creates the digital twin and the output will be used by the Internet of Things and after the interactions, it will be processed and generated into Big data and the flow continues.

In Manufacturing sectors, WoT will decrease the person's mistakes and increase the production of the output product. In Farming, WoT helps in the improvisation of entire food fecundity. In the area of fitness, the Internet of things helps the health industry in many ways. The fitness wearable will identify the particular health issues, detect the vitals and the experienced interconnected appliances.

---

[1]vichitra.vichi05@gmail.com, [2]drmangaiprabu@gmail.com

DOI: 10.1201/9781003428466-8

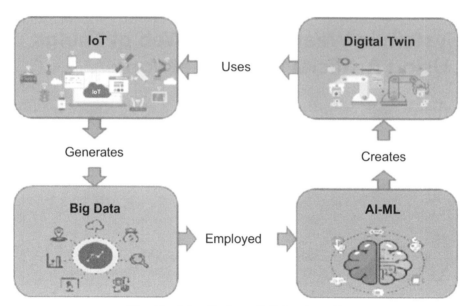

Fig. 8.1 Scientific flow of IoT, Big data, AI

Source: https://www.researchgate.net/figure/Relationship-between-IoT-big-data-AI-ML-and-digital-twins_fig4_349508978

2. Web of Things in India

Our Indian government is using the WoT as a portion of the programmed India. The Automated interactions strategy was initiated in the year 2018 to grow and appeal to the Internet of Things.

The unit of IT and voltaic both produced a Cheque rule for WoT. Our Indian government has set an aim of 15 billion dollars in the period of 2020.

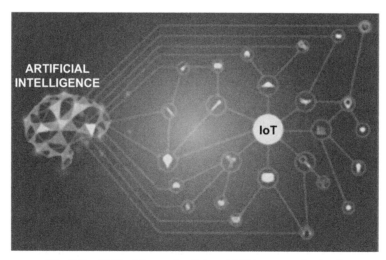

Fig. 8.2 AI (decision authority) and IoT (connect devices)

https://www.softwebsolutions.com/resources/advantages-of-ai-powered-iot.html

Fig. 8.2 states that, The Internet of Things will connect all the devices whereas the functioning of the devices, decision authority will be the Artificial Intelligence.

3. Drawbacks Faced in Web of Things (WoT)

People started losing their jobs because they were all replaced with the machineries. Robotization will automatically make a job mislaying.

Security and seclusion is impossible in this field. There will be a problem in vending private values without the end user's knowledge. Due to the online transactions lots of bank accounts were hacked and so on.

The directive and policy substructure is required to monitor the appliances connected with the internet.

4. Implementation and Result of Wearable Using Artificial Neural Network

The data collected using wearable are taken as A_1, A_2 and A_3. Heart rate will be taken as A_1, Respiratory count will be taken as A_2 and Blood pressure will be taken as A_3. The normal heart rate, respiratory count and blood pressure will be considered as value 0 and abnormal health rates are taken as 1. The data were received by using the smart watch.

Table 8.1 The values collected from the people by using smart watch

Person	A_1	A_2	A_3	Output
Person 1	0	1	1	0
Person 2	1	1	0	0
Person 3	1	0	0	-1
Person 4	1	0	1	0
Person 5	0	0	1	-1
Person 6	0	1	0	-1

Source: Collected the real time data from the people using the smart watch such as heartbeat, respiratory count and blood pressure as A1, A2 and A3.

$A_1 \to 0.3$

| INPUT |
| NODE |

$A_2 \to 0.3$

| INPUT |
| NODE |

$A_3 \to 0.3$

| INPUT |
| NODE |

The Strength between the units and output node is assigned as 0.3. The output node is ∫ and it is referred to as $Y^\wedge$. T is a bios factor, which is assigned to a value 0.4.

$Y^\wedge = 1\{0.3 A_1 + 0.3 A_2 + 0.3 A_3 - T > 0$
$\quad -1\{0.3 A_1 + 0.3 A_2 + 0.3 A_3 - T < 0$

$Y^\wedge = Sign(W_d A_d + W_d\text{-}1 A_d\text{-}1 + \ldots\ldots + W_2 A_2 + W_1 A_1 - t)$

$Y^\wedge = Sign(W_d A_d + W_d\text{-}1 A_d\text{-}1 + \ldots\ldots + W_2 A_2 + W_1 A_1 + W_0 A_0)$

Sign is the activation function. Assigning the values $W_0 = --T$, $A_0 = 1$;

Table 8.2 Obtained Result for $Y^\wedge$ after applying artificial neural network

A1	A2	A3	$Y^\wedge$
0	1	1	−1.2
1	1	0	−1.2
1	0	0	−0.8
1	0	1	−1.2
0	0	1	−0.8
0	1	0	−0.8

Source: The collected data is then evaluated and the result obtained is mentioned for $Y^\wedge$ after applying artificial neural network algorithm

When we have too much data for detection with accuracy the neural networks can be used. Table 8.2 states that, the person 1,2 and 4 is at stress by using the values collected. The persons 3,5 and 6 were in a cool state was detected by using the heart rate, respiratory count and blood pressure.

5. Conclusion

Wearable is an important constituent in the upcoming IT methodologies. Lots of drawbacks are unconsidered in accordance with assembling of data, data operations, safety and so on. The goal of this work was to provide a clear review of wearable in multiple fields. The play of artificial intelligence in improvement of wearables have been explored. In ensuing sessions, the AI in wearables to refine the standards of life.

References

1. M. Choi and J. Kim, "Electrical characteristics and signal transmission characteristics of hybrid structure yarns for smart wearable devices," *Fibers and Polymers*, vol. 17, no. 12, pp. 2055–2061, 2016.
2. H. Thapliyal, V. Khalus, and C. Labrado, "Stress detection and management: a survey of wearable smart health devices," *IEEE Consumer Electronics Magazine*, vol. 6, no. 4, pp. 64–69, 2017.
3. Y. Chang, J. Zuo, H. Zhang, and X. Duan, "State-of-the-art and recent developments in micro/nanoscale pressure sensors for smart wearable devices and health monitoring systems," *Nanotechnology and Precision Engineering*, vol. 3, no. 1, pp. 43–52, 2020.
4. M. Khatib and G. Ahmed, "Management of artificial intelligence enabled smart wearable devices for early diagnosis and continuous monitoring of CVDS," *International Journal of Innovative Technology and Exploring Engineering*, vol. 9, no. 1, pp. 1211–1215, 2019.
5. J. A. Naslund, K. A. Aschbrenner, and S. J. Bartels, "Wearable devices and smartphones for activity tracking among people with serious mental illness," *Mental Health and Physical Activity*, vol. 10, no. 8, pp. 10–17, 2016.
6. Narasimhan R, Parlikar T, Verghese G, McConnell M V. Finger-Wearable Blood Pressure Monitor. In: 2018 40th Annual International Conference of the IEEE Engineering in Medicine and Biology Society (EMBC). 2018. p. 3792–5.
7. Zhang G, Cottrell AC, Henry IC, McCombie DB. Assessment of pre-ejection period in ambulatory subjects using seismocardiogram in a wearable blood pressure monitor. In: 2016 38th annual international conference of the IEEE Engineering in Medicine and Biology Society (EMBC). 2016. p. 3386–9.
8. Dai M, Xiao X, Chen X, Lin H, Wu W, Chen S. A low-power and miniaturized electrocardiograph data collection system with smart textile electrodes for monitoring cardiac function. Australas Phys Eng Sci Med. 2016;39(4):1029–40.
9. Nakajo K, Takahashi S, Shiraishi Y, Komori Y, Motegi K, Miyashita H. Pressure transfer function for aorta model in cardiovascular simulator: Feasibility study of wearable central blood-pressure gauge. In: 2016 International Conference on Medical Engineering, Health Informatics and Technology (MediTec). 2016. p. 1–4.
10. Sasidharan P, Rajalakshmi T, Snehalatha U. Wearable cardiorespiratory monitoring device for heart attack prediction. In: 2019 International Conference on Communication and Signal Processing (ICCSP). 2019. p. 54–7. 10. Arif NH, Surantha N. IoT Cloud Platform Based on Asynchronous Processing for Reliable Multi-user Health Monitoring. In: Advances in Intelligent Systems and Computing. Springer Verlag; 2020. p. 317–30.
11. Prawiro EAPJ, Yeh C-I, Chou N-K, Lee M-W, Lin Y-H. Integrated wearable system for monitoring heart rate and step during physical activity. Mob Inf Syst. 2016;2016.
12. Huen D, Liu J, Lo B. An integrated wearable robot for tremor suppression with context aware sensing. In: 2016 IEEE 13th International conference on wearable and implantable body sensor networks (BSN). 2016. p. 312–7.

Human Machine Interaction in the Digital Era – Prof. J. Dhilipan et al. (eds)
© 2024 Taylor & Francis Group, London, ISBN 978-1-032-54998-9

Cognitive Based Learning in Feed Forward MLP for Classification of Neurodegenerative Disorders

9

B. Mahalakshmi[1]

Research Scholar, Department of Computer Science, Vels Institute of Science,
Technology & Advanced Studies (VISTAS)

A. Thirumurthi Raja

Research Supervisor & Assistant Professor, Department of Computer Science,
Vels Institute of Science, Technology & Advanced Studies (VISTAS)

Abstract Neurodegenerative illnesses are global epidemics that disrupt the lives of millions of people in every part of the world. These disorders are known as neurodegenerative diseases. There is currently no medicine that has been proved to be capable the progression of the disease itself. In this paper, we develop a cognitive based learning in Feed forward MLP (Multi-Layer Perceptron) for classification of neurodegenerative disorders. The dataset is split into training and test data, where the FFMLP model is trained using the pre-processed datasets. Then the trained MLP model is tested using the test dataset and finally the classified instances are depicted. The simulation is conducted to test the efficacy of the FFMLP model over various input images. The results of simulation show that the proposed method achieves higher accuracy, precision, recall and f-measure than existing methods.

Keywords Cognitive, Feed forward multi-layer perceptron, Classification, Neurodegenerative disorders

1. Introduction

Neurodegenerative disorders are a global epidemic that impacts the lives of millions of people in every region of the world. Neurodegenerative diseases are those that reveal themselves whenever neurons in the brain gradually lose their function and eventually die. These diseases are known as neurodegenerative diseases. There is currently no medication that has been shown to be capable of arresting the progression of the disease itself [1]-[5]. This is because there is no cure for the sickness. On the other hand, there is a possibility that there are methods that can help reduce some of the mental or physical anguish that is caused by neurodegenerative conditions. There is a wide variety of neurodegenerative illnesses, and each one is placed into a distinct category based on the area of the brain that is impacted by the death of nerve cells and the symptoms that is brought on by the condition [6].

In addition, if a correct evaluation of the severity of the patient disease is performed, the right drug can be given to the patient at the right time. Disagreements between a patient report and a physician assessment have the potential to cloud the true gravity of a disease prognosis. This is because both parties are biased in their perspectives [11]-[15]. At this point in time, a system supported by AI and which depends solely on gait data has the potential to deliver a more objective assessment at a lesser cost than the approaches that are now being used.

[1]runali.16@gmail.com, [2]dr.a.thirumurthiraja@gmail.com

DOI: 10.1201/9781003428466-9

In order to classify neurodegenerative diseases into appropriate categories, we have integrated into Feed forward MLP a learning technique that is cognitively based (Multi-Layer Perceptron). First thing that we do when we want to train the FFMLP model is take the dataset and divide it into the training set and the test set. After that, we put the training set to use. After the trained MLP model has been evaluated on the test dataset, the examples that have been detected are shown to the user.

2. Literature Review

One of the contributions made by Sadeeh et al. [16] was the demonstration of a sensor prototype for fall detection and prediction. In the event that a fall is anticipated, the patient would be warned to take measures, and the device would alert medical care in the event that a fall really takes place. Clouds have the ability to notify medical staff when an elderly patient has fallen, allowing the team to immediately assist the patient.

Researchers Du et al. [17] used longitudinal support vector regression (SVR) to determine whether or not demographic parameters may correctly predict the severity of ALS. They propose employing an innovative approach to machine learning in order to accomplish the task of collecting longitudinal data. In order to keep track of the participant behaviour throughout the length of the study, this technique calculates new weighted parameters at a variety of intervals throughout the research process. These include the characteristics of the SVR hyper plane as well as the parameters of the trend in time over a given period of time. Because of this, they were inspired to develop LSVR, which combines longitudinal modelling with the paradigms of machine learning, in order to satisfy a demand that was obvious in the business. In doing so, LSVR was able to mix longitudinal modelling with the paradigms of machine learning.

The mono triaxial accelerometer sensor that was developed by Sadeeh et al. [18] is a device that is worn on the thigh and is utilized to both anticipate and recognize instances of people falling. After that, the information is saved so that a medical professional can do a comprehensive follow-up on the patient. In the event that the patient has a fall, a notification will be transmitted over the internet to the relevant medical personnel. The patient will also be informed in the event that a fall is anticipated.

Benassar et al. [19] wanted to establish a system that was low-cost, objective, and automated for measuring the impairment of upper limb mobility caused by HD. They did this by utilizing techniques from machine learning and signal processing. This was supposed to be their end goal. They offer a method for evaluating the motor dysfunction of HD patients in an objective and continuous manner as the patients carry out a novel task involving their upper limbs. This method was developed by the researchers at the University of Washington. An impairment score that reflected the extent of mobility impairment was automatically constructed from the recorded accelerometer signals by applying methods from signal processing and machine learning. This score was derived from the obtained data. The data from the accelerometer were used to generate this score.

Gaßner et al. [22] decided to perform their investigation utilizing a sensor-based gait analysis because they wanted to objectively define the characteristics of the gait in HD patients. Using mobile sensor-based gait analysis, the purpose of this study was to compare and contrast the gait parameters of HD patients with those of controls who were the same age and were of the same gender. To determine whether or not these objective measures had any clinical relevance, in particular, we investigated the correlation between gait measurements and total motor score (TMS) and total functional capacity (FRC). The GRF data that was gathered from these walk sensors was fed into a machine learning model that was presented by Aşurolu et al. [24].

One of the services that they offered was regression analysis that was based on the numerical properties of the generated signal. They referred to the model that they used as a hybrid and gave it the name Locally Weighted Random Forest. Another one of the services that they offered was analysis of the correlation between two variables (LWRF). In specifically, they provide a computational method for assessing the severity of motor symptoms in patients who are diagnosed with Parkinson disease (PD).

3. Proposed Method

In this section, the framework illustrated in Figure 1 shows how the process of training and testing is carried out to classify well the neurodegenerative disorders using FFMLP.

3.1 FFMLP

An example of a neural network that employs the back-propagation algorithm for the purpose of supervised learning is the Fully-Fed Multilayer Perceptron, which is also sometimes referred to as the FFMLP MLP.

MLP functions most well with a three-layer design consisting of an input layer, a hidden layer or levels, and an output layer or layers. In addition to this, in this architecture, every neuron is connected to each and every neuron in the layer that lies below it. It has been argued that the MLP is successful when utilized in the context of solving non-linear issues.

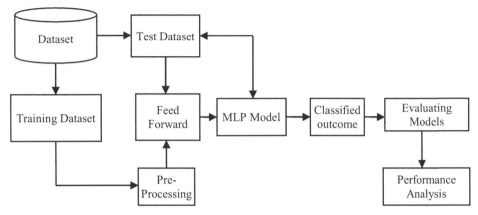

Fig. 9.1   Proposed cognitive ML framework

The MLP design can be utilized to determine the input variables, the bias values, and the output values, as shown by Equation (1):

$$S_i = \sum_{i=1}^{n} w_{ij} I_i + \beta_i \tag{1}$$

Where I - input layer, I_i - input variable i, n - Total inputs, β_j - bias value, and ω_{ij} - connections weight at level j.

In the multilayer perceptron, the sigmoid function, which can be obtained by solving Equation (2), is frequently utilized as an activation function.

$$f_j = \frac{1}{1 + e^{-S_j}} \tag{2}$$

Where, S - Activation function.

Equation (3) may be utilized to find out what the final action of neuron j will be:

$$y_i = f_i \left(\sum_{i=1}^{n} w_{ij} I_i + \beta_i \right) \tag{3}$$

In order to evaluate the efficacy of a model, it is necessary to compare the output of the model, which is denoted by y, to values that have been established. This is required in order to do so. 70% of the information was used to teach MLP, and then later on, the model the training dataset in a random order. During the time that the predictor model was being trained, we varied the total number of neurons that were present in the hidden layer within a four-point range. This allowed us to determine the architecture that produced the most accurate results for the model. Tan h(x) was chosen to act as the activation function in place of the other potential choices since it exhibited higher performance when compared to the alternatives.

3.2 Cognitive Learning

In order to facilitate supervised learning, the input format settings have been standardized at 335×335, the image resolution has also been standardized at 335×335, the add and sub trace maps have been labelled, and the training set accounts for 80% of the total data while the validation set accounts for 20% of the total data. When repeated on several occasions, the results of a set of examinations should remain unchanged. Because of this, it is essential to make use of seeds that produce the exact same random number each and every time. Because of this, it is much simpler to enhance the accuracy of models, which in turn makes it easier for them to be optimized.

A random value with a standard deviation of 0.05 is added to the weight, which results in the addition of an offset of 0.05 to the total. In order to improve the classification performance of the convolution neural network and to more closely match the data, this step is taken. Additionally, this contributes to an overall improvement in the appropriateness of the network. The RMS Prop optimizer and the gradient descent method developed by Momentum are the two approaches that are combined to form the Adaptive Moment Estimation methodology. This technique is built by merging the two methods. The configurations of the network are updated on a regular basis by an automated system that makes adjustments.

In order for training to converge on a superior performance setting with an initial learning rate of $1*10^{-4}$, Adam excellent computational efficiency and small memory demand are both extremely helpful. As a result of the linear non-saturation features possessed by the RELU function, the Sigmoid and tan functions converge at a significantly slower rate than the RELU function. By using RELU as the activation function of the convolution neural network during training, the gradient disappearance problem can be overcome in an effective manner. This improves the convolution neural network ability to sparsely express data, which ultimately results in a more successful resolution of the issue. The simulation was run for a total of 800 cycles, with 32 iterations making up each batch. The function known as softmax is what is utilized in order to map the input of neurons onto a scale ranging from 0 to 1.

The ideal objective function, often referred to as the loss function, should serve as the basis for the choice of the cross-entropy function that is most appropriate. The cross-entropy function is what used to determine the degree to which the two values differ from one another, and the softmax function is what responsible for computing the cross-entropy loss. Let make the assumption that, for the purpose of iteratively training the network, p is the probability distribution of the desired output, q is the distribution of the observed output, and H (p, q) is the cross entropy. This will allow us to say that q is the distribution of the observed output. Because of this, we will be able to assert that the cross entropy is H. (p, q). The following is the questionable formula:

$$H(p,q) = -\sum_x \left(p(x) \log q(x) + \left(1 - p(x) \log\left(1 - q(x)\right)\right)\right) \qquad (4)$$

4. Results and Discussions

In this section, the proposed FFMLP is validated at both training and testing phases over various other methods. The simulation is conducted in python that runs on a i5 core processor with 16GB RAM. The proposed model is tested in terms of accuracy, precision, recall, f-measure and classification loss during training and testing phase. In order to evaluate the performance of the neural network, we made use of the following measurements:

Sensitivity = (TP)/(TP + FN), Specificity = (TN)/(TP + FP), Accuracy = (TP + FP)/(TP + FP + TN + FN)

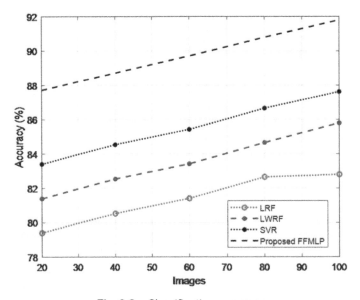

Fig. 9.2 Classification accuracy

Figure 9.2 shows the classification accuracy between the proposed FFMLP and existing LRF, LWRF and SVR. The results of simulation shows that the proposed FFMLP has higher rate of training classification accuracy than the existing classification models.

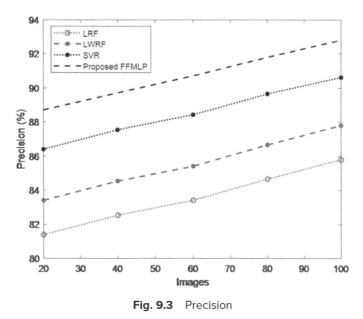

Fig. 9.3 Precision

Figure 9.3 shows the precision between the proposed FFMLP and existing LRF, LWRF and SVR. The results of simulation show that the proposed FFMLP has higher rate of precision than the existing classification models.

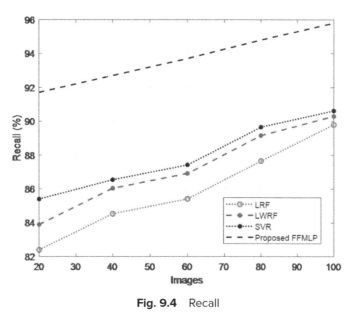

Fig. 9.4 Recall

Figure 9.4 shows the recall between the proposed FFMLP and existing LRF, LWRF and SVR. The results of simulation shows that the proposed FFMLP has higher rate of recall than the existing classification models.

Figure 9.5 shows the F-measure between the proposed FFMLP and existing LRF, LWRF and SVR. The results of simulation show that the proposed FFMLP has higher rate of f-measure than the existing classification models.

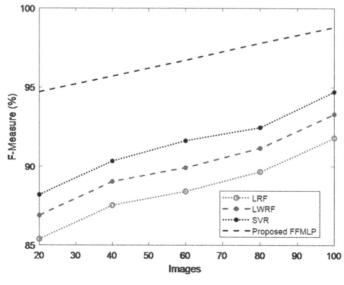

Fig. 9.5 F-Measure

5. Conclusion

In order to classify neurodegenerative diseases into appropriate categories, we have integrated into Feed forward MLP a learning technique that is cognitively based (Multi-Layer Perceptron). First thing that we do when we want to train the FFMLP model is take the dataset and divide it into the training set and the test set. After that, we put the training set to use. The examples that have been discovered are not displayed until after the trained MLP model has been evaluated on the test dataset. The goal of the simulation is to perform the evaluation necessary to determine how well the FFMLP model operates with the many different types of input photographs, and the purpose of the simulation is to do this. The results of simulations indicate that the proposed method outperforms solutions that are currently considered to be state-of-the-art in terms of accuracy, precision, recall, and f-measure.

References

1. Vieira, S., Pinaya, W. H., & Mechelli, A. (2017). Using deep learning to investigate the neuroimaging correlates of psychiatric and neurological disorders: Methods and applications. *Neuroscience & Biobehavioral Reviews, 74*, 58–75.
2. Noor, M. B. T., Zenia, N. Z., Kaiser, M. S., Mahmud, M., & Mamun, S. A. (2019, December). Detecting neurodegenerative disease from MRI: a brief review on a deep learning perspective. In *International conference on brain informatics* (pp. 115-125). Springer, Cham.
3. Beyrami, S. M. G., & Ghaderyan, P. (2020). A robust, cost-effective and non-invasive computer-aided method for diagnosis three types of neurodegenerative diseases with gait signal analysis. *Measurement, 156*, 107579.
4. Lins, A. J. C. C., Muniz, M. T. C., Garcia, A. N. M., Gomes, A. V., Cabral, R. M., & Bastos-Filho, C. J. (2017). Using artificial neural networks to select the parameters for the prognostic of mild cognitive impairment and dementia in elderly individuals. *Computer methods and programs in biomedicine, 152*, 93–104.
5. Lee, G. G., Huang, P. W., Xie, Y. R., & Pai, M. C. (2019, October). Classification of Alzheimer's disease, mild cognitive impairment, and cognitively normal based on neuropsychological data via supervised learning. In TENCON 2019-2019 IEEE Region 10 Conference (TENCON) (pp. 1808–1812). IEEE.
6. Duch, W. (2007). Computational models of dementia and neurological problems. *Neuroinformatics*, 305-336.
7. Sarić, R., Jokić, D., Beganović, N., Pokvić, L. G., & Badnjević, A. (2020). FPGA-based real-time epileptic seizure classification using Artificial Neural Network. *Biomedical Signal Processing and Control, 62*, 102106.
8. Berke Erdaş, Ç., Sümer, E., & Kibaroğlu, S. (2022). CNN-based severity prediction of neurodegenerative diseases using gait data. *Digital Health, 8*, 20552076221075147.
9. Lauraitis, A., Maskeliūnas, R., & Damaševičius, R. (2018). ANN and fuzzy logic based model to evaluate huntington disease symptoms. *Journal of Healthcare Engineering, 2018*.
10. Munteanu, C. R., Fernandez-Lozano, C., Abad, V. M., Fernández, S. P., Álvarez-Linera, J., Hernández-Tamames, J. A., & Pazos, A. (2015). Classification of mild cognitive impairment and Alzheimer's Disease with machine-learning techniques using 1H Magnetic Resonance Spectroscopy data. *Expert Systems with Applications, 42*(15-16), 6205–6214.

11. Pahuja, G., & Nagabhushan, T. N. (2021). A comparative study of existing machine learning approaches for Parkinson's disease detection. *IETE Journal of Research*, *67*(1), 4–14.

12. Ren, P., Tang, S., Fang, F., Luo, L., Xu, L., Bringas-Vega, M. L., ... & Valdes-Sosa, P. A. (2016). Gait rhythm fluctuation analysis for neurodegenerative diseases by empirical mode decomposition. *IEEE Transactions on Biomedical Engineering*, *64*(1), 52–60.

13. Ieracitano, C., Mammone, N., Bramanti, A., Marino, S., Hussain, A., & Morabito, F. C. (2019, July). A time-frequency based machine learning system for brain states classification via eeg signal processing. In *2019 International Joint Conference on Neural Networks (IJCNN)* (pp. 1–8). IEEE.

14. Ren, P., Zhao, W., Zhao, Z., Bringas-Vega, M. L., Valdes-Sosa, P. A., & Kendrick, K. M. (2015). Analysis of gait rhythm fluctuations for neurodegenerative diseases by phase synchronization and conditional entropy. *IEEE Transactions on Neural Systems and Rehabilitation Engineering*, *24*(2), 291–299.

15. Saadeh, W., Butt, S. A., & Altaf, M. A. B. (2019). A patient-specific single sensor IoT-based wearable fall prediction and detection system. *IEEE transactions on neural systems and rehabilitation engineering*, *27*(5), 995–1003.

16. Du, W., Cheung, H., Johnson, C. A., Goldberg, I., Thambisetty, M., & Becker, K. (2015, November). A longitudinal support vector regression for prediction of ALS score. In *2015 IEEE International Conference on Bioinformatics and Biomedicine (BIBM)* (pp. 1586–1590). IEEE.

17. Saadeh, W., Altaf, M. A. B., & Altaf, M. S. B. (2017, February). A high accuracy and low latency patient-specific wearable fall detection system. In *2017 IEEE EMBS International Conference on Biomedical & Health Informatics (BHI)* (pp. 441–444). IEEE.

Note: All the figures in this chapter were made by the Authors

Human Machine Interaction in the Digital Era – Prof. J. Dhilipan et al. (eds)
© *2024 Taylor & Francis Group, London, ISBN 978-1-032-54998-9*

Prediction of Lung Cancer Disease in Human Beings from Biomedical Text Dataset Using Transfer Learning

10

K. Jabir[1]

Research Scholar, Department of Computer Science,
Vels Institute of Science, Technology & Advanced Studies(VISTAS)

A. Thirumurthi Raja

Research Supervisor & Assistant Professor, Department of Computer Science,
Vels Institute of Science, Technology & Advanced Studies(VISTAS)

Abstract It has been determined that lung cancer is the key factor responsible for the increased cancer mortality rate seen by this generation. In order to successfully treat lung cancer, early diagnosis of symptoms is an imperative necessity. In order to establish a sustainable prototype model for the treatment of lung cancer that does not have a negative impact on the natural environment, the most recent technology breakthroughs, such as the Internet of Things and computational intelligence, may be applied. As a result, there will a less time and effort throughout the operation, and you will have access to a more expedient method of diagnosing lung cancer that will require the assistance of a lesser number of people. In this paper, we present a transfer learning-based algorithm to classify the lung cancer from the input real time data based on the training of the classifier by UCI datasets. The method involves pre-processing, feature extraction and classification using transfer learning. The simulation is conducted in python to test the efficacy of the model against UCI repository datasets. The results of simulation shows that the proposed method achieves higher classification accuracy, precision, recall and f-measure than the existing methods.

Keywords Lung cancer, Mortality rate, Transfer learning, Classification

1. Introduction

In cities where the population is growing at a rapid rate, it is more challenging to provide proper medical care to the residents. Raw data is being generated in the realm of healthcare in a continuous and steady stream, and its origins can be broken down into a huge number of different sources [1]. A large amount of information about patients is amassed as a direct result of the routine collecting of such information. The process of disease diagnosis, on the other hand, is significantly influenced by a number of factors, including as a shortage of knowledgeable people, delays in functions, and outmoded manual procedures [2].

In this kind of environment, making a diagnosis of a chronic illness can be an extremely difficult task. If medical professionals have a better understanding of the physiological and genetic aspects of their patients, they will be better equipped to diagnose patients who are afflicted with chronic diseases [3]. In this respect, computational intelligence can be of significant assistance in creating models for forecasting the risk of illness that take into consideration a broad variety of different parameters. This can be done more accurately and more reliably than ever before [4].

---
[1]jabu.jaf@gmail.com, [2]dr.a.thirumurthiraja@gmail.com

DOI: 10.1201/9781003428466-10

The method of disease diagnosis has benefited significantly from these innovative algorithms, which have brought about major advancements. Among these are the establishment of an accurate diagnosis at an early stage, the identification of risk factors, and the restricting and scheduling of healthcare unit visits in line with the requirements of the particular patient being treated [5]. An example of a computational intelligence model for the diagnosis of disease is the patient medical histories serve as the foundation for a model that might potentially provide predictive diagnoses of disease [6]. Disease risk factors and symptoms, as well as textual and image-based samples from the maintenance of healthcare records, are all examples of diverse sorts of medical data.

The ability to recognize patterns in previous data is one of the many applications for computational intelligence, which can subsequently be used to make forecasts about what will happen in the future [7]. Researching historical information allows one to achieve this goal. The integration of AI and IoT gives businesses the chance to perform predictive analysis on test cases, which, in turn, boosts their capability to automate processes [8].

The ultimate goal of artificial intelligence is to tackle challenging issues by conceiving of solutions that are viable. The possibilities present a combination of outcomes that are not as easily implementable as the others [10]. Within the context of this scenario, the search algorithm plays the role of the artificial intelligence agent. It conceals itself and works in the background, studying the many paths forward and pointing you in the right direction while remaining completely out of sight. Without these algorithms, an AI system cannot be able to decide which action to take that will yield the best results [11].

2. Related Works

Lung cancer, which is the most common form of the disease in humans, originates in the tissues of the organ and then spreads to other parts of the body. There is a correlation between the use of tobacco products and the development of about 85% of all lung cancer cases. Lung cancer is caused by the unchecked growth of abnormal cells in the lung, particularly in the region around the line of air division. This is the primary factor that contributes to the progression of the disease [12]. The chances of life for diabetic people are exceedingly unpredictable. According to the findings, those with diabetes who need a greater amount of insulin are at a greater risk for having complications than those who require a lower amount [13].

Diabetes mellitus has been hypothesized to have a part in the development of lung cancer. This is according to a study that was conducted by Mishra et al. Those who smoke cigarettes on a regular basis are at a greater risk of developing insulin resistance and have a poorer insulin sensitivity than those who don't smoke [14]. Stopping smoking is an absolute prerequisite for bringing one diabetes under control and lowering one vulnerability to complications associated with the disease [15]. By utilizing electronic health records (EHRs), patients can receive assistance with the administration of their individualized health plans. These records can also be utilized to coordinate the delivery of medical treatment to patients [16].

Previous studies that have been conducted on the treatment of lung cancer have made use of a wide array of computer tools for their analysis. Setio et al. [17] created a model of a 3D convolutional neural network (CNN) for the goal of FP minimization in the classification of lung nodules. This model was created for the aim of reducing the number of false positives in the process of identifying lung nodules. Using a multi-patches method in conjunction with the Fragni filter, Hongyanget al. [18] successfully develop a community-oriented pulmonary nodule identification model with improved capability for prediction.

Kattan and Bach [19] advocated doing a multifactoral study to analyze the phenomena of the observed shift in lung cancer risk anomaly among smokers. It was shown that young people who were addicted to smoking had a greater risk of developing lung cancer compared to young people who did not have an addiction to smoking. These young folks ranged in age from 18 to 28 years old, on average.

Knowledge mining methods were utilized by Ramachandran et al. [20] to construct a preventative model for a variety of disease risks, including hepatitis and lung cancer. Classifiers such as Bayes trees and decision trees were utilized by Thangaraju et al. [21] to make predictions regarding cardiovascular risk variables utilizing a large-sample heart disease dataset. This was done in order to make predictions regarding cardiovascular risk variables.

An exhaustive study on the detection and categorization of lung cancer using the application of computer prediction models. The majority of the existing models, in our opinion, make use of machine learning algorithms for classification that are on the more fundamental end of the spectrum. Even after carrying out a variety of processes for optimizing the qualities, the data samples still contain characteristics. There was an increase in the reliability of the lung cancer detection process. The examination of lung cancer does not typically make use of the transfer learning technique, which is another aspect of this disease.

3. Proposed Method

A Transfer learning classifier is applied with the intention of classifying patients who have lung cancer into the right category depending on the symptoms that they present with. Within the framework of the approach that has been proposed, the phase of pre-processing comes before the step of data collection. The chosen classifiers are then put through a training phase, followed by two independent testing phases on the benchmark dataset, which are carried out in accordance with the usual method of 10-fold cross-validation.

In order to find the screening approach that is most successful for lung cancer, the information is analyzed and broken down into its component parts. The general strategy that will be implemented may be seen depicted in given Fig. 10.1.

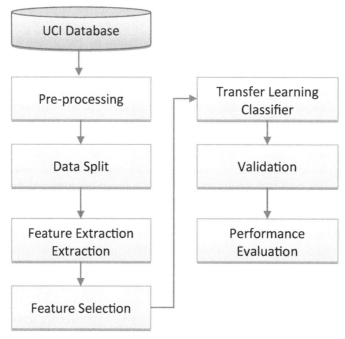

Fig. 10.1 Proposed design

The data that was supplied is shown here in the same order in which it was supplied in the original query. These inputs are dependent on a wide variety of data kinds, each of which, in turn, requires a significant number of inquiries to be conducted before the necessary information can be received. These preprocessing techniques are used to divide the entering information into more manageable bits before any additional processing is carried out on the data.

3.1 Data Acquisition

The lung cancer dataset is available online in a repository that is managed by the University of California, Irvine, and can be viewed through that platform. The dataset has a total of 32 samples, each of which has its very own unique collection of 57 attributes and a single class attribute. In total, there are 58 attributes and one class attribute. The fundamental reason behind our proposed line of inquiry is the ultimate goal of analyzing and comparing the efficacy of SVM and other methods that are comparable in nature. This purpose will be accomplished by conducting a series of tests.

3.2 Data Preprocessing

The first thing that is done in the process of identifying lung cancer is called preprocessing. During this phase of the process, the dataset will be finished by removing unnecessary data and completing missing values. Therefore, in order to improve the general trustworthiness of the entire dataset, missing values are inferred using a method termed nearest neighbor, paired with three other values that are located in close proximity. It is essential to make sure that you have training as well as testing samples.

3.3 Training and Testing Samples

A transfer learning, a form of artificial intelligence, is the principal instrument that is used both for training purposes and for testing the samples of input data after training has been completed. At the outset, the weights of the neural network are derived from the input data in a completely arbitrary fashion. After having been trained on a dataset that is intended to be an accurate representation of the whole, the neural networks are next put to the test using the same dataset. During the process of classification, data weighting is used to calculate the mistake rate as well as the error frequency; reweighting the dataset fixes the errors that were introduced by the initial weighting.

3.4 Feature Extraction

After a dataset has been segmented, it is possible to extract a large number of characteristics from the dataset, which makes the process of detecting lung cancer significantly simpler. This type of diagnosis makes it easier to surgically remove the lung tumor that was caused by the proliferation of cancer cells. The tumor was caused when the cancer cells multiplied. This type of diagnosis makes it possible to surgically remove the lung tumor that was caused by the proliferation of cancer cells. The tumor was brought on because of the multiplication of cancer cells.

The Ant Colony Optimization (ACO) is used to carry out this method of feature extraction so that the desired results can be achieved. For the purpose of classifying data, support vector machines (SVMs) are typically put to use, as this is standard procedure. The process of obtaining the principal qualities that are more significant and nonredundant from the input data is referred to as feature extraction, and it is a component of the pattern recognition algorithms that are used.

3.5 Classification Model

A classification model known as BERT, which is an abbreviation that stands for Bidirectional Encoder Representations from Transformers, has been pre-trained to construct word representations based on context that has been learned from both sides of the conversation.

The BERT consists of a Bertrange network with 24 layers has 1024 hidden units and 16 attention heads, whereas a Bertbase network with 12 levels only has 768 hidden units and 12 attention heads. A Bertrange network with 24 layers is referred to as a Bertrange network.

Six encoder layers, 512 hidden unit feed forward layers, and eight attention heads make up the base version. The sophistication level of the base variation is far higher than that of either of these variants. There are two distinct layers of abstraction present inside of each and every encoder.

The following formula is going to be used for the purpose of computing the output of the Z layer, which is of self-attention:

$$Z = \text{softmax}\left(\frac{QK^T}{\sqrt{d_k}}\right)V \tag{1}$$

The significant number of attention heads into the model makes it possible for the model to shift its focus and make use of a variety of representation subspaces, each of which possesses its own Z matrix. This is made possible by the fact that the model incorporates a large number of attention heads. This is made feasible by the fact that the model is able to flip between focusing on different things (12 for BASE and 24 for LARGE). The matrices that are sent from the attention heads are combined into a single matrix that is Z. This is necessary since the feedforward layer will only take a single Z matrix. A weight matrix known as W^O is applied to this Z matrix, and the result is multiplied.

$$Z = \text{LayerNorm}(X + Z) \tag{2}$$

The first encoder of a BERT stack provides a comprehensive look at the structure of the stack from the inside. The foundation of BERT training was the interdependence of two distinct obligations, each of which played a unique role. The first task is known as next sentence prediction (NSP), and it needs the input of two sentences into a model.

4. Results and Discussions

The dataset contains 32 samples with 57 features and a hypothetical range of 0–3 for all predictive parameters. The data was gathered from the machine learning repository at UCI. The process of simplifying data analysis, data transformation involves converting data from nominal attributes and class labels into binary form. In the field of data analysis, the approach that has shown to be the method that is the most often used and accepted technique is the transformation of nominal data into binary form.

The analysis of the data needs to be carried out with considerable caution because the dataset contains some unaccounted-for numbers. On a scale from severe to moderate to mild, the severity of the label can be scaled, with severe representing the most extreme case. The information that was supplied is deficient in a considerable number of essential pieces of data. Therefore, when cleaning up the data, it is recommended to replace any missing values with the value that appears in that column the majority of the time. This is the value that is most frequently observed.

The implementation analysis made use of a variety of different performance metrics, all of which are generated from the confusion matrix. One of the goals of the analysis was to determine how well the implementation was carried out. You may see the performance parameters that we used in our analysis stated here. The total amount of time needed to train and evaluate a computational intelligence model is referred to as the execution time delay.

Classifiers take historical data and convert it into a format that is more appropriate for being categorized, so enabling the data to be used. In order to put the classifier through its paces, a total of ten different cross-validation techniques are used. It is a strong type of data analysis that enables reliable predictions to be made using 10 times as much data as is possible with conventional techniques of approach. This is a significant improvement over the previous state of affairs, which was only possible with conventional approaches. When compared to how things were before, this is a major step in the right direction. The Fig. 10.2 to Fig. 10.5 shows the result obtained by performing several metrics.

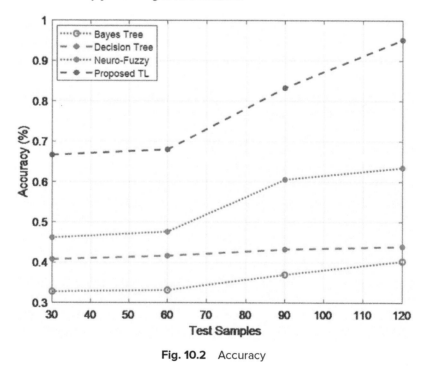

Fig. 10.2 Accuracy

- The accuracy rate is defined as the percentage of correct disease predictions in relation to the total number of forecasts, and it is shown. When compared to the total number of predictions, this results in a lower accuracy.
- The sensitivity of a decision in recognizing incidences of the disease in a population is referred to as its sensitivity, and the word is employed here.
- Specificity is referring to the ability of a prediction decision to correctly identify persons who do not pose a risk for a disease.
- F-score is a representation of the harmonic mean average of the values of specificity and sensitivity.

Every cancer dataset was taken into consideration when figuring out the ideal number of attributes to apply.In order to evaluate the effectiveness of the model, many distinct cancer datasets were utilized at various points in the process.

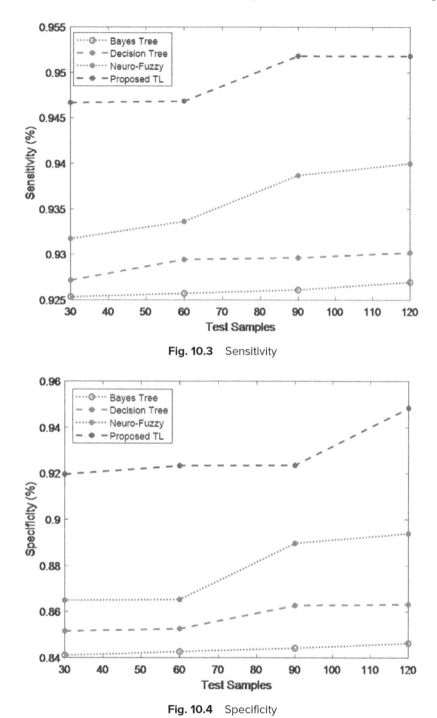

Fig. 10.3 Sensitivity

Fig. 10.4 Specificity

The percentage of accurate predictions that a forecast generates in comparison to the total number of forecasts that it creates can be used as a metric to determine how well accurate the classification accuracy of a forecast is. The results of the experiment will have an effect on the values of these variables, causing them to change in a manner that is appropriate to the new circumstances.

5. Conclusions

In this article, I present an approach for detecting lung cancer from real-time data input using a classifier that has been trained using data from the University of California, Irvine. The classifier was trained by using data from the University of California,

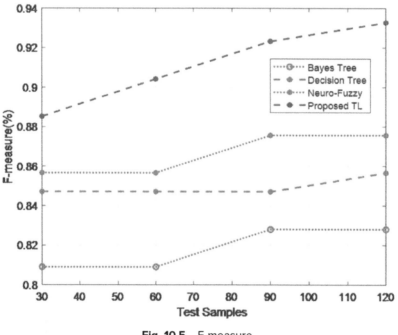

Fig. 10.5 F-measure

Irvine. This method, which utilizes transfer learning as its foundation and was developed as part of this research, is presented below. The procedure begins with the pre-processing of the data, then moves on to the extraction of features, and finally, the categorization based on transfer learning brings the whole thing to a close. By implementing the transfer learning classifier, it was possible to gain improved classification precision. One of the many performance measures that were utilized in order to evaluate the efficiency of the hybrid heuristic classifier model that was proposed was accuracy in classification. This was just one of the many performance measures that were utilized. The classification technique that was proposed was one that was based on transfer learning, and the results that it provided were consistent, for the most part, across all of the datasets.

References

1. Bogere, N., Bongomin, F., Katende, A., Omaido, B. A., Namukwaya, E., Mayanja-Kizza, H., &Walusansa, V. (2022). A 10-year retrospective study of lung cancer in Uganda. *BMC cancer, 22*(1), 1–8.
2. Patra, R. (2020, March). Prediction of lung cancer using machine learning classifier. In *International Conference on Computing Science, Communication and Security* (pp. 132–142). Springer, Singapore.
3. Pokkuluri, K. S., Usha Devi, N. S. S. S. N., &Mangalampalli, S. (2022). DLCP: A Robust Deep Learning with Non-linear CA Mechanism for Lung Cancer Prediction. In *Innovations in Computer Science and Engineering* (pp. 299-305). Springer, Singapore.
4. Kibudde, S., Kirenga, B. J., Nabwana, M., Okuku, F., Walusansa, V., & Orem, J. (2021). Clinical profile and initial treatment of non-small cell lung cancer: a retrospective cohort study at the Uganda Cancer Institute. *African Health Sciences, 21*(4), 1739–45.
5. Prasad, S. K., Johnson, A., & Kumar, S. M. (2021). Lung Cancer Detection with Prediction Employing Machine Learning Algorithms: A Recent Study. *New Approaches in Engineering Research Vol. 14*, 99–109.
6. Dritsas, E., &Trigka, M. (2022). Lung Cancer Risk Prediction with Machine Learning Models. *Big Data and Cognitive Computing, 6*(4), 139.
7. Varsini, V. R., &Mohanasundari, M. (2021). Lung Cancer Prediction and Classification Using Recurrent Neural Network. *International Journal of Research in Engineering, Science and Management, 4*(11), 8–10.
8. Manju, B. R., Athira, V., & Rajendran, A. (2021). Efficient multi-level lung cancer prediction model using support vector machine classifier. In *IOP Conference Series: Materials Science and Engineering* (Vol. 1012, No. 1, p. 012034). IOP Publishing.
9. Tejaswini, C., Nagabushanam, P., Rajasegaran, P., Johnson, P. R., & Radha, S. (2022, April). CNN Architecture for Lung Cancer Detection. In *2022 IEEE 11th International Conference on Communication Systems and Network Technologies (CSNT)* (pp. 346–350). IEEE.
10. Anil Kumar, C., Harish, S., Ravi, P., Svn, M., Kumar, B. P., Mohanavel, V., ... & Asfaw, A. K. (2022). Lung Cancer Prediction from Text Datasets Using Machine Learning. *BioMed Research International, 2022.*

11. Anil Kumar, C., Harish, S., Ravi, P., Svn, M., Kumar, B. P., Mohanavel, V., ... & Asfaw, A. K. (2022). Lung Cancer Prediction from Text Datasets Using Machine Learning. *BioMed Research International*, *2022*.

12. Manikandan, R., Sara, S. B. V., Yuvaraj, N., Chaturvedi, A., Priscila, S. S., & Ramkumar, M. (2022, May). Sequential pattern mining on chemical bonding database in the bioinformatics field. In AIP Conference Proceedings (Vol. 2393, No. 1, p. 020050). AIP Publishing LLC.

13. Mishra, S., Mallick, P. K., Jena, L., & Chae, G. S. (2020). Optimization of skewed data using sampling-based preprocessing approach. *Frontiers in Public Health*, *8*, 274.

14. Mishra, S., Tripathy, H. K., Mallick, P. K., Bhoi, A. K., &Barsocchi, P. (2020). EAGA-MLP—an enhanced and adaptive hybrid classification model for diabetes diagnosis. *Sensors*, *20*(14), 4036.

15. Mishra, S., Mallick, P. K., Tripathy, H. K., Bhoi, A. K., & González-Briones, A. (2020). Performance evaluation of a proposed machine learning model for chronic disease datasets using an integrated attribute evaluator and an improved decision tree classifier. *Applied Sciences*, *10*(22), 8137.

16. Ahmad, T., & Chen, H. (2020). A review on machine learning forecasting growth trends and their real-time applications in different energy systems. *Sustainable Cities and Society*, *54*, 102010.

Note: All the figures in this chapter were made by the Authors

Human Machine Interaction in the Digital Era – Prof. J. Dhilipan et al. (eds)
© 2024 Taylor & Francis Group, London, ISBN 978-1-032-54998-9

An End-to-End Machine Learning Model to Predict the First Innings Score in IPL

11

Helen Josephine V.L.[1]
Associate Professor, Business Analytics, School of Business and Management,
Christ University, Bengaluru, India

Mansurali A.[2]
Assistant Professor, Business Analytics, School of Business and Management,
Christ University, Bengaluru, India

Aishwarya Sathyan[3]**, Sujay S. Hedge**[4]
MBA Student, Business Analytics, School of Business and Management,
Christ University, Bengaluru, India

Abstract Sports analytics is a very popular segment. Every modern day sport uses a form of analytics to analyze either the player performance, training effect, match statistics or match result predictability. As huge volumes of data is being generated due to Indian premier league (IPL), match result predicting websites like Dream11, CricBuzz, etc. have gained popularity. Such companies have shifted from basic player analysis to accurate match result prediction. This paper deals with important game factors to be considered to develop a match result predicting system. We employ popular machine learning techniques like linear regression, random forest, ridge modeling and others to develop such a system. The machine learning model has been tested with game conducted in 2021 April between Royal Challengers Bangalore and Mumbai Indians. The proposed model predicts a score range of 152 - 167 when the actual first innings score was 159, hence the model is capable of making the closest prediction with the help of lasso regression.

Keywords Regression model, Machine learning regularization, IPL prediction, Lasso regression, Ridge regression, Random forest, Linear regression

1. Introduction

The Twenty20 cricket league known as the Indian Premier League (IPL) is played in India. Every year, it is often performed in April and May. The game's title sponsor as of 2021 is Vivo, and it is known as the Vivo IPL 2021. Board of Control for Cricket India (BCCI) established the league in 2008. The shortest variation of cricket is called T20.   It gets completed in 3-4 hours. There are two innings one for each team. Each innings consists of 20 overs (6 legal balls to bowl) in an over. T20 version has gained popularity in the last 15 years, thanks to India's inaugural T20 world cup win in 2007, which led to the formation of IPL. The data can be scraped and transformed from Cricsheet.org and IPL T20 - Official Website or espncricinfo.com, the above-mentioned sites have ball-by-ball details and it contained summary of all the matches by season and time. Hence the data from 2008 till 2021 can be scraped from the same.

[1]helen.josephine@christuniversity.in, [2]mansuralia@christuniversity.in, [3]aishwaryas@mba.christuniversity.in, [4]sujay.hegde@mba.christuniversity.in

DOI: 10.1201/9781003428466-11

The IPL event has had 14 seasons, and owing to the COVID-19 epidemic, the 2020 season will be played in the UAE. The IPL game can significantly increase the GDP because it manages to increase scale, branding, fan support, and tremendous fan following globally, which has led to a continuous increase in the GDP of the nation. The survey report indicates that the tournament has brought in close to 11.5 billion USD, and the BCCI has appointed the KPMG sports advisory group to handle the surveys. It has also made a major financial contribution to the nation's tourist industry, increasing travel-related earnings by roughly 30%. India especially is ranked 34th out of the world's most appealing tourism destinations and 11th in the Asia-Pacific area. Additionally, it generates a significant number of employment possibilities, which has helped the Indian economy develop successfully.

2. Literature Review

Cricket is a dynamic sport that evolves ball by ball. The goal of the study in this paper is to look into machine learning technologies that can be used to the issue of forecasting cricket match outcomes based on historical match data. [1]. In order to create predictive models, factors are discovered using filter-based techniques. Once the features were chosen, the proper machine learning algorithm was then used. In comparison, tree-based models like random forest performed better in terms of accuracy, precision, and recall metrics.

The IPL is a team game and not an individual sport, which makes the winning chances to be dependent on each and every factor of all the players within the team. This study has mainly focused upon building a robust prediction model, which cab make it easier to classify players and help in predicting the best team that has the highest chances of winning the match, in this study modelling algorithms like decision tree, support vector machine and random forest are used in order to make the predictions for the same[2]. The study showed 95.78% accuracy for random forest model, which showed that the model is efficient in predicting the right selection of players to the team.

When modelling the dataset, a multivariate regression-based solution is proposed to calculate the points for each team member and the overall weight of the team based on past performance. A cricket match is dependent on a number of factors, and these ultimately have a significant impact on how it turns out. Players' performance has a relatively huge impact. Lamsal et al. identified 7 factors and trained the models for outcome prediction; it was observed that nearly three of the trained models demonstrated a correct prediction with multilayer perception and an accuracy of 72%; it was also observed that the multilayer perception classifier outperformed the other classifiers by correctly predicting 43 out of 60 matches[3]. The analysis took into account the five characteristics of an IPL career and the five characteristics of an international 120 career for both bowlers and batsmen. Deep et al. gave idea to construct mathematical model to make the prediction of match results. Future research opportunities will be demonstrated by using more work features that can be created and taken into account, more data that is generated and gathered, better model learning and classification techniques, and various other types of data analytics techniques[4].

This paper widely discusses the winner prediction using machine learning and data analytics. In order to test and train the data set of various sizes, different machine learning algorithms have been applied. Some of the algorithms are random forest SVM, naive bayes logistic regression, and decision tree. The paper researchers various features that can be used to analyse and predict the match winner of the game. The cricketing boards can use this to assess team strength and conduct a more thorough examination of cricket. This may be applied to gambling and media coverage of sporting events, which is a blessing in disguise. Mago Vistro et al.using a decision tree model, the model's performance was enhanced when the parameters were adjusted with an accuracy of 76%. The model's performance was improved from 76% to 94 percentage, and positive results were obtained[5]. The study finds that in order to produce superior analyses, a wide range of data must be analysed, the computation must run quickly, and it must satisfy business systems and problems.

3. Data and Variables

3.1 Study Period and Sample

The data from various sites like cricbuzz, cricsheet, Indian Premier league cricket from 2008 to 2020, etc. can be gathered, which can include two forms of data. This gives information regarding the matches and the ball-by-ball information. The matches dataset will contain information with regard to the location, teams, umpires and results over the period of time, ball-by-ball dataset will have the information on the batting and bowling teams, batsman, bowler, non-striker, striker, runs scored, wickets which can be scraped and collated from the respective online resources, which can be later converted to useable format.

This data is used to build a model which provides prediction of the first innings score of the batting team in IPL[5]. To identify the important areas of data comprehension, model building, data dependency, exploration of the data relationships, data descriptions, data dictionary metadata and data linkage.

4. Methodology and Findings

4.1 Problem Statement

Today, Indian Premier league means version 2.0 of the cricket, and the level of game has only risen with time the bar is so high that any other cricket league seems to be at a domestic level, with huge fan following, excellent involvement in social media to being one of the world's most profitable leagues. The main factors that lead to winning of a team is what every stakeholder wish they knew, since a lot of money is invested on the games during auctions, hence a model can be built in order to identify the most significant factors in determining the winning probability of a team and the factors can be focused upon by the stakeholders while making business decisions regarding the same.

4.2 Exploratory Data Analysis

As a part of the EDA, the summary statistics are checked, the absence of a nan value does not imply that our data is accurate. Below is a scatter plot, which is plotted against all the teams runs that has played in the IPL right from the first season up to recently in 2021, along with the overs, since the correlation matrix helped identify that the overs and the runs of the data are highly correlated, an analysis of the same via a scatter plot has been as shown below.

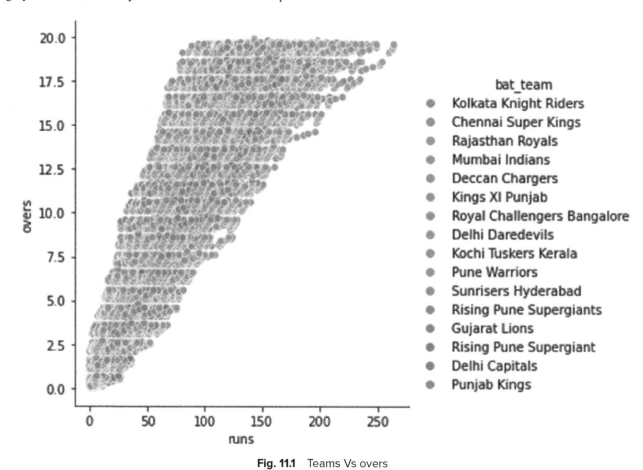

Fig. 11.1 Teams Vs overs

4.3 Data Preparation

In this project for Data Cleaning several steps were carried out because the data was huge consisting of ball-by-ball statistics of all the matches from 2008 2021 IPL hence the information gathered had to be properly formatted in order to make it ready for analysis hence the Data Cleaning included the following steps:

On checking the information on the number of rows and columns in the data set before cleaning the data it was found to be 16 columns and 103881 rows. The data has to be checked for any form of null values which was checked with a function is null and it was found to be 0 hence the data did not contain any null values. Now the data had some columns that were not necessary for the analysis hence unwanted columns like mid, season, venue, batsman, bowler, striker and non-striker were removed to go ahead with further analysis. Once the data was cleaned off of the unnecessary columns the number of columns reduced to 9. Now to go ahead with the further analysis the most consistent teams of IPL had to be identified which was done using the unique function. Since the first 5 overs information had to be collected during the deployment of web app the first 5 overs column had to be removed from the data set. Now the date column required to be converted to object format since it was in string format. All of these cleaning the data was check for duplicate values and null values once again. Finally, the correlation Matrix was generated which revealed a high correlation between the runs and the overs of the dataset.

4.4 Modeling and Evaluation

Feature importance had to be identified to start off with building the model. Extratreesregressor was used as a meta estimator in order to fit a number of randomised decision trees on various some samples within the data set and utilise the averaging form in order to maintain a better accuracy on prediction and to keep the overfitting under control. In this project 4 different algorithms are deployed as a part of model building which is linear regression, lasso regression, random forest and ridge regression[6].

4.5 Linear Regression

Linear Regression model is built using the Sklearn and the prediction is made and matrices MSE, RMSE and MAE are used to identify the fitness of the model and the score of the model is also predicted in order to understand the model better and below is a plot of the prediction.

4.6 Random Forest

The random forest model is also built using the Sklearn and the random forest regressor is used in order to make the regression happen and fit the multiple decision trees randomly on the extracted subsets and to average the prediction that is made across the test data[7].

4.7 Ridge Regression

The ridge regression is also performed by using scale, randomised searchCV and gridsearch CV are used to fit the parameters and to identify the lambda value and the negative means squared error is used to identify the score of the algorithm and below is density graph of the same.

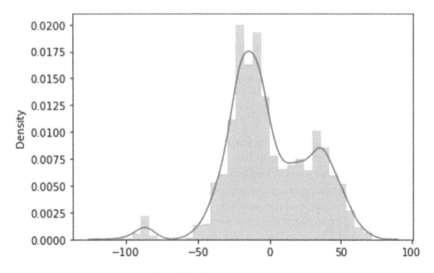

Fig. 11.2   Ridge regression

4.8 Lasso Regression

The lasso regression algorithm is also built using the Sklearn and the gridsearchCV to identify the Alpha value and the negative means squared error is used to get the score of the algorithm and the model is predicted using the laws of regression and below is density graph of the same[8].

The respective models are built by importing the models from Sklearn. And in order to interpret the best fit model for the prediction of this project several metrics can be utilised however three different metrics which is MSE, MAE and RMSE are used in order to pick the best algorithm to make the prediction.

(a) MAE: Also known as the Mean Absolute Error, is the value the denotes the average of the absolute difference between the actual values and the predicted values within the dataset, and it is used to measure the average of all the residuals within the dataset.

(b) MSE: Also known as the Mean Squared Error, is the value the denotes the average of the squared difference between the original values and the predicted values within the dataset, and it is used to measure the variance of all the residuals within the dataset.

(c) RMSE: Also known as the root mean squared error is the square root of mean squared error, it is something that measures the standard deviation of all the residuals within the dataset.

Table 11.1 Model Metrics Comparison

Model	MAE	MSE	RMSE
Linear Regression	24.68133723	910.8728591	30.18067029
Random Forest	25.86872274	1053.227113	32.45346072
Ridge	24.67406353	910.4989473	30.1744751
Lasso	**23.80395399**	**863.3588529**	**29.38296876**

From the above Table 11.1 for comparison, we can see that the lowest values for MAE, MSE and RMSE are for the LASSO REGRESSION, hence the lasso regression model is used to make the first innings score prediction in this project.

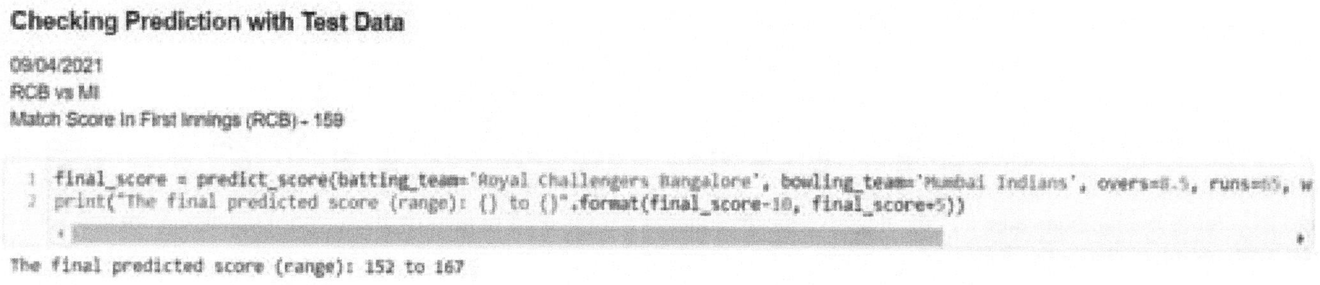

Checking Prediction with Test Data

09/04/2021
RCB vs MI
Match Score in First innings (RCB) - 159

```
1 final_score = predict_score(batting_team='Royal Challengers Bangalore', bowling_team='Mumbai Indians', overs=5.5, runs=55, w
2 print("The final predicted score (range): {} to {}".format(final_score-10, final_score+5))
```

```
The final predicted score (range): 152 to 167
```

Fig. 11.3 Lasso model testing

The above snapshot shows the model predicting the scores of one of the games conducted in 2021 April between Royal Challengers Bangalore and Mumbai Indians, the model predicts a score range of 152 - 167 when the actual first innings score was 159, hence the model is capable of making the closest prediction with the help of lasso regression.

5. Conclusion

Lasso regression was found to be the model that had the highest accuracy among the models and it did a good job in making the future prediction effective, appropriate and close to the range. The factors like runs, wickets, overs and the batting and the bowling team, along with the total runs plays a major role in predicting the score of the first innings. Factors like runs and wickets in the last 5 overs matter the most in case of prediction as the game's fate is almost sealed after the completion of the first 5 overs, hence the state of the game in the previous 5 overs play a major role. The venue of the match, was not considered in building the model; however, its significance is not identified, hence this aspect opens window for future research to explore more factors in predicting the score of the match. IPL has 8 consistent teams based on the no. of games played, being an established franchise and the data proves their consistency based on the average score of these teams as well. A crucial aspect of cricket is the use of machine learning algorithms to forecast the results of a match. By analysing player performance data and other related information, models that predict the winning team can be developed. The worth of a victory manifests itself in a variety of ways, including trickle-down effects to the fans, broadcast contracts, fan store items, parking, concessions, sponsorships, enrolment, and retention.

References

1. Kapadia, K., Abdel-Jaber, H., Thabtah, F., & Hadi, W. (2020). Sport analytics for cricket game results using machine learning: An experimental study. Applied Computing and Informatics, ahead-of-print(ahead-of-print).
2. Balasundaram, A., Ashokkumar, S., Jayashree, D., & Magesh Kumar, S. (2020). Data mining based Classification of Players in Game of Cricket. Proceedings - International Conference on Smart Electronics and Communication, ICOSEC 2020, 271–275. https://doi.org/10.1109/ICOSEC49089.2020.9215413
3. Lamsal, R., & Choudhary, A. (2018). Predicting Outcome of Indian Premier League (IPL) Matches Using Machine Learning. http://arxiv.org/abs/1809.09813
4. Deep, C., Patvardhan, C., & Singh, S. (2016). A new Machine Learning based Deep Performance Index for Ranking IPL T20 Cricketers. International Journal of Computer Applications, 137(10), 42–49. https://doi.org/10.5120/ijca2016908903
5. Rushe, S. (2009-2022). Cricsheet. Retrieved from Cricsheet: https://cricsheet.org/matches/
6. Basit, A., Alvi, M. B., Jaskani, F. H., Alvi, M., Memon, K. H., & Shah, R. A. (2020). ICC T20 Cricket World Cup 2020 Winner Prediction Using Machine Learning Techniques. 2020 IEEE 23rd International Multitopic Conference (INMIC), 1–6. https://doi.org/10.1109/INMIC50486.2020.9318077
7. Ishi, M. S., & Patil, J. B. (2021). A Study on Machine Learning Methods Used for Team Formation and Winner Prediction in Cricket. Lecture Notes in Networks and Systems, 173 LNNS, 143–156. https://doi.org/10.1007/978-981-33-4305-4_12
8. Jain, P. K., Quamer, W., & Pamula, R. (2021). Sports result prediction using data mining techniques in comparison with base line model. OPSEARCH, 58(1), 54–70. https://doi.org/10.1007/s12597-020-00470-9.

Note: All the figures in this chapter were made by the Author

Human Machine Interaction in the Digital Era – Prof. J. Dhilipan et al. (eds)
© 2024 Taylor & Francis Group, London, ISBN 978-1-032-54998-9

Real Time Intrusion Detection Network Using Artificial Neural Network

12

Deepa M.[1]
Research Scholar, Department of Computer Science and Applications,
SRMIST, Ramapuram Campus, Chennai, India

J. Dhilipan[2]
Professor & Head, Department of Computer Science and Applications,
SRMIST, Ramapuram Campus, Chennai, India

Abstract The fortification of networks and databases has become increasingly dynamic in recent years, and intrusion detection systems are essential to this process[1][2]. An IDS can provide real-time statistics on network traffic in the event that an attack is identified, inform users, and/or halt doubtful activity. The indefinite features continue to limit the effectiveness and accuracy of ML-based IDS, despite the fact that they have been shown to be efficient at monitoring real-time traffic.

The challenge with machine learning is how the system becomes accustomed to experiencing accumulation to automatically develop performance and is compatible with the idea of detecting the attacks by making the machine learning against intrusion to increase the rate of detection and to decrease the false positive rate[2].

A standard feature set may not be suitable for identifying different forms of intrusion attacks since some attributes may be unnecessary or unrelated and slow down the functionality of machine learning. In order to increase a detection system's accuracy, it is important to research the ideal qualities. If an attack is discovered, an IDS can offer real-time statistics on network traffic and immediately alert users or stop questionable activity. The accuracy and effectiveness of ML-based IDS are still hampered by their unspecified features, despite the fact that they have confirmed strength in real-time traffic monitoring. The principal task is to focus on topology and hyperparameter configurations that are near to the ideal configuration, which are used to identify the smallest topologies that require the least computational resources.

Keywords Intrusion attack, Security, Machine learning, Neural network

1. Introduction

More defence systems have been created and implemented, including firewalls, antiviruses, and malware detection software. However, it has been demonstrated that these solutions fall short of providing complete protection against attacks in the current network environments. Intrusion detection systems have become increasingly important for ensuring network and database security in recent years. Network intrusion detection [3]systems have been created using ML approaches in a variety of ways (NIDSs).

[1]dm8027@srmist.edu.in, [2]hod.mca.rmp@srmist.edu.in

DOI: 10.1201/9781003428466-12

Real-time statistics can be provided for intrusion detection system in the event that an attack is identified, inform users, and/or halt suspicious activity. Although ML-based IDS have demonstrated success in monitoring actual time traffic, their effectiveness and accuracy are constrained by the inaccurate characteristics.

The variety of cybersecurity infractions has sparked study into a wide spectrum of different detection techniques. The working of signature-based intrusion detection systems (IDS) relies on a database of known assaults, but other approaches like anomaly-based approaches create a model of "normal" traffic and raises an alarm if the model deviates from it.

Each and every automated machine learning system has hyperparameters, and the most fundamental objective is to automatically set these hyperparameters to maximize performance. Particularly contemporary deep neural networks heavily rely on a variety of hyperparameter decisions regarding the architecture, regularisation, and optimization of the neural network. The use cases for automated hyperparameter optimization (HPO) are numerous and significant.

Often, research on the use of neural networks to intrusion detection presents setups for hyperparameters and topologies that are picked at random, oblivious to the significant impact that these parameters might have on the accuracy that is attained. The initial goal of this research is to draw attention to the potential that exists in frequently untested situations. Second, the ideal configuration benchmarks that are applied for detecting the intrusions is desired. Then to locate the smallest topologies that demand the minimum resources, topology and hyperparameter mixtures that are near to the optimal arrangement are identified. A strategy for hyperparameter tuning is provided and used to accomplish the aforementioned goals. Grid search is a method e producing maximum candidates for the setup of a network using neurons and ranking it in accordance with one selected scoring function while searching through hyperparameter space. The results of the function is collected, scheduling the results with different hyperparameter combination has been designed for a given topology. The results that are obtained highlight the optimized topology and hyperparameters for data in the security domain. Grid search was employed in this study because of its transparency and ability to manually direct tests to the most promising areas.

2. Related Work

An artificial neural network is made up of a different layers of neurons, which are known as units for processing, that are coordinated and connected in the desired topology (ANN). A wide range of data-intensive applications have effectively been applied using ANN. Because there are so many possible attack vectors, cybersecurity is a very broad subject[3][4].

In Pruning deletion of input or hidden layers of neural nodes takes place. As a result, the ANN runs more quickly because less computations need to be performed. In [5],by using a feed-forward ANN, authors have tried to give solutions for the problems like overfitting, excessive memory usage. A two-layered neural network topology was suggested. In particular, a 2-layered neural network with neurons based configuration was suggested. A training function and corresponding dataset were used to address the aforementioned issues. According to the authors, although having less computing cost, their method produces outcomes comparable to those of conventional methods. KDD'99 dataset was used and the technique was tested. Less data is preferable, according to the paper's findings, because the machine needs more time to process it. The method was examined using the benchmark KDD'99 dataset. In [6], according to the paper's conclusions, data for testing should be less as it takes the machine longer to process. In [7], as a feature extractor Principal component analysis is used before the data is sent for training and testing the data. Thereby considerably reducing the memory needs of the approach and the amount of training time required, as the paper demonstrates. In terms of accuracy, the two assessed approaches produced comparable results. This makes using PCA the obvious better choice. The authors of [8] compare three types of efficient algorithms - Support Vector Machine, Naive Bayes, and C4.5 algorithm with a neural network implemented using single hidden layer. The 3 layered network framework's simple configuration allows the network to perform malware detection at a level that is comparable to or higher than other examined techniques while utilizing fewer calculations. The tests were conducted using the NSL-KDD dataset, the most recent KDD'99's replacement.

The authors of [9] want to discover a solution by tackling the issue of large dimensionality. To do this, they have combined two types of networks: deep belief networks (DBN) and Feature-weighted support vector machines (SVM) (WSVM). The machine is trained using a feature extractor known as an adjustable learning rate. A WSVN that has been optimized for particle swarms is then fed the characteristics. The solution was tested on NSL-KDD, and its binary classification accuracy was 85.73%.

3. Proposed System

The tool known as artificial neural networks (ANN) has several applications. Natural language processing, biometrics, calculating the roots of polynomials, and detecting attacks[10] are the various fields that are spanned by this methodology.

An network which uses neurons imbibes the learning capacities of a typical natural neural network that is seen in brains of human beings while being extremely effective . There are input , hidden layers and an output layer in a neural network with neurons present in each layer. Through the first layer which is the input layer, data enters the network which are connected by different neurons, where it is refined inside the hidden layers before being available for retrieval in the last layer which is an output layer. The following diagram depicts an normal neural network model with all the three layers:

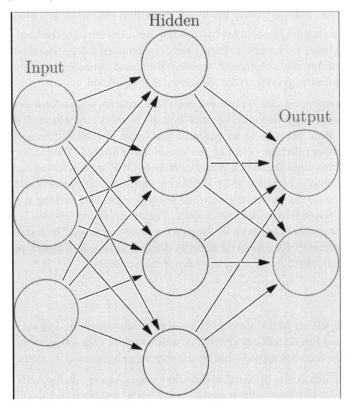

Fig. 12.1 An overall schema of ANN with n layers, neural nodes, and output layer

Source: Adapted and Redesigned from https://blog.knoldus.com/architecture-of-artificial-neural-network/

One neuron in the network in this kind of configuration can accept any number of inputs, n, and each input has a weight assigned to it. "x0" refers to the value which is called bias which is added to the input (activation functions input). If the inputs are X_i are the neuron's inputs, and W_i is weighted, b is biased, its output is determined by the equation (1).

$$a = f(\sum_{t=0}^{n} wixi + b) \tag{1}$$

where f is the function that is employed to obtain the output of that layer and then send the output to the next successive layer[6]. Learning patterns in machine learning have two major categories: Unsupervised and supervised learning. In the supervised technique, the output is labeled as the first phase, and the network knows what should be accepted as a response. In unsupervised learning, a network that uses neurons examines the patterns and brings out the features related to the input properties. As a result less data because the ML algorithm needs to train and test the data.

The ANN's modeling process in recognizing the patterns is due to the remarkable flexibility in adapting to new input. This enormous approximation capability is extremely important when working with real-time data, where there is more information but the patterns that are hidden are not identified. The results of the configuration generated may be significantly influenced by how well it is optimized. By adjusting weights with progressively larger sets of datasets, an ANN learns. The learning algorithm is capable of recognizing the connections between variables and generalizing in a way that enables the successful application of novel, unforeseen data. In most cases, it works in the same way for fitting a line, or plane with a set, in most cases. The ANN's surprising modeling prowess in pattern recognition is due to its remarkable malleability when it comes to adapting to new input. The results of the configuration generated can be significantly influenced by how well it is optimized. This algorithm is capable of recognizing the connections between variables and generalizing in a way that enables successful application to novel, unforeseen data [11]. Practically speaking, it's equivalent to fitting a line with a set [12]. An ANN called a perceptron has only one processing layer and input, output computational layers. Data points are fed into the input layer and then transferred to the computational layer. The output neuron calculates W X 14 Pdi141wixi) from the dnodes in the input layer that stand in for the d features represented by X 14 12x1... xd and the weight edges by W 14 12w1... wd. When using a perceptron, the binary is used for the forecast. Incorporating bias can help with unequal distribution. The prediction of y is obtained as:

$$y = sign\{W.X + b\} = sign\left[\sum_{i=1}^{d} wixi + b\right] \tag{2}$$

The sign is represented in the equation by the activation functions Uv. Different functions for activating can be applied in neural networks combined with many layers that are hidden. For effective training in multi-layered networks, the Rectified Linear Unit (ReLU) is widely utilized. Regression error is represented as the difference between the actual value that is used for testing and the value that is predicted. If the found-out error is more than zero, the weights need to be changed. The objective of the perceptron is to predict the least-squares difference of y for every point in dataset D. This goal is mentioned by the following function:

$$\sum_{(X,y)} (y - sign\{w \cdot X\} \tag{3}$$

The function is specified throughout the complete dataset X, and the learning rate is updated with weights, and then the entire dataset is used in each iteration of the algorithm until convergence. This algorithm is also known as stochastic gradient-descent and is expressed as follows:

$$W = W + \alpha E(X)X \tag{4}$$

A many layer network which uses neurons is constructed by employing numerous complex layers, also termed as the layers that are hidden . As stated by the title, "black box" alludes to the fact that the computations underlying such layers are concealed from the user. The points of the data are transmitted starting from the input layer to following levels up to the output layer, with computations occurring at each stage. The mentioned technique is coined as a feed and forward network with neurons. Usually, the count of the nodes in the input layer and the computational layer on the topmost are not equal. The accurate number of neurons and layers that are hidden will be calculated by the complexity of the necessary model and data[13].Utilizing hidden layers with fewer neurons than the input often leads in a loss of representation while still boosting the performance of the network, even though using a fully-connected layer is typical in some exceptional cases. This is most likely what happened when the noise in the data was removed.

The undesirable behaviour that a network with an excessive number of neurons may display is overfitting. This circumstance arises if a network that uses neurons fits the patterns exactly in the training dataset that the algorithm struggles to function well on unexpected input because the expected data is not sufficient [14]. The work which has been mentioned has looked at the way the hidden layers are used in the network and the neurons in those layers alter the quality of ANNs (along with other hyperparameters). These characteristics, like topology, and structure of the network, can also be referred to as hyperparameters.

3.1 The Application of Backpropagation

The loss function of a single-layer perceptron depends on the weights, making training it straightforward. When there are additional levels, the process is more challenging since the weights in the various layers interact. By using backpropagation, the error gradient is determined along several pathways to the output node[14]. Both the forward phase and the backward phase of the algorithm are present. The data points are delivered to the input nodes during the forward phase, where the outcomes at subsequent layers are computed one at a time using the current weights. The outcome of this prediction is evaluated against the training example. The loss function gradient for each weight is shown by the backward phase. The gradients step backward from the output layer all the way to the first layer to update the weights. Each repetition of this weight update procedure over the training data is referred to as a "epoch," and ANNs frequently require thousands of iterations to for convergence.

3.2 Technologies

TensorFlow, which is open-source library and has a high-performance, is made available by the Google Brain team's developers and engineers, and has been utilised in this article. It is now used in a variety of scientific and commercial applications as a capable support for machine learning as well as deep learning[15]. Incomparable user experience is combined with an incredible speed of experimentation using Keras, that is trained to run on TensorFlow top layer and various other libraries that come under machine learning. Modular, extendable design is used to achieve this. Instead of being an independent library, Keras was created primarily as an interface. The TensorFlow library now fully supports Keras, allowing for the straightforward coding of methodologies and its supporting operations[16]

3.3 Optimising the Hyperparameters of the Chosen Algorithms

The activation function's role in the creation of an ANN is one of its most crucial aspects because it directly affects the outcomes that can be obtained. Different sorts of activation functions can be supported by the network. Particularly in multi-layer networks, where each layer may have its own non-linear activation function, the choice of an activation function's type is critical[17]. Every separate function has the potential to have a unique impact on the ANN's outcomes, convergence, and overall network coverage. The four activation functions that have appeared the most frequently in recent research were chosen from a large pool of activation functions.

a)_Sigmoid b)-Hard Sigmoid c)_Rectified Linear Unit (ReLU) d) _Hyperbolic Tangent (tanh)

A grid search process is used, that finishes a thorough search over the space of the hyperparameters, and the best network configuration is identified. The criteria for the grid search included:

- Batch size
- Activation function
- Optimiser
- Epoch count
- Number of neuron nodes
- Count of layers that are hidden

3.4 Dimensionality Reduction

Dimensionality reduction refers to the pursuit of a feature vector that conveys the essence of the dataset while omitting the need to express each individual feature.

The topic is how to build an n-dimensional projection that can describe the information in a k-dimensional space. This process has clear computational advantages. Preventing the "curse of dimensionality" is one of the additional advantages. Consequently, when dataset dimensions increase, the ML classifiers fail to produce the desired results [18]. The algorithms must exponentially increase the number of samples to make up for this flaw. There are many different methods for reducing dimensionality .Feature selection is typically used to obtain the most relevant feature set because insignificant features frequently produce noise. However, there are other ways to get a more focused feature set that fully describes the data. Linear Discriminant Analysis is a different strategy that can be used (LDA). The two main aim of the mentioned strategy is the using the feature of maximisation between the selected class centroids and the variation that results from minimisation within those classes. In other words, LDA aids in establishing boundaries across class clusters.

Principal Component Analysis is another technique for dimensionality reduction that is extensively utilized in academic literature (PCA). PCA seeks out the data view with the greatest amount of variance. According to an example given in [28], if the data form a line of input, conducting the process of Analysis using the Principal component strategy would definitely show that the dimension is equal to 0 across all variances. Therefore, since the characteristics of those dimensions are unnecessary, they can be excluded. Data collection may indicate a strong signal in one direction, but noise will likely be present in a large number of aspects. The evoked projection with the largest value of variance is definite to capture the value and meaning of the data, given that the signal outweighs the noise to a sufficient degree.

4. Experimental Setup

The NSL-KDD data collection was developed to address issues with the malware and data that comes inside as intruded data was frequently brought up in the literature. Even though it still exhibits some undesirable traits, it is now a recognized benchmark dataset. NSL-KDD is nevertheless a trustworthy option for detecting malware detection research due to the missing values of accessible IDS datasets and the difficulties in gathering required data. The collection has about 5,000,000 records, which distinguishes it as acceptable for machine learning while also not being excessively big to make researchers have to choose random samples from the set. The outcomes are so easy to compare. In comparison to the KDD'99 dataset, the NSL-KDD has been cleaned of redundant data to avoid bias in ML algorithms. The shortage of trustworthy and sufficiently updated cybersecurity datasets led to the creation of the Intrusion Detection Evaluation Dataset - CICIDS2017 . The IDS data that are used for the researchers typically have a number of drawbacks of their own, such as a lack of attack variety, lack of traffic diversity, and a lack of appropriate characteristics. By studying 25 users behaviour with a wide variety of protocols, CICIDS2017 provide data with real network traffic. Then the data was collected over a period of five days, of which four were occupied by attacks on the model setup.

In addition to being one of the newest datasets accessible to academics, CICIDS2017 includes more than 80 network flow features. Figure 12.1

To limit the number of features to 50, PCA is used. Based on preliminary setup tests, an arbitrary number of features was extracted. The feature and the dataset delivered to the first layer that is the input of the neural network contains feature vector. In Fig. 12.2, the process's pipeline is depicted.

Fig. 12.2   Process in the form of pipeline with dataset

Source: Adapted and Redesigned from https://hazelcast.com/glossary/data-pipeline/

4.1 Cross Validation

Implementations of ML and many other aspects of AI come with a few unique problems. The overfitting of some of those issues is one of those issues. If the algorithms face these difficulties, they may perform brilliantly in a test setting but fall short when put to use on actual data. The models go through a process known as k-fold cross validation, also known as rotation estimation, to help alleviate this problem. The data that is used for training is divided into parts labelled as k, out of which k-1 data are used for training and the last component is utilised for testing. The process is then carried out k times while altering the portion of the test The data folds are randomly sampled and 10-fold cross-validation was applied in this study.

5. Results and Discussion

Multiple topological setups were examined, starting with 1-10(3 layers which is hidden mapped with ten neurons (Table 12.1) and progressing to 7 layers hidden mapped with 25 neurons (Table 12.2), ten layers of hidden with 25 neurons, four levels mapped with 25 neurons, and ten layers mapped with 25 neurons (Table 12.3).

Table 12.1   3 Layers with 10 neurons with NSL KDD dataset

Accuracy	Activation	Optimizer	Batch size
0.998981	Sigmoid	Adam	10
0.987871	Hard sigmoid	Adam	100
0.998768	Hard_sigmoid	Adam	10
0.987898	Tanh	Rmsprop	10
0.998789	Sigmoid	Adam	100
0.989879	Tanh	Adam	100
0.989888	Sigmoid	Sgd	100
0.998787	Hard Sigmoid	Sgd	100
0.998787	Tanh	Sgd	100
0.965459	Sigmoid	Sgd	100

Source: Self simulation using KDD dataset

Table 12.2 7 layers,25 neurons,NSL-KDD dataset

Accuracy	Activation	Optimizer	Batch-size
0.98787	tanh	Rms prop	10
0.96689	tanh	Adam	10
0.96453	Hard_sigmoid	Adam	10
0.96835	tanh	Adam	10
0.95567	tanh	Sgd	100
0.98456	tanh	Adam	10
0.78976	hard Sigmoid	Sgd	10
0.89786	sigmoid	Sgd	10
0.45689	sigmoid	Sgd	100
0.87612	tanh	Adam	10

Source: Self simulation using NSL KDD dataset

Table 12.3 10 layers,25 neurons, NSL-KDD dataset

Accuracy	Activation	Optimizer	Batch size	Epochs
0.98898	Tanh	Adam	10	100
0.97888	Tanh	Adam	10	100
0.96678	Tanh	Ms prop	10	100
0.85876	sigmoid	Ms prop	10	100
0.85879	hard_sigmoid	Adam	10	100
0.84767	sigmoid	Adam	10	30
0.49898	sigmoid	S g d	10	30
0.49768	Sigmoid	S g d	10	100
0.49787	Hard-sigmoid	S g d	10	30
0.49234	Hard Sigmoid	S g d	10	100

Source: Self simulation for NSL KDD dataset used for intrusion detection and security

Tables demonstrating the accuracy levels attained by various hyperparameters arrangement of various complexity that increase steadily are used to present the findings of the tests that were conducted. Although the trials were not carried out in this order, this method of presentation has been chosen to make things more clear. Accuracy attribute was selected as it is simple as well as effective for any learning algorithms that are used for detecting the attacks in network environment. It offers several confusion-matrix based metrics that have the capacity to describe precisely the classifiers capabilities .

Accuracy is also one of the metrics that IDS research uses the most frequently. This work uses a balancing data process that reduces the impact of datasets with substantial class imbalance, despite the fact that this method has drawbacks for these datasets. Precision and recall are used to further amplify the information that accuracy offers. Tables are arranged in order of accuracy. To avoid placing a simple condition on the models, early stopping was not used.

The achieved accuracy variation in the smallest of the examined topologies is shown in the first table (Table 12.1). A batch size of 10, and 300 epochs and sigmoid activation function were combined to get the best performance, which achieved nearly 99.9%. Although the trials were not carried out in this order, this method of presentation has been chosen to make things more clear. It is also one of the metrics that IDS research uses the most frequently. This work uses a data balancing process to reduce the impact of datasets with substantial class imbalance, despite the fact that this method has drawbacks for these datasets. Table 12.4 shows the Accuracy and Activation function along with the Batch size for CICIDS2017 which are arranged in order of accuracy. To avoid placing a smoothness constraint on the models, early stopping was not used.

The hyperparameter and accuracy for one hidden layer with 25 neural nodes are shown in the second table (Table 12.2). There are 300 epochs in total. This table's best accuracy, which is also the best accuracy obtained in all of the studies, exceeds 99.9%. (Table 12.4 and 12.5). The authors found that this particular combination of hyperparameters was interesting. Modern neural

Table 12.4   CICIDS2017 results

Accuracy	Activation	Batch Size	Layers	Neurons
0.99912	tanh	100	1	25
0.9909	tanh	10	2	25
0.9989	sigmoid	10	4	10
0.9708	sigmoid	10	7	25
0.9682	tanh	1001	3	25
0.9649	tanh	10	2	25
0.9640	tanh	10	7	10

Source: Self simulation results for CICIDS2017 dataset

Table 12.5   Balanced CICIDS2017 results (2)

	Precision	Recall	F1-Score
0	0.99	0.96	0.98
1	1.00	1.00	1.00
2	0.92	0.95	0.95
Micro avg	0.98	0.98	0.98
Macro avg	0.97	0.98	0.98
Weighted avg	0.98	0.98	0.98

Source: Self-simulation results for CICIDS2017 dataset

networks frequently omit the sigmoidal function known as the hyperbolic tangent activation function (with _R e L U being the go-to function). Modern deep learning models frequently employ micro batch sizes of 1 or 2. Tanh, ADAM, and a batch size of 100 are thus a very intriguing combination, especially for a tiny topology. The hyperparameter mixes for a topology with 2 hidden layers with 25 neurons each as in (Table 12.3). It is noteworthy that all of the results in this table, including the eight best ones, exceed 99.9% accuracy, and that all of the results in the two tables before it (Tables 12.1 and 12.2) exceed 98.9%. In Table 12.4, the highest accuracy was greater than 99.89%. Few accuracy values in this table dip below 90%, with the lowest performer going below 6%. For more complicated topologies, this trend will persist. The best accuracy in Table 12.4 was 96.83%, with a batch size of 10 and a combination of tanh and rms prop. Only 50% of the configurations achieved 90%. The hyperbolic tangent tops the list of factors that contribute to accuracy once more. The tanh function can be found in six out of ten alternative configurations (Table 12.4).

The output data where accuracy is highlighted is shown in Table 12.6 shows that increasing the neural network depth has no further advantages. Tanh and ADAM were the setups with the highest scores. The highest-scoring topologies among those with the best performance in Table 6 contain a collection of hyperparameter configurations. As stated in earlier sentences the tanh function which is used for activation and the optimizer contributes to the best outcomes than those of their rivals.

Surprisingly, the results of the smaller settings outperform those of the deeper structures. The statistical significance component of the research will go into greater detail about these findings. The algorithm has been used on the CICIDS2017 dataset in a number of trials. The preliminary findings in Table 12.5 were highly positive, with an accuracy rate of more than 99%. (0.9936). There may be a balancing issue in the dataset, as shown in one of the classes.. The accuracy of the method outperformed the test set by 97%, and the 10-fold cross-mean validation's accuracy was 95%.

5.1 Comparison of Performance of ML Algorithms

This section focuses on the performance of ML approaches. Tests were run on an SVM with a gaussian kernel, the Naive Bayes classifier, and ADA Boost to put the performance of the illustrated methods into perspective. The variations in the outcomes the classifiers have produced using the CICIDS2017 data are shown in Table 12.6 lists the precision, recall, and f1-score of each class found in the dataset along with the results the Naive Bayes classifier was able to achieve on the entire CICIDS2017 dataset. The table is further expanded to display the outcomes from additional methodologies across the entire CICIDS2017 dataset.

5.2 Hyperparameter Optimization

Each of the settings has its hyperparameters optimised. The selected hyperparameters are evaluated using the grid search approach in all of their conceivable permutations. In order to attain the best accuracy, the batch size, the activation function, the optimiser, and the used epochs count are sequentially permutated as shown in the Tables 12.1–12.6. It is obvious that, as was the case with NSL-KDD, the results obtained with various parameter configurations vary significantly.

5.3 Statistical Analysis

To assess the statistical significance of this study, the tried-and-true Wilcoxon Signed-Rank Test was applied. Table 5 compiles the NSL-KDD setups that perform the best. All of those configurations were evaluated in comparison to the top-ranking configuration, but the test failed to demonstrate statistical significance which led to the conclusion that, if the hyperparameters are properly set for this dataset, the topology of the neural network has no discernible influence on accuracy.

Table 12.6 4 hidden layers, 25 neurons each, 30 Epochs CICIDS2017 dataset

Accuracy	Activation	Optimizer	Batch size
0.97698	R e l u	Ms prop	50
0.97888	R e l u	Ms prop	100
0.97425	R e l u	Adam	50
0.97192	R e l u	S g d	100
0.97051	Hard_sigmoid	Ms prop	50
0.96978	Hard_sigmoid	Ms prop	100
0.96911	Hard_sigmoid	S g d	100
0.96557	sigmoid	Ms prop	50
0.95818	R e l u	S g d	100
0.93022	sigmoid	S g d	100
0.70845	sigmoid	S g d	50
0.58551	Hard_sigmoid	Ms prop	50
0.49982	Hard_sigmoid	S g d	100

Source: Self simulation results for CICIDS2017 dataset

The acquired data and analysis of the statistical significance of the findings attained by specific setups showed that, past a certain point, neither deepening network under neural environment nor increasing the neurons numbered layers, ANN accuracy can be increased for the base dataset under evaluation. In this study, the proposed method was thoroughly explained and its performance on two benchmark datasets was demonstrated, demonstrating its usefulness by displaying equivalent findings. The numbered layers that are hidden and the numbered neurons on those layers were varied on testing a vast range of neural based topologies. The effect that different hyperparameter choices have on the metrics attained was estimated. Threats to the legitimacy of the suggested approach were explored, as well as the issue of data imbalance. Two standard sets of datasets which are always used for testing and training are NSL-KDD and CICIDS2017 respectively. The tests showed that employing the incorrect combination of hyperparameters might result in radically different results while applying the same topology over the same dataset. In one instance, increasing the batch size from 10 to 100 causes the accuracy to decrease from 0.9498 to 0.8592. When using SGD instead of rms prop optimizer, accuracy decreases from 0.9545 to 0.8592. When using tanh instead of R e L U activation, accuracy plummets startlingly from 0.9545 to 0.0564. However, the accuracy was increased to 0.9990 when the sigmoid activation function was used in the identical arrangement. The experiments that employed CICIDS2017 show a similar outcome. The accuracy falls from 0.9582 to 0.5850 this time. As demonstrated in the previous paragraph, this work offers experimental evidence that even a little change in configuration may have a large impact on the metric to be calculated and used. The ideal amalgamation of the examined standards was discovered. Additional settings that are comparable to the quality setup (for the value p = 0.05) were also used for NSLKDD.

6. Conclusion

The proposed work has successfully implemented and simulated hyperparameter optimization in Artificial Neural Networks (ANN) to achieve a efficient network intrusion detection.

References

1. Michał Choras , Marek Pawlicki,"Intrusion detection approach based on optimised artificial neural network",Elsevier Publishing,Neurocomputing 452(2021) 705-715
2. M. Feurer, F. Hutter, "Hyperparameter Optimization", Springer International Publishing, Cham, 2019, pp. 3–33. doi:10.1007/978-3-030-05318-5_1.
3. Roopa M and Dr.Selvakumaraja, "Intelligent Intrusion Detection and Prevention System using Smart Multi-Instance Multi- Learning Protocol for Tactical Mobile Ad hoc Networks", KSII Transactions on Internet and Information Systems, VOL. 12, NO. 6, 2895-2921, Jun. 2018. ISSN : 1976-7277 .
4. M. Choras´, R. Kozik "Machine learning techniques applied to detect cyber attacks on web applications", Logic J. IGPL 23 (1) (2015) 45–56, https://doi.org/10.1093/jigpal/jzu038.
5. Seungwon Lee, Changbae Mun and Ook Lee, "A Study of Neural Network Based IoT Device Information Security System", Journal of Theoretical and applied Information Technology, Volume 96, Issue 22, ISSN: 1992-8645, E-ISSN: 1817-3195,2018.
6. I. Mukhopadhyay, et.al , "Back propagation neural network approach to intrusion detection system" ,International Conference on Recent Trends in Information Systems (2011) 303–308, https://doi.org/10.1109/ReTIS.2011.6146886.
7. F. Haddadi, S. Khanchi, M. Shetabi, V. Derhami, "Intrusion detection and attack classification using feed-forward neural network" , Second International Conference on Computer and Network Technology,262–266, 2010 https://doi.org/10.1109/ICCNT.2010.28
8. W. Gong, W. Fu, L. Cai, "A neural network based intrusion detection data fusion model", Third International Joint Conference on Computational Science and Optimization, vol. 2, pp. 410–414. doi:10.1109/CSO.2010.
9. H.A. Sonawane, T.M. Pattewar, "A comparative performance evaluation of intrusion detection based on neural network and pca, ", International Conference on Communications and Signal Processing (ICCSP) ,0841–0845, 2015,https://doi.org/10.1109/ICCSP.2015.7322612.
10. B. Subba, S. Biswas, S. Karmakar, " A neural network based system for intrusion detection and attack classification, Twenty Second National Conference on Communication (NCC) 2016 (2016) 1–6, https://doi.org/10.1109/NCC.2016.7561088
11. F. Haddadi, S. Khanchi, M. Shetabi, V. Derhami, " Intrusion detection and attack classification using feed-forward neural network", Second International Conference on Computer and Network Technology 262–266, 2010 ,https://doi.org/10.1109/ICCNT.2010.28.
12. Y. Goldberg, "A primer on neural network models for natural language processing", J. Artif. Intell. Res. 57 (2016) 345–420.
13. D.-S. Huang, H.H.-S. Ip, K.C.K. Law, Z. Chi, "Zeroing polynomials using modified constrained neural network approach", IEEE Trans. Neural Networks 16 (3) 721–732, 2005.
14. O. Maimon, L. Rokach, "Data Mining and Knowledge Discovery Handbook",second ed., 2010.
15. C.C. Aggarwal, "Neural Networks and Deep Learning", a Textbook,.doi:10.1007/978-3-319-94463-0, 2018
16. O. Maimon, L. Rokach, " Data Mining and Knowledge Discovery Handbook", second ed., 2010.
17. G. James, D. Witten, T. Hastie, R. Tibshirani, "An introduction to statistical learning", Cluster Computing, 2018.
18. P. Branco, L. Torgo, R. Ribeiro, "Relevance-based evaluation metrics for multiclass imbalanced domains", in: Pacific-Asia Conference on Knowledge Discovery and Data Mining, Springer, 2017, pp. 698–710.

Human Machine Interaction in the Digital Era – Prof. J. Dhilipan et al. (eds)
© *2024 Taylor & Francis Group, London, ISBN 978-1-032-54998-9*

Detection of People or Animals Using Deep-learning in Railway Tracks

13

V. Dinesh[1]
Assistant Professor, Easwari Engineering College
(Anna University) Chennai India

G. Mahalaxmi[2], M. Manjari[3]
Student, Easwari Engineering College
(Anna University) Chennai India

Abstract The Indian Railways one of the world's largest rail networks, stretches more than 1,15,000 kilometers across the entirety of the country. Deep neural networks, as well as other machine learning techniques, have received a lot of interest recently when applied to railroad transportation data in order to detect the animals or people in the railway tracks, Convolution Neural Networks (CNN) is used. They provide a lot of advantages over other techniques, resulting in the season they are most frequently utilized in pattern and image identification tasks. This paper employs Keras python library as the backend as well as Tensor flow, which able to precisely forecast the classification report. For classifier images, the convolutional neural network technique is used. Given that this classifier is a deep learning model, the accuracy will increase as you train it. As an object is detected in the railway tracks, the buzzer will go off near the loco pilot. This system will aid in preventing accidents caused by creatures on rail tracks.

Keywords Keras, CNN, Buzzer, Tensorflow

1. Introduction

Railway line tracking system in forest area is used to minimize accidents on the railway tracks. The system monitors the tracks for the presence of any animals or people. A significant improvement in this area will have a major impact on the nation. There is a mechanism in place to adequately monitor and maintain the rails. Because of the enormous size, bad maintenance will lead to accidents on the tracks. The absence of carelessness has an impact on many lives. This paper puts in place a system that can stop many accidents from occurring in order to avoid this with rails. This technique largely concentrates on areas where wildlife is commonly seen on railroad tracks. Images are recognized using image processing in the proposed system. Here, this paper employs Keras as a backend as well as Tensor flow and is able to precisely forecast the classification report [15]. It detects an object in the image in a fraction of a second, and the application immediately generates an alert message and alerts the loco pilot. Serious accidents include those involving trains carrying people that result in fatalities, serious injuries to one or more passengers, serious damage to railroad property worth more than Rs. 25,000, or any other accident that, in the judgment of the "CHIEF COMMISSIONER OF RAILWAY SAFETY OR COMMISSIONER OF RAILWAY SAFETY," the commissioner of railway conducting an investigation.

[1]dinesh.v@eec.srmrmp.edu.in, [2]mahauma2000@gmail.com, [3]mmanjari2002@gmail.com

DOI: 10.1201/9781003428466-13

2. Literature Review

Enhanced YOLOv4 for Pedestrian and Locomotive Signal Light Detection on Railway Tracks

A method for detecting organisms on railroad tracks is called the Railway Track Tracing Technique. The system consists of a camera, a train status system, and an alert application. At periodic intervals, the camera takes pictures of the track, which are later identified by image processing.

Using the Enhanced YOLOv4, a Technique for Detecting and Recognizing Traffic Lights

The enhanced YOLO V4 algorithm is examined using the deep feature enhancement and boundary box uncertainty prediction mechanisms. To extract features from the network and increase its capability to detect small items and color resolution, the shallow feature enhancement technique combines two shallow characteristics at different phases with the rising semantic features generated after rounds of upscaling

An IoT based Staple Food Endowment System and Waste Management System for Foster Care using Arduino and Blockchain

Utilizing an Arduino Uno microcontroller, a railway track fracture detection system, the primary focus of this work is on the dynamic integration of a tracking system, a GSM module for alert message transmission and location's GPS coordinates for sensors used in railway line break detection. This device's operations are guided and coordinated by an Arduino microcontroller

3. Methodology and Model Specifications

Neural network-based method for measuring and detecting railway deterioration. Train accidents are frequently caused by railway problems and deficiencies. An approach that uses selective search to extract just 2000 areas from the image and call them region proposals was presented to get over the issue of picking a large number of regions. Corrupted images will prevent the system from operating correctly. Low Precision due to Poor Lighting The suggested approach not only resolves these issues but also boosts precision.

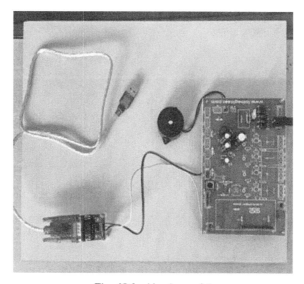

Fig. 13.1   Hardware kit

3.1 Convolutional Neural Network

A typical method of processing images is called convolution, which modifies a pixel's brightness to reflect the brightness of pixels nearby. Image filters are a frequent application for convolution.

3.2 Convolution Layer

When a neuron is connected to a small area in the input volume, the Convolution layer will estimate the outputs of that neuron. Each neuron will compute the dot product between its weights and that area.

3.3 Relu Layer

The elementwise activation function of the RELU layer will be applied, including the max(0,x) thresholding at 0. The volume remains the same size as a result.

3.4 Pooling Layer

The Pooling layer will do down sampling all along dimensions (width, height).

3.5 Fully Connected Layer

A FC layer will calculate the class scores, producing a volume of size [1x1xN], in which each of the N numbers represents a class score, such as within the N categories.

Module 1 Image classification

- Resize, orient, and colour adjustments fall within this category
- The pace of model inference and model training may both be accelerated by image pre-processing.

Module 2 image extraction

- The dimensionality reduction procedure, which divides and condenses a starting collection of raw data into smaller, easier-to-manage groupings, includes feature extraction.
- It will therefore be simpler for you to process it when you want to.

Module 3 Detection

- Finding huge animals before they enter the track and alerting drivers to their presence while on or close to it.

3.6 Block Diagram

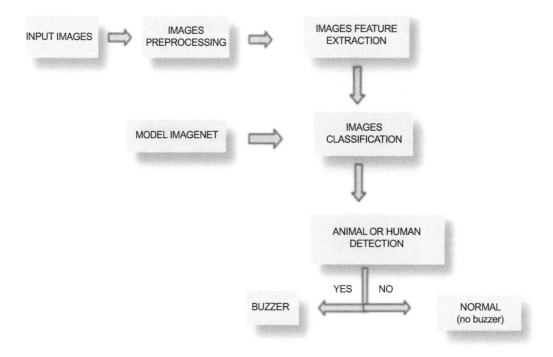

Fig. 13.2 Block diagram

3.7 Analysis Result

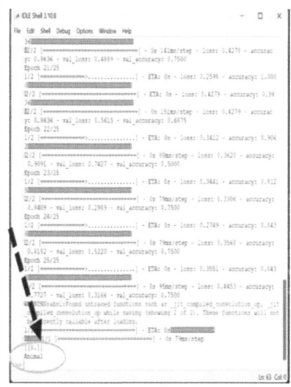 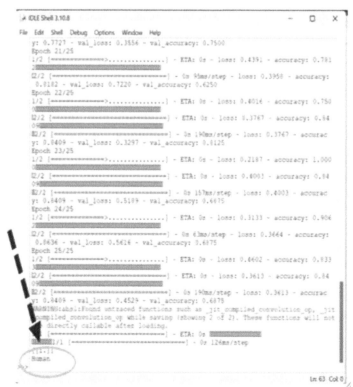

Fig. 13.3 Animal and human detected

4. Conclusion

The railroad is the most practical and easy form of transportation. But several accidents are happening as a result of carelessness. This paper, therefore, introduces the proposed method, which makes use of the image processing approach, to prevent such accidents. The photos are stored in the database after being captured. After capturing, the comparing process started, and the image was found. When a creature or object was found, there was a recapturing process that required further detection and verification that the creature or object was still on the track. The loco-pilot can use this information to halt the train and prevent incidents that endanger the critter on the track. When a creature is found, it makes a buzzing sound.

5. Acknowledgement

Firstly, we want to express our gratitude to Dr. T. R. Paarivendhar, Founder Chairman, SRM Group of Institutions, without whose support our project could not have been completed successfully. We are very much grateful to Dr. M. Devaraju, Head of the Department, Electronics and Communication Engineering for helping us without attendance and providing continuous encouragement for the completion of the project. We take the extreme privilege in extending our hearty thanks and sincere gratitude to our supervisor Mr. V. Dinesh, Assistant Professor, Electronics and Communication Engineering, for her intellectual guidance and valuable suggestions in making this project a successful one. Finally We would extend our hearty thanks to all our parents for their support in completing this project.

References

1. Bhojwani, N & Ansari, A & Jirge, S & Baviskar, M & Pawar, D. (2021). RAILWAY TRACK – CRACK DETECTION SYSTEM BY USING ARDUINO MICROCONTROLLER. International Journal of Engineering Applied Sciences and Technology. 6. 10.33564/IJEAST. 2021.v06i05.044.
2. H. Wang, H. Pei and J. Zhang, "Detection of Locomotive Signal Lights and Pedestrians on Railway Tracks Using Improved YOLOv4," in IEEE Access, vol.10, pp. 15495–15505, 2022, 10.1109/ACCESS.2022.3148182.
3. Clark, G. L. and Wójcik, D. (2005). Financial valuation of the German model: the negative relationship between ownership concentration and stock market returns, 1997–2001. Econ. Geogr. 81(1): 11–29.
4. Iliev, Rosen, and Angel Genchev. "Possibilities for Using Unmanned Aerial Vehicles to Obtain Sensory Information for Environmental Analysis." Information & Security: An International Journal 46, no. 2 (2020): 127–140.
5. Deepa, M & Raji, C & Ajina, VA & Ashla, & Azra, Afsal & Susanna, George. (2021). Railway track tracer system for creature detection. IOP Conference Series: Materials Science and Engineering. 1055. 012041.10.1088/1757- 899X/1055/1/012041.Gompers, P. A. and Metrick, A. (2001). Institutional investors and equity prices. Q. J. Econ. 116(1): 229–259.
6. B. Malarvizhi, V. Dinesh, M. Janani, R. Gunaseeli, B. Abarna, "An IoT based Staple Food Endowment System and Waste Management System for Foster Care using Arduino and Blockchain," 2020 International Conference on Recent Trends in Computer Science & Information technology" ICRCSIT - 20, ISBN No.978-93-80831-66-4.
7. P. Lad and M. Pawar, "Evolution of railway track crack detection system," 2016 2nd IEEE International Symposium on Robotics and Manufacturing Automation (ROMA),2016, pp. 1-6, doi: 10.1109/ROMA.2016.7847816.
8. Wang, Qingyuan et al. "Traffic Lights Detection and Recognition Method Based on the Improved YOLOv4 Algorithm." Sensors (Basel, Switzerland) vol. 22, 1 200. 28 Dec. 2021, doi:10.3390/s22010200
9. Alzubaidi, L., Zhang, J., Humaidi, A.J. et al. Review of deep learning: concepts, CNN architectures, challenges,applications, future directions. J Big Data 8, 53 (2021).
10. B. Malarvizhi V. Dinesh M. Janani, R. Gunaseeli, B. Abarna, "IoT based Staple Food Endowment System and Waste Management System for Foster Care," 2020 Sambodhi- UGC Care Journal, ISSN: 2249-6661, Volume-43, UGC Care
11. S. Albawi, T. A. Mohammed and S. Al-Zawi, "Understanding of a convolutional neural network," 2017 International Conference on Engineering and Technology
12. M. -Y. Lee, J. -H. Lee, J. -K. Kim, B. -J. Kim and J. -Y. Kim, "The Sparsity and Activation Analysis of Compressed CNN Networks in a HW CNN Accelerator Model," 2019 International SoC Design Conference (ISOCC), 2019, pp. 255–256, doi: 10.1109/ISOCC47750.2019.9027643.

Note: All the figures in this chapter were made by the Author

Human Machine Interaction in the Digital Era – Prof. J. Dhilipan et al. (eds)
© 2024 Taylor & Francis Group, London, ISBN 978-1-032-54998-9

Confronting Floating Plastic Debris as Pollutants in Marine Using Technological Solutions Deep Learning Approach

14

S. Belina V. J. Sara[1]

PDF Research Scholar, Srinavas university, Karnataka, India
Assistant Professor, Department of Computer Science, SRMIST Ramapuram, Tamilnadu, India.

A. Jayanthiladevi[2]

Professor-Institute of Computer Science and Information Science,
Srinivas University,Karnataka,India

Abstract Estimating how much macro-plastic is present in the world's oceans is one of the most critical environmental challenges of our day. The most popular techniques for counting the amount of floating plastic trash take a long time and only work in a relatively narrow region. To overcome the shortcomings, an automatic identification method based on a deep learning framework was developed to find and quantify the marine trash. As a result, it is suggested to utilize the FMA-YOLOv5 approach, an improved version of the YOLOv5 technique that uses a feature map attention (FMA) layer at the end of the backbone for detection and classification. The deep learning-based segmentation model that was used by the image classifier allowed it to quantify the marine debris in the sample. The quantity of coastal debris objects predicted by the suggested strategy was compared with the estimation of the manual monitoring method in order to demonstrate the differences in predicting the marine debris standing-stock. Apparently, the proposed method verified on seven categories of items yielded a mean average precision of 0.90. The outcomes offer vital data for formulating efficient marine trash control programs and regulations.

Keywords Debris, Marine, Aquatic, Pollution, Yolo, Garbage, Plastics

1. Introduction

Arcangeli et al., (2017) Marine debris, particularly pollution from plastics, is one of the largest environmental issues currently endangering our ecosystems. Ryan et al., Gall et al., Jambeck et al (2008) Marine debris is present in all marine sectors worldwide and is described as any durable, made, or treated solid mass disposed, discarded of, neglected, or misplaced in the coastal and marine environment. Garcia-Garin et al., (2020) Marine debris pollution of coastlines has a terrible effect on ecosystems, human life, and marine life. Because it can entangle any type of marine species (including fishes, turtles, and marine animals) and because it can be consumed by marine life, particularly big filter-feeding species, floating marine debris is particularly damaging. Nnaji et al., (2015) The amount of plastic generated worldwide has surpassed 500 million tons, and predictions show that 30% of this plastic will be dumped in the oceans. Clapp et al., (2012) Inside the Central Pacific Gyre, plastic trash has multiplied five times, and studies have indicated that it now outpaces natural plankton six to one in abundance. About 80% of the marine plastic we see comes from land-based sources: Most frequently in the form of packing supplies and food storage containers like plastic bags and bottles. The remaining 20% comes from discharges from transport vessels

[1]belinavs@srmist.edu.in, [2]Drjayanthila@srinivasuniversity.edu.in

DOI: 10.1201/9781003428466-14

and abandoned commercial fishing gear. According to Carlton et al., (2017) studies, clearing the oceans of plastic will have a huge positive impact on ecosystem health. Royer et al., (2018) includes stopping invasive species from spreading between regions, stopping it from degrading into microplastics, and reducing greenhouse gas emissions (thereby slowing down climate change).Barnes et al., (2009) Determining the amount of plastic marine debris (PMD) is crucial for conducting efficient beach cleaning efforts and evaluating the harmful effects of marine litter on the environment. Boat surveys are the primary technique for gathering information on plastics lying on (or around) the water's surface, including its form and thickness. Wang et al., (2015) Vessel surveys are time-consuming, costly, and have a restricted geographic scope notwithstanding the possibility that human error would taint observations. This is because of the subsequent examination of the content and data that are gathered. However, survey techniques have not yet been developed that can objectively and precisely determine the quantity of PMD that has washed ashore. To detect floating trash, some researchers have recommended employing a single component with an attached camera. However, using an automated system based on machine learning for identifying and categorizing marine debris has some additional benefits. A technique like this could speed up the identification of large-size plastic garbage and reduce expedition costs while cutting down on search time. A machine learning approach can produce very accurate results for the automation of the macro-plastic identification procedure.

On numerous computer vision tasks, such as object localization and detection Bochkovskiy et al., (2020) classification, and semantic segmentation Tao et al., (2020) deep learning techniques have been successfully applied. Unlike normal machine learning approaches, which demand domain-specific as well as humanly designed feature extraction, deep-learning algorithms autonomously uncover latent data characterizations out from source data and provide end-to-end training processes. Fukushima et al(2014) Deep learning techniques can be used to develop quicker and more precise tools for locating and classifying floating trash in the scope of marine debris study. When it comes to picture and video data, convolutional neural networks (CNNs) are among the top techniques advancing the success of deep learning. CNNs are created in a way that is influenced by the receptive field patterns seen in the animal visual system in order to retrieve features hierarchical order by integrating low level information into high level ones.

The single-stage single shot multi-box detector (SSD) as well as YOLO series algorithm and the two-stage region-convolutional neural network (R-CNN) series algorithm are two subsets of algorithms for deep learning. The most recent studies demonstrate that YOLO series techniques can be used in a variety of detecting applications. The YOLOv5 method has been selected to be used in the network model for the spotting of floating debris.YOLOv5 has a more adaptable network structure than YOLOv2, YOLOv3, and YOLOv4, which makes it easier to change the network architecture of various jobs. The four versions of YOLOv5 are YOLOv5s, YOLOv5m, YOLOv5l, and YOLOv5x. The model's depth and width, as well as its number of parameters, are growing from YOLOv5s to YOLOv5l.Even though accuracy will grow as model complexity increases, detection speed is also significantly impacted. Due of their numerous parameters, which would significantly slow down detection, YOLOv5l as well as YOLOv5x are not included.

This paper's research intends to address the demand for a quick and accurate estimate of floating plastic. These are the primary contributions to this paper:

- To create a model for the detection, classification, and quantification of various types of marine trash by autonomous means;
- A feature map attention (FMA) layer was placed in between backbone and neck of the YOLOv5 network, as seen in Figure 1. The input feature map for this layer was adjusted to improve the YOLOv5 backbone's lighter object recognition's ability to extract features.
- To estimate the coastal debris, U-Net, segmentation technique based on deep learning was employed.
- The proposed methodology is utilized to quantitatively estimate and classify waste into six categories: plastic, glass, fishing gear, thermopolis, rubber, metal and undefined.
- This investigation also demonstrated the difference between the suggested automatic method and established techniques for estimating and predicting beach waste.

The remaining sections of the work are structured as follows. Related works are reviewed in section 2. Materials and techniques utilized in this study are described in Section 3:dataset (Section 3.1) and implemented deep convolutional architectures (Section 3.2).Section 4 presents and discusses the experimental findings. Conclusions and suggestions for further research are provided in Section 5.

2. Literature Survey

Understanding and combating the expanding advent and latest developments in the dispersion and accumulation of plastic particles depend on the identification and detection of macroplastic waste in aquatic settings. Fieldwork studies and ocular methods are now the most widely used strategies for gathering data, although since 2018 remote sensing and artificial intelligence-based methods have advanced. Despite their shown effectiveness, speed, and broad usability, there remain challenges to be solved, particularly when considering the accessibility and availability of data. As a result, Gnann N et al., (2022) review provides an overview of the most recent studies on the visual identification and recognition of various macroplastic types. The emphasis is on methods for both gathering data and evaluating it, such as machine learning and deep learning, although outcomes and publicly available might also be taken into account. The goal is to provide a critical outlook and viewpoint at a period when this field of study is developing swiftly. This research reveals that the bulk of machine learning and deep learning algorithms remain in their immaturity in terms of accuracy and precision comparing to visual surveillance, although having highly promising results.

In order to design a protocol that citizens can use to check shorelines for marine debris, Papakonstantinou et al., (2021) framework blends drone technology with advancements in artificial intelligence for machine vision applications. In the current study, a customer-grade off-the-shelf drone was used to capture really high resolution aerial photographs from a beach with a complicated background. In order to detect marine litter in the coastal area and produce its density maps, these photos were employed as input data in deep learning models. The input data was annotated into the litter and no litter classifications using the Zooniverse citizen science tool. With the aid of volunteers, the annotation procedure was quickly put into place, increasing its effectiveness and efficiency. In order to recognize marine litter from very high quality photos gathered from beaches with complicated backdrops, five deep learning algorithms were looked at and trained. By contrasting these density maps with those generated by a manual screening categorization, it was demonstrated that the method could be applied geographically to new and undiscovered beaches.

Li P et al., (2022) suggests using picture enhancement and YOLOv5S to detect marine biological items. To address the issues of underwater image distortion and blur, contrast-limited adaptive histogram equalization (CLAHE) is used. They also present an upgraded YOLOv5S to increase object detection's accuracy and real-time functionality. The enhanced YOLOv5S incorporates coordinated attention and dynamic spatial feature fusion to precisely detect the object of attention and fully integrate the features of different dimensions. Soft non-maximum suppression has been adopted in favor of non-maximum suppression to improve the capacity to detect overlapping items. The results of the experiment show that the CLAHE method can greatly improve the accuracy and quality of underwater picture detection.

To better understand the instantaneous drift and distribution of marine trash and to enhance the management of marine waste, Sun et al. (2020) perform a numerical analysis on the marine debris drifting in the western Bohai sea. The impact of important variables on the drift trajectory and range, including wind velocity and direction, and trash source, has been thoroughly investigated and debated. The program for treating marine trash is also examined. It has been discovered that the topographical characteristics of the area where the ocean trash oirgin is located have an impact on the particle's diffusion as well as its eventual route of drift. The collection of marine debris will be triggered by an increase in wind speed, and the drift distance and diffusion zone of the particles will be periodically impacted by changes in wind direction.

In this study by Kylili et al., (2020) an image classifier was built using an artificial intelligence application using a CNN design and the bottleneck approach. The trained bottleneck approach classifier was used to divide plastic packs, containers, baskets, food wrappers, straws, abandoned netting, fishes, and other objects into eight distinct groups. These plastics were found either at the coastline or floating at the surface of the ocean. With a 90% rate of success in effectively differentiating objects, the recommended deep learning technique is a move to the intelligent detection of plastic products at the seashore and in the ocean. The results of training and testing for losses and accuracy for different epochs and batches have given the proposed method more confidence. Results from such a resolution sensitivity examination showed that even with reduced image quality, the prediction system can still successfully identify plastics.

The process of visually identifying trash in indigenous underwater ecosystems has been carried out by Bajaj R et al. (2021) using a deep learning based method in order to deploy autonomous underwater vehicles (AUVs) to examine, monitor, and eliminate that trash. For training, the InceptionResNetV2 architecture of deep learning is employed. A huge and freely accessible dataset of real trash in diverse locations is labeled using data from the JAMSTEC e-library about deep-sea trash. After that, the certified network is put to the test on a variety of images from various datasets, offering knowledge about ways to improve an AUV's detection capabilities for removing undersea trash. The trained system uses bounding boxes and target class maps to find interesting features in test images. It is expected that the findings will advance efforts to use AUVs to autonomously examine, detect, and collect aquatic garbage in underwater environments.

Marine trash has a negative impact on the marine ecology and marine life's ability to exist. The identification of deep-sea debris utilizing deep learning techniques is established in this study by Xue et al., (2021). In order to facilitate future study, a true deep-sea trash identification dataset (3D dataset) is first established. The dataset includes 7 different forms of trash. Second, the ResNet50-YOLOV3 one-stage deep-sea trash detection system is suggested. Eight other sophisticated detection algorithms are also utilized in the search for deep-sea debris. Finally, experiments are used to validate ResNet50-YOLOV3's performance.

This work by Winans et al., (2021) assesses the effectiveness of deep learning centered object identification to autonomously identify, track down, and categorize abandoned marine debris along several kilometers of the diverse Hawaiian shoreline. The two most popular object detection techniques, SSD with MobileNetV2 (SS-MN), which is the faster model, and faster RCNN with Inception-ResNet-v2 (FR-IR), which has a greater computational cost, were put to the test. These two models provide a trade-off between identification accuracy and detection speed, with one model operating five times as quickly. When thoroughly studied, the results of the object identification and deep learning classifiers clearly show the best methods for the development of deep learning for the management of coastline strand maritime rubbish at the broad, regional level.

3. Materials and Methods

This work uses a segmentation technique based on deep learning to identify and quantify the residual stock of marine debris. The purpose of this proposal is to identify and quantify the standing stock of coastal trash using deep learning-based segmentation algorithms. The implementation of real-time floating debris identification and classification employs a method based on FMA-YOLOv5. Here, using the image segmentation method, the total number of debris objects along the coast is approximated. Additionally, this study demonstrated the discrepancy between automatic and manual approaches for predicting the standing-stock of beach waste.

3.1 Data Acquisition

A Galaxy S20 super camera with highest resolution was utilized to examine photographs of beach trash. The field of view and focal length are 25 mm and 79 degrees, respectively. A 10 m-wide transect was used to gather images of the entire Dasa-port seashore, which is situated in the West Sea, Korea (2021). For the training and test sets, 110 pictures were gathered. Although every effort was made to prevent duplication, some debris pieces did overlap. Therefore, the overlapped photos were shaded so that they couldn't be spotted before deep learning. Additionally, 736 close-up photos were gathered solely for the data set. Under the specified circumstances, close-up photos were captured at a height of just 0.5 m. 1500 items were taken for individual item labeling from 736 pictures.

The litter on the entire beach was personally counted and categorized to make comparisons of the results of the deep learning method utilized in this work. Because of the small size, beach debris measuring >2.5 cm was used to determine the residual stock. Each piece of beach trash was categorized into one of seven categories, including plastic (containing PET bottles, vinyl ropes, and synthetic fiber), glass, fishing gear, thermopolis, rubber, metal and undefined. The undefined objects includes paper, fabric, manufactured wood pieces etc.. Following that, each class's items were separately counted.

3.2 Implementation of FMA-YOLOv5 Object Detection Algorithm

Using the acquired pictures, a network architecture based on YOLOv5—You Only Look Once—was utilized to identify and classify the debris. As a 1-stage detector, YOLO was created to expedite the image processing procedure. The 1-stage detector simultaneously performs region proposal and classification. To improve the ability of network feature extraction, an enhanced YOLOv5 (FMA-YOLOv5) technique is developed. This method extends the backbone by adding a feature map attention (FMA) layer. The input feature map's size is left unchanged by the FMA layer, allowing for flexible and versatile addition to any network configuration. Figure 14.1 illustrates the network training and detection procedures that make up the method.

YOLOv5 Network Structure

As demonstrated in Fig. 14.2, this approach uses the YOLOv5 network structure for both its training and identification networks, with the Cross Stage Partial Network (CSPNet) acting as the network's central node. The issue of needing numerous inference calculations is solved with CSPNet. By combining the cross-stage hierarchical structure, CSPNet separates the base layer's

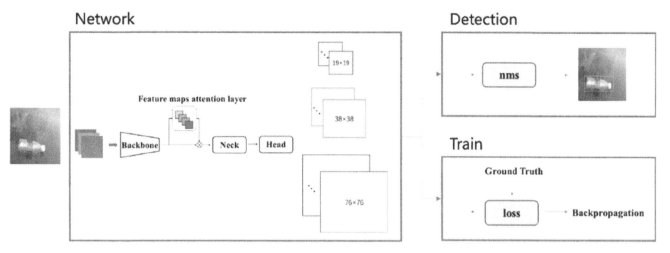

Fig. 14.1 FMA-YOLOv5s algorithm's general framework

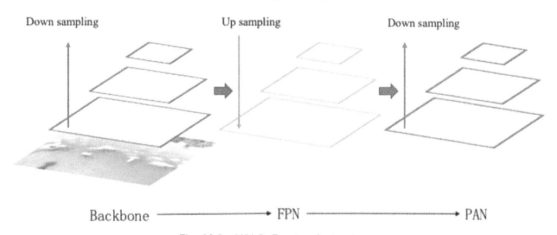

Fig. 14.2 YOLOv5 network structure

feature map into two portions before extracting the picture features. This approach has the benefit of decreasing the amount of repetitive gradient data, lowering the number of computations, speeding up the equipment, and not impacting the model's correctness. YOLOv5 additionally uses the Feature Pyramid network topology to fully utilize the feature information retrieved from various levels (FPN). The new feature map is constructed by combining the source mapping on the left with the level-appropriate feature maps that were created from the downsampling of the input image, following a process of upsampling from top to bottom. By integrating shallow data with deep data, this structure improves the effect while also enhancing the magnitude of the produced feature map. Following the FPN feature combinations, the Path Aggregation Network (PAN) framework is built on top of this base. After convolutional downsampling, the merged bottom map is combined with the scale-matched feature map from the left FPN structure to create 3 output feature maps of varying magnitudes. With this combination, the model's feature extraction will be more accurate and strong location features will be communicated from the ground up.

Feature Map Attention Layer Structure

There are numerous convolutional layers in YOLOv5. Each component's convolution has a unique depth and width. The depth and width coefficients of YOLOv5s are 0.33 and 0.5, respectively. By controlling the number of stacking layers, the depth index controls the depth of the network. The width coefficient, which affects both the network's width and the quantity of convolution output channels, controls this. By limiting the depth and width, the total quantity of variables in the network can be controlled. It is recommended to add an FMA layer in between backbone as well as neck to improve the network structure that depends on YOLOv5s, as shown in Fig. 14.3. The attention feature map, which is shown in Fig. 14.3 as having four channels, and the input feature map are both weighted in the FMA layer to obtain the features for the four groups. Then, 1 x 1 convolution is used

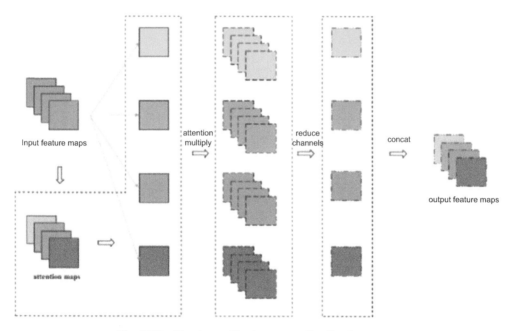

Input feature maps

attention
multiply

reduce
channels

concat

output feature maps

attention maps

Fig. 14.3 Structure of feature map attention layer

to downsample the quantity of channels with in feature map for every group. After channel downsampling, the feature maps from the four groups are finally concatenated to create a new output feature map. Since the attention layer of said feature maps' channel downsampling multiplicity is the same as that of the attention feature map, the output and the input feature maps are of the same size.

Detection Process

The detection procedure follows the following flow:

Step 1: Using the test image as input, extract the image's features using the backbone.

Step 2: Get the feature maps of the different depths of the backbone network.

Step 3: In order to combine features, the retrieved multiscale feature maps are fed into the FPN structure. Bilinear interpolation is used as an improvement to the feature map upsampling.

Step 4: The multi-scale feature maps are supplied into the PAN framework for reliable feature localization following FPN fusion. The outcomes of the identification of 3 feature maps with different scales are then obtained.

Step 5: The final results will be generated and the detecting boxes and classifications in the original input photographs will be labeled once nms has processed all feature map detecting results.

Step 6: To finish identifying each frame of the movie, redo steps 1 through 5 and extract the next frame of the visual to be identified, as illustrated in Fig. 14.4.

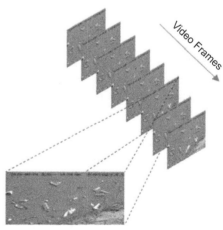

Detection result

Fig. 14.4 Frame by frame detection

3.3 Quantification of Coastal Debris

The beach debris identification and classification performed well using the object detection model, but it was unable to predict the types of debris present in the covered region. To solve the issue, the debris classes of the research region were estimated using the image segmentation method. The predicted area of the debris items is referred to as the covered area. The segmentation model was trained using the patch pictures created by the survey's extraction of images from a movie. A commercial geographic information system (GIS) tool was utilized to label the debris pieces. A polygon was produced in each patch image by joining the vertices along the outer edge of the debris object. In order to set each polygon apart from the background in the images, it was designated as a piece of trash. The tagged items were divided into seven groups for the segmentation model's training, comprising Plastic, glass, fishing gear, Thermopolis, rubber, metal and undefined. A pair of photos composed of the source patch picture and a labeled picture were also uploaded as one dataset. Utilizing the picture segmentation model U-Net, which is based on a fully convolutional network, the trash classes of the enclosed region were estimated (FCN). In the U-Net, a highly symmetric U-shaped design, skip connections are utilized to instantly link each level's output from the encoder to the corresponding level of the decoder. The model communicates the combined feature map to the appropriate decoders after concatenating the decoder feature maps which was upsampled *with* it. Even with minimal amounts of training data, the U-Net model performed well.

4. Results and Discussion

On the testing set, the model's performance was assessed employing the intersection over union (IoU) and mean average precision (mAP). The network model's predictions and the actual survey findings for the entire Dasa-port beach were compared for confirmation.

- The IoU assesses the accuracy with which projected bounding boxes correspond to the position of an item, which is defined as,

$$IoU = \frac{Area\ of\ Overlap}{Area\ of\ Union}$$
(1)

As the IoU increases, so does the overlapping among the two bounding boxes. 0.5 IoU thresholds were used to examine mAP and evaluate the performance of the model. IoU = 0.5 is true when the overlapping area of the two boxes exceeds 50% of the total area, and any value less than this is considered false.

The mAP is calculated by calculating the Average Precision (AP) for each class and then averaging it across numerous courses.

$$mAP = \frac{1}{N}\sum_{i=1}^{N}AP_i$$
(2)

The mAP takes into account both false positives (FP) and false negatives (FN), and accounts for the trade-off between precision and recall (FN). Due to this characteristic, the majority of detecting applications can use mAP as a measure.

The network model achieved a mAP score of 0.86 upon that training set having the IoU threshold set at 0.5. (Fig. 14.5). The average accuracy (AP) scores of plastic, Thermopolis, wood and rubber tubes were all greater than the mAP rating of total coastal debris, with values of 0.91, 0.90, 0.89, and 0.87, respectively. Glass (0.83), Fibre (0.83), and Unspecified (0.79) were the three classes with AP values that were lower than the mAP. Plastic (0.91) is predicted by the network model with the best precision, while Unspecified (0.79) is predicted with the lowest precision. The following Table 14.1 and Fig. 14.5 represents precision rate in predicting different kind of debris.

Table 14.1 Precision Rate in predicting different kind of debris

Debris	Precision Rate
Plastic	0.91
Thermopolis	0.90
Wood	0.89
Rubber Tubes	0.87
Glass	0.83
Fibre	0.83
Unspecified	0.79

As stated in Section 3.1, there are seven categories used to categorize the beach trash standing stock: plastic, thermopolis, wood, rubber tubes, glass, fibre and unspecified. There were 950 items in all that were actually surveyed. More counts were collected

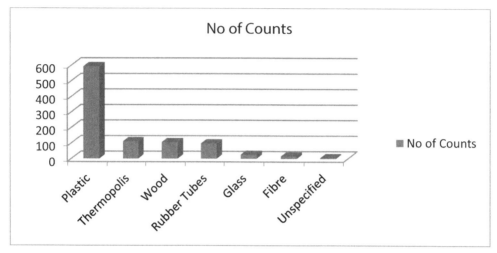

Fig. 14.5 The network model's mAP (IoU threshold = 0.5) for classifying things on beach trash images

for the class of "Plastic" than all other classes combined (apart from its own). From the research totally 950 items have been found. In that 593 pieces of plastics, 110 pieces of thermopolis, 105 pieces of wood, 97 pieces of rubber tubes, glass of 25, Fibre of 15 and unspecified was 5. The following Table 14.2 and Fig. 14.6 represents number of counts of debris by proposed algorithm.

Table 14.2 No of counts in predicting different kind of debris by proposed FMA-YOLOv5

Debris	No of Counts
Plastic	593
Thermopolis	110
Wood	105
Rubber Tubes	97
Glass	25
Fibre	15
Unspecified	5

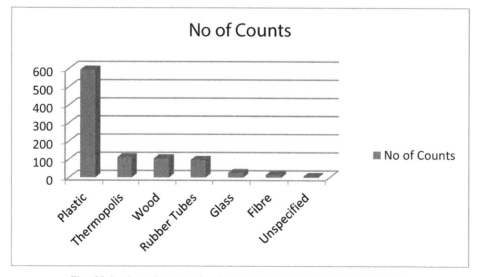

Fig. 14.6 Actual count of debris items by proposed FMA-YOLOv5

5. Conclusion

A deep learning approach was employed in this study to develop an automated method for finding and measuring marine rubbish. This study suggests the FMA-YOLOv5s network architecture for object identification and classification. A feature map attention (FMA) layer is introduced at the end of the backbone based on YOLOv5s to improvise feature extraction. U-Net, a segmentation model built using a deep learning methodology, was also employed to calculate the quantity of trash. With a precision of 0.90 mAP, the network model created for this investigation categorised the items. Then, the proposed automatic approach's prediction of marine waste standing-stock on the entire beach was compared to a statistical calculation based on a manual method. This method is capable of detecting floating items in water bodies with sufficient speed and precision in real-time. In the long term, greater focus can be placed on enhancing the recognition of blurry and dense items and supporting new categories of floating debris. Furthermore, the deep learning methodology developed in this work might be integrated with drones or unmanned airplane, allowing for the future creation and implementation of an automated monitoring system at national marine debris monitoring stations.

References

1. Arcangeli, Antonella & Campana, Ilaria &Angeletti, Dario & Atzori et al., (2017). Amount, composition, and spatial distribution of floating macro litter along fixed trans-border transects in the Mediterranean basin. Marine Pollution Bulletin. 129. 10.1016/j.marpolbul.2017.10.028.
2. Ryan, Peter & Moore, Charles & Van Franeker, Jan & Moloney, Coleen. (2009). Monitoring the abundance of plastic debris in the marine environment. Philosophical transactions of the Royal Society of London. Series B, Biological sciences. 364. 1999–2012. 10.1098/rstb.2008.0207.
3. Gall, Sarah & Thompson, R.C.. (2015). The impact of debris on marine life. Marine pollution bulletin. 92. 10.1016/j.marpolbul.2014.12.041.
4. Jambeck, Jenna & Geyer, Roland & Wilcox, Chris & Siegler et al., (2015). Marine pollution. Plastic waste inputs from land into the ocean. Science (New York, N.Y.). 347. 768-771. 10.1126/science.1260352.
5. Garcia-Garin, Odei& Sala, Berta & Aguilar, Alex &Vighi et al., (2020). Organophosphate contaminants in North Atlantic fin whales. Science of The Total Environment. 721. 137768. 10.1016/j.scitotenv.2020.137768.
6. Nnaji, Chidozie. (2015). The Status Of Municipal Solid Waste Generation And Disposal In Nigeria. Management of Environmental Quality An International Journal. 26. 10.1108/MEQ-08-2013-0092.
7. Clapp, Jennifer. (2012). The rising tide against plastic waste: Unpacking industry attempts to influence the debate.In book: Histories of the Dustheap: Waste, Material Cultures, Social Justice. 199–225.
8. Carlton, James & Chapman, John & Geller, Jonathan & Miller, et al., (2017). Tsunami-driven rafting: Transoceanic species dispersal and implications for marine biogeography. Science. 357. 1402-1406. 10.1126/science.aao1498.
9. Royer, Sarah-Jeanne &Ferrón, Sara & Wilson, Samuel & Karl, David. (2018). Production of methane and ethylene from plastic in the environment. PLOS ONE. 13. e0200574. 10.1371/journal.pone.0200574.
10. Barnes, David &Galgani, François & Thompson, Richard &Barlaz, Morton. (2009). Accumulation and fragmentation of plastic debris in global environments. Philosophical transactions of the Royal Society of London. Series B, Biological sciences. 364. 1985-98. 10.1098/rstb.2008.0205.
11. Wang, Yong & Wang, Dianhong& Lu, Qian & Luo, Dapeng& Fang, Wu. (2015). Aquatic Debris Detection Using Embedded Camera Sensors. Sensors (Basel, Switzerland). 15. 3116-37. 10.3390/s150203116.
12. Bochkovskiy, A.; Wang, C.Y.; Liao, H.Y.M. Yolov4: (2020) Optimal speed and accuracy of object detection. arXiv 2020, arXiv:2004.10934.
13. Tao, Andrew &Sapra, Karan & Catanzaro, Bryan. (2020). Hierarchical Multi-Scale Attention for Semantic Segmentation.arXiv 2020, arXiv:2005.10821.
14. Fukushima, Kunihiko. (2014). Modeling Vision with the Neocognitron. 10.1007/978-3-642-30574-0_44.
15. Gnann N, Baschek B, Ternes TA. (2022) Close-range remote sensing-based detection and identification of macroplastics on water assisted by artificial intelligence: A review. Water Res. 2022 Aug 15;222:118902. doi: 10.1016/j.watres.2022.118902. Epub 2022 Jul 30. PMID: 35944407.
16. Papakonstantinou, Apostolos &Batsaris, Marios&Spondylidis, Spiros &Topouzelis, Konstantinos. (2021). A Citizen Science Unmanned Aerial System Data Acquisition Protocol and Deep Learning Techniques for the Automatic Detection and Mapping of Marine Litter Concentrations in the Coastal Zone. Drones. 6. 10.3390/drones5010006.
17. Li, Peng, Yibing Fan, Zhengyang Cai, ZhiyuLyu, and Weijie Ren. (2022). "Detection Method of Marine Biological Objects Based on Image Enhancement and Improved YOLOv5S" Journal of Marine Science and Engineering 10, no. 10: 1503. https://doi.org/10.3390/jmse10101503

18. Sun, Junkai& Yao, Zhenfeng& Zhao, Enjin. (2020). Prediction of ocean debris drift trajectory and recovery range using a coupled numerical model.International Ocean and Polar Engineering ConferenceAt: Shanghai, China

19. Kylili, Kyriaki&Hadjistassou, Constantinos&Artusi, Alessandro. (2020). An intelligent way for discerning plastics at the shorelines and the seas. Environmental Science and Pollution Research. 10.1007/s11356-020-10105-7.

20. R. Bajaj, S. Garg, N. Kulkarni and R. Raut, 2021, "Sea Debris Detection Using Deep Learning : Diving Deep into the Sea," IEEE 4th International Conference on Computing, Power and Communication Technologies (GUCON), 2021, pp. 1–6, doi: 10.1109/GUCON50781.2021.9573722.

21. Xue, Bing & Huang, Baoxiang& Wei, Weibo & Ge, Chen & Li, Haitao& Zhao, Nan & Zhang, Hongfeng. (2021). An Efficient Deep-Sea Debris Detection Method Using Deep Neural Networks. IEEE Journal of Selected Topics in Applied Earth Observations and Remote Sensing. PP. 1–1. 10.1109/JSTARS.2021.3130238.

22. Winans, W. Ross & Chen, Qi & Franklin, Erik &Qiang, Yi. (2021). Automatic detection of Hawai'i's shoreline stranded mega-debris using deep learning-based object detection. 10.13140/RG.2.2.23593.72807.

23. Song, Kyounghwan& Jung, Jung-Yeul& Lee, Seung & Park, Sanghyun. (2021). A comparative study of deep learning-based network model and conventional method to assess beach debris standing-stock. Marine Pollution Bulletin. 168. 112466. 10.1016/j.marpolbul.2021.112466.

Note: All the figures and tables in this chapter were made by the authors.

Human Machine Interaction in the Digital Era – Prof. J. Dhilipan et al. (eds)
© 2024 Taylor & Francis Group, London, ISBN 978-1-032-54998-9

Recommender Model for e-Learning Environment Using Enhanced Support Vector Space Model (ESVM)

15

Manikandan Rajagopal[1]

Associate Professor, CHRIST (Deemed to be University), Bangalore

Surekha R Gondkar[2]

Associate professor, Department of Electronics and Communication Engineering,
BMS Institute of Technology and Management, Bengaluru, Karnataka

Ramkumar Sivasakthivel[3], Gobinath Ramar[4]

Associate Professor, CHRIST (Deemed to be University), Bangalore

Abstract One of the most intriguing studies in the world of education is custom online learning based on a recommender system. The majority of academics created a variety of recommender e-learning approaches that use recommendation methods in mining of educational data particularly for the determination of learning choices of students. But it doesn't produce satisfactory outcomes. The suggested system designed an Enhanced Vector Space Model (ESVM) based recommendation for the Protus system in order to improve suggestion accuracy and reduce query processing time. The learning styles of students are initially derived from server blogs. Similarity computation and recommendations are carried out when preprocessing is finished. The system first creates a suggestion list using content-based filtering. Based on the adjusted cosine similarity of their content, the results are ranked. In order to accurately classify the most active participant in the group, we use a collaborative method. The experimental results prove that the designed EVSM scheme performs better than the old system in terms of query processing speed, MAE, and correctness.

Keywords Document frequency, Adjusted cosine, Recommendation system, and Vector space model (VSM)

1. Introduction

E-learning scenarios are becoming more and more commonplace in educational settings. A significant change has been brought in conventional methods of learning owing to the advanced pedagogy through e-learning, with a new situation (understudies) [1-3]. In order to stimulate student input and direct students' learning processes, teachers need a programmed strategy on their end. On the student's side, an e-learning framework that could organically guide the student's exercises and intelligently produce and suggest learning resources would be incredibly beneficial [4-7]. Finding learning resources pertaining to student preference is difficult due to the abundance of learning resources available online. A group of things are expected to be prescribed to students by e-learning recommender frameworks, i.e., the most effective or practical approaches within a wide range of learning resources to achieve a specific fitness[8-9]. Additionally, selecting the most effective teaching method for each student and putting it into practise in a real classroom is quite difficult for educators, and the present eLearning frameworks do not provide

[1]manikandan.rajagopal@christuniversity.in, [2]surekha.r.gondkar@bmsit.in, [3]ramkumar.s@christuniversity.in, [4]gobinath.r@christuniversity.in

DOI: 10.1201/9781003428466-15

a good way to monitor students' progress. Many academic and industry researchers have already tried to address the challenges of RS for improving on the quality of recommendations which are supplied to the learners. The learner centric approach is what expected from the online learners [10-11]. The level in which the learners are motivated and due interest are also of more importance and one of the key metric in measuring the outcome of learning. This motivates the researchers to come out with more efficient recommendation models for improving the quality of learning outcome through recommendations. According to certain researchers, advancing the execution of personalized learning can be done by taking into account the students' dimensions of information [12-13]. Along these lines, personalization is strongly impacted by a student's capacity. According to the theory of item responses that gives a tailor made learning environment that largely depend on the materialistic parameters, B-Spoke framework for enhanced e-learning is introduced in[14]. A personalized framework for learning English vocabulary was developed based on the memory cycles in learning pattern was introduced in [15]. The framework recommends suitable and appropriate content for mastering vocabulary based in individual participants' capacity and memory based cycles [16]. A multi-specialist tailored e-learning framework based on IRT and Artificial Neural Network (ANN) was developed [17].

2. Proposed Methodology

Proposed demonstrate a recommendation system for Protus with a module that can automatically adapt to student needs and skill levels. By assessing students' learning styles and mining their server logs, this system detects various examples of learning style and students' propensities. In a case where the student is new to the learning platform, the proposed module for recommender understands the present level of aptitude of the students and encourages them to take basic level courses and an personal profile is built on basis of how the learner is performing. Once the course is completed by taking up a test at final, the recommender framework develops a recommendation set for the future courses which the student is likely to prefer based on the results. When a student engages with the framework, information mining algorithms employ data about the student's learning preferences to create a student profile and intelligent recommendation. Examples of this data include navigational examples, preferences, accessible contents, and bookmarks. The intended system's flow diagram is displayed in Fig. 15.1.

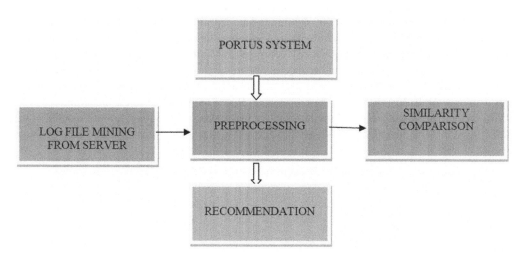

Fig. 15.1 Flow diagram of the proposed system

2.1 Recommendation Process

The vector space model is then used foe indicting the recommended basic level course, the articles and the relevant materials. For co-relating the numerical functions on the documents, the VSM is implemented for representing the appropriate reports in a multi-dimensional view. The reports are seen as vector here. Every item is represented as a point in the vector space, and the closer things are assumed to be the most pertinent or relevant. The significant sub vectors of a course, article, or video are compared to one another, and comparability is calculated using the adjusted Cosine Similarity and TF-IDF loads. Incorporate the comparable Vector Space Model into our exploration in this way. The frequency of a related term (t) within a specific record (d) is known as the time frequency (TF), which may be written as $tftd$. The following is the TF weight equation.

$$W_{t,d} = \begin{cases} 1 + \log_{10} tf_{t,d} & \text{if } tf_{t,d} > 0 \\ 0 & \text{otherwise} \end{cases} \tag{1}$$

Document Frequency (DF), abbreviated as *dft*, provides the number of records that include a specific phrase *t*. On the other hand, Inverse Document Frequency (IDF) increases the relevance of less often used terms while decreasing the importance of frequently used terms.

$$idf_t = \log_{10}\left(\frac{N}{df_t}\right) \qquad (2)$$

The term specific weight of the scheme is determined by multiplying the TF and IDF weights. In order to get adjusted cosine similarity, this value is used.

$$W_{t,d} = (1 + \log tf_{t,d}) \times \log_{10}\left(\frac{N}{df_t}\right) \qquad (3)$$

The similarity between two vectors is used to calculate cosine similarity. It can be used by the system to compare two courses based on a corresponding feature *p*. The adjusted cosine similarity is computed by the system using TF-IDF weights.

The cosine similarity measure does not take into account the situation where a certain user uses a different rating system. The average rating of user u for all the things user u has evaluated is subtracted by adjusted cosine similarity.

$$\text{Adjusted Cosine Similarity} = \frac{\sum_{i \in I}(r_{u,i} - \overline{r_u})(r_{v,i} - \overline{r_v})}{\sum_{i \in I}(r_{u,i} - \overline{r_u})^2 + \sum_{i \in I}(r_{v,i} - \overline{r_v})^2} \qquad (4)$$

Where, *i*-target item (materials) and *u* and *v*-Users. The system hence can determine the similarity index using the TF-IDF approach for weighting through comparison of each of the course for the initial inclination with all other courses in the similar domain or the list.

3. Experimental Results

3.1 A comparison of Query Processing Times

The proposed EVSM-based recommendation and the current VSM-based recommendation scheme are compared in terms of query processing time in Fig. 15.2. The *y*-axis represents query processing time, and the *x*-axis represents the number of learners. The experimental results demonstrate that the new system produced better results than the existing system.

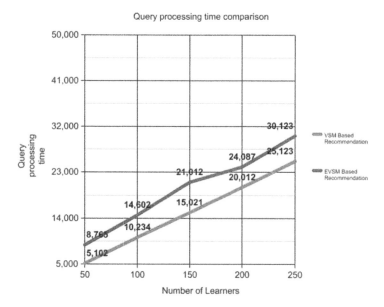

Fig. 15.2 Comparison of query processing time

3.2 Mean Absolute Error Metric (MAE)

The MAE comparison of the proposed EVSM and the current VSM-based recommendation approach is shown in Fig. 15.3. The results of the inquiry demonstrate that, in comparison to the prior approach, the proposed method achieves the smallest MAE value.

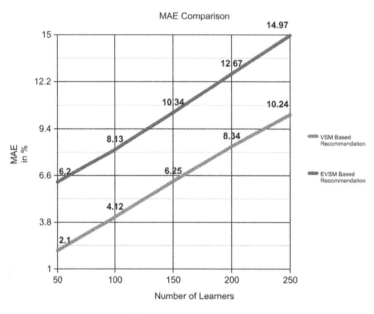

Fig. 15.3 Comparison of MAE

3.3 Comparison of Precision

The accuracy performance for the proposed EVSM-based recommendation and the current VSM-based recommendation approach are compared in Fig. 15.4. The experimental results demonstrate that the designed system produced better results than the existing system.

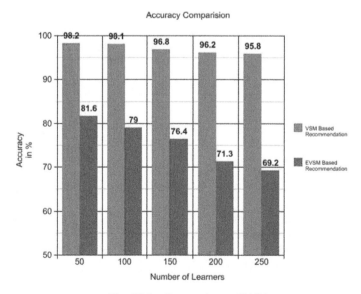

Fig. 15.4 Comparison of MAE

4. Conclusion

This study has described the creation of a recommendation method for an online learning system based on an Enhanced Vector Space Model (EVSM). This strategy combined content-based filtering with collaborative filtering techniques to boost the effectiveness of the tailored system. The adjusted cosine similarity measure is offered in this proposed study to assess the similarity between learners. The analysis's findings indicate that, in terms of query processing time, MAE, and accuracy, the suggested system outperforms the current methodology. Future recommendations will be more accurate thanks to the use of various optimization methods like the Firefly Algorithm (FA) and Particle Swarm Optimization (PSO) algorithms.

References

1. Brockman, P., French, D. and Tamm, C. (2014).REIT organizational structure, institutional ownership, and stock performance. J. Real Estate Portf. Manag. 20(1): 2136.
2. Cella, C. (2009).Institutional investors and corporate investment. United Sates: Indiana University, Kelley School of Business.
3. Chuang, H. (2020). The impacts of institutional ownership on stock returns. Empir. Econ. 58(2): 507533.
4. Clark, G.L. and Wójcik, D. (2005). Financial valuation of the German model: the negative relationship between ownership concentration and stock market returns, 1997–2001.Econ. Geogr. 81(1): 1129.
5. Dasgupta, A., Prat, A. and Verardo, M. (2011). Institutional trade persistence and long-term equity returns. J. Finance. 66(2): 635653.
6. Demsetz, H. and Lehn, K. (1985). The structure of corporate ownership: causes and consequences. J. Polit. Econ. 93(6): 11551177.
7. Dyakov, T. and Wipplinger, E. (2020). Institutional ownership and future stock returns: an international perspective. Int. Rev. Finance. 20(1): 235245.
8. Gompers, P. A. and Metrick, A. (2001).Institutional investors and equity prices. Q. J. Econ. 116(1): 229259.
9. Han, K. C. and Suk, D. Y. (1998). The effect of ownership structure on firm performance: Additional evidence. Rev. Financ. Econ.7(2): 143155.
10. Kennedy, P. (1985). *A Guide to Econometrics*, MIT Press, Cambridge.
11. La Porta, R., Lopez-de-Silanes, F., and Shleifer, A. (1999). Corporate ownership around the world. J. Finance. 54(2): 471517.
12. Manawaduge, A.S., Zoysa, A., and Rudkin, K. M. (2009).Performance implication of ownership structure and ownership concentration: Evidence from Sri Lankan firms. Paper presented at the Performance Management Association Conference. Dunedin, New Zealand.
13. McNulty, T. and Nordberg, D. (2016). Ownership, activism and engagement: institutional investors as active owners. Corp. Gov.: Int. Rev. 24(3): 346358.
14. Othman, R., Arshad, R., Ahmad, C.S. and Hamzah, N.A.A. (2010). The impact of ownership structure on stock returns, In 2010 International Conference on Science and Social Research (CSSR 2010) (pp. 217–221). IEEE.
15. Ovtcharova, G. (2003). Institutional Ownership and Long-Term Stock Returns. Working Papers Series. Available at SSRN: https://ssrn.com/abstract=410560.
16. Shleifer, A. and Vishny, R.W. (1986).Large shareholders and corporate control. J. Polit. Econ. 94(3): 461488.
17. Sikorski, D. (2011).The global financial crisis. *The Impact of the Global Financial Crisis on Emerging Financial Markets. Contemporary Studies in Economic and Financial Analysis*,ed. A. Jonathan Batten, and G Peter Szilagyi, 93:1790. United Kingdome: Bingley: Emerald Group Publishing.
18. Singh, A. and Singh, M. (2016). Cross country co-movement in equity markets after the US financial crisis: India and major economic giants. J. Indian Bus.

Note: All the figures in this chapter were made by the Authors

Human Machine Interaction in the Digital Era – Prof. J. Dhilipan et al. (eds)
© 2024 Taylor & Francis Group, London, ISBN 978-1-032-54998-9

Ripeness Detection in Watermelon Using CNN

16

A. T. Madhavi[1]

Assistant Professor, Department of Electronics and Communication Engineering,
Easwari Engineering College, Chennai, India

K. Rahimunnisa[2]

Associate Professor, Department of Electronics and Communication Engineering,
Easwari Engineering College, Chennai, India

E. Thilagavathy[3]

Student, Department of Electronics and Communication Engineering,
Easwari Engineering College, Chennai, India

N. Vithiya[4]

Student, Department of Electronics and Communication Engineering,
Easwari Engineering College, Chennai, India

Abstract Watermelon is a valuable fruit with a long storage period, and its ripeness is crucial for its taste, quality, and preservation. Traditional methods of assessing ripeness are time-consuming and prone to errors. This study used CNN models to categorise watermelon ripeness into unripe, ripe, and overripe, using Jupyter and several libraries. The MobileNetV2 network performed the best, with an accuracy rate of 95.97%.

Keywords CNN-Convolutional neural network, Watermelon, MobileNetV2

1. Introduction

The watermelon is a huge fruit with a maximum diameter of 25 cm and a maximum weight of 15 kg. Its form is oval or spherical, and its smooth, dark-green rind occasionally sports errant pale-green patches. It has high water content and little protein or fat, making it low in calories. People have difficulty determining ripeness with their naked eyes and even by touching. Convolutional neural networks (CNN) are a popular method for doing this because of their exceptional effectiveness in extracting useful information from raw images without the need for humans. This study concentrated on using deep learning to capture images of watermelon fruit and analyse them in order to detect its ripeness stages using non-destructive testing techniques.

2. Literature Review

[Yinghao Zhang et al., 2018] comparison of ELM and SVM is performed in relation to the classification results, and an accuracy of 92% is achieved. [DengfeiJie et al., 2019] used VIS/NIR transmittance technology and created a spectral characteristic

---

[1]madhavi.t@eec.srmrmp.edu.in, [2]rahimunnisa.k@eec.srmrmp.edu.in, [3]darshudisha2404@gmail.com, [4]vithiyarajeswari@gmail.com

DOI: 10.1201/9781003428466-16

analysis classification technique using the ratio of peak1 to peak2 intensity (RPP) and the normalised difference intensity of peak (NDIP) approaches to identify the mature phases of watermelon.

[Edwin R. Arboleda, et al., 2020] employed near-infrared spectroscopy (NIRS) to determine the ripeness level of watermelon. [Ketsarin et al., 2021] Proposed a non-destructive method that uses auditory signals, image processing, weight, rind designs, sound signals, and ML techniques to detect watermelon sweetness and achieved an accuracy of 92%.

[Anthony B. Villa, et al., 2022]studied the technology of using an Android-Based Application on Watermelon Fruit Recognition using a deep learning CNN model to classify a watermelon's ripeness.

3. Data and Variables

Datasets were created by capturing images using an Android phone's camera at a vertical distance of around 40 to 50 cm from the top of the sample. The top, bottom, and lying sides were all taken into consideration as datasets. 300 images used in this study were split into (100 for unripe, 100 for ripe, and 100 for overripe) which can be seen in watermelon from various angles. 80% of the image dataset was used to train the model, while 20% was used to test it.

4. Proposed Methodology

In order to categorise the watermelon's ripeness using the non-destructive method, CNN is used. By comparing and analysing CNN models that provide high accuracy, low loss, and less compilation time among various models, the best model for watermelon ripeness detection is chosen.

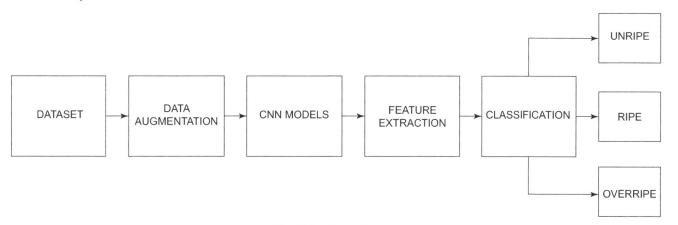

Fig. 16.1 Block diagram

Before using the image dataset for training, the background of each was manually removed using an online background removal tool, and the resulting image was given as the input to the CNN. For training, the image datasets are reduced in size to 224 by 224 pixels.

Convolutional neural networks (CNNs) employ the data augmentation approach to artificially increase the size of a training dataset. This helps to improve the generalisation of the model and avoid overfitting. Figure 16.2 shows the data augmentation techniques. The Keras deep learning library offers data augmentation to train a model with the ImageDataGenerator class.

The four types of CNN models, such as AlexNet, ResNet50, VGG19, MobileNetV2, and DenseNet121, have been studied. Deeper layers of a CNN detect more complicated features such as textures and patterns, while early layers often only detect simple features like edges and corners. Figure 16.3 shows the ripeness stages of watermelon. Based on the classification, it would predict the ripeness stages of watermelon.

5. Results and Discussion

The performance of the deep learning model used for classifying watermelon ripeness was evaluated using training accuracy, validation accuracy, training loss, validation loss, and compilation time at each epoch. There were 20 epochs in total for each experiment. In Fig. 16.4, the training and validation accuracy, testing and validation loss, for each experimental run, are represented graphically, and Table 16.1 shows the compilation time for different CNN.

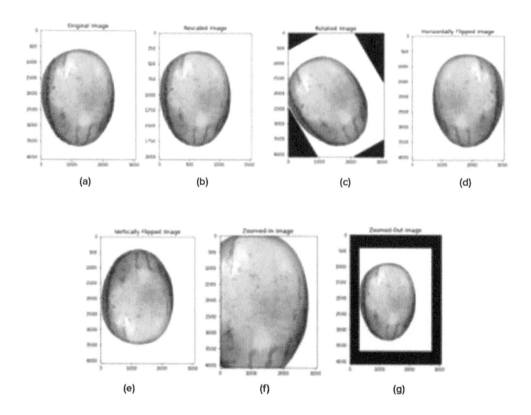

Fig. 16.2 Data augmentation techniques

UNRIPE   RIPE   OVERRIPE

Fig. 16.3 Ripeness stages of watermelon

5.1 Accuracy

The models' classification performance was evaluated based on their training and validation accuracy values. Figure 16.4(a) shows the accuracy result for various CNN models. Among them, MobileNetV2 achieved the highest accuracy for both training and validation, with 95% and 90%, respectively. VGG19 had the second-highest accuracy, with a training accuracy of 88% and a validation accuracy of 81%. ResNet50, AlexNet, and DenseNet121 had lower accuracy values.

5.2 Loss

The loss value measures a model's performance after each iteration, with less loss indicating better performance, except for over-fitting. Figure 16.4(b) shows the loss rates of different models during training and testing. As the number of epochs increased, the rate of loss generally decreased. MobileNetv2 had the lowest loss rates of 0.528% and 0.7928% for training and validation, respectively. VGG19 had the second-lowest loss rates. AlexNet, ResNet50, and DenseNet121 had higher loss rates.

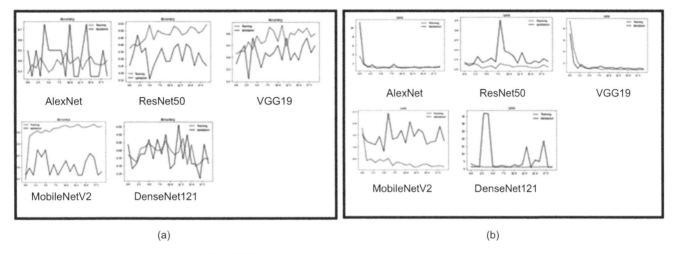

(a)   (b)

Fig. 16.4   Comparison of accuracy (a) and Loss (b)

5.3 Compilation Time

A model's effectiveness depends on both its accuracy and completion time. Table 16.1 shows the compilation time in minutes and seconds. DenseNet121 had the longest completion time of 31 minutes and 86 seconds, while MobileNetv2 had a shorter time of 3 minutes and 51 seconds. MobileNetV2 had the best prediction results, with the highest accuracy and lowest compilation time.

Table 16.1   Compilation time of different models

Models	Compilation time
AlexNet	5 min 53 s
ResNet50	13 min 3 s
VGG19	14 min
MobileNetV2	3 min 51 s
DenseNet121	31 min 86 s

6. Conclusion

The study shows that non-destructive methods can be used to determine whether a watermelon is unripe, ripe, or overripe. This study presents a comparison of different CNN-based models for classifying watermelon ripeness. Models evaluated include AlexNet, ResNet50, VGG19, MobileNetV2, and DenseNet 121. A dataset made up of 300 images of watermelons was used in these models. Performance analysis was performed by comparing three performance metrics: accuracy, loss, and classification time. Overall, the MobileNetV2 showed the best performance in watermelon ripeness classification, with 95.97% accuracy, the lowest loss, and the least compilation time.

References

1. YinghaoZhang,XiaoyanDeng,Zhou Xu, PeipeiYuan,"Watermelon Ripeness Detection via Extreme LearningMachine with Kernel Principal ComponentAnalysis Based on Acoustic Signals", November 2018International Journal of Pattern Recognition and Artificial Intelligence doi: 10.1142/S0218001419510029
2. Dengfei Jie, Wanhuai Zhou, Xuan Wei, "Nondestructive detection of maturity of watermelon by spectral characteristic using NIR diffuse transmittance technique",https://doi.org/10.1016/j.scienta.2019.108718

3. Wacharawish Daosawang, Komsan Wongkalasin, Natthapong Katewongsa, "A Study Sound Absorption for Ripeness and Unripe Classification of Watermelon", Published in 2020 8th International Electrical Engineering Congress (iEECON) doi: 10.1109/iEECON48109.2020.229521

4. Ketsarin Chawgien, Supaporn Kiattisin,"Machine learning techniques for classifying the sweetness of watermelon using acoustic signal and image processing",https://doi.org/10.1016/j.compag.2020.105938

5. Joe Garvin et al.,"Microwave imaging for watermelon maturity determination", December 2022 Current doi:10.1016/j.crfs.2022.100412

6. Edwin R. Arboleda, Kimberly M. Parazo, Christle M. Pareja, "Watermelon ripeness detector using near-infrared spectroscopy", October 2020Jurnal Teknologi dan Sistem Komputer 8(4):317-322 doi:10.14710/jtsiskom.2020.13744

7. Pavadharini T, Anita HB, "Classification of Watermelon using Sound Processing", Published By: Blue Eyes Intelligence Engineering & Sciences Publication, doi: 10.35940/ijeat.D8498.049420

8. Anthony B. Villa et al.,"Determination of Citrullus Lanatus "Sweet-16" Ripeness Using Android-Based Application", Proc. of the 3rd International Conference on Electrical, Communication and Computer Engineering (ICECCE) 12-13 June 2021, Kuala Lumpur, Malaysia

9. Az-AimanFariq Mat Saat, Nur Anida Jumadi,"Development of Watermelon Ripeness Grading System Based on Colour Histogram", Conference Paper · December 2020 doi: 10.30880/eeee.2020.01.01.030, ResearchGate, 2021

10. N. Ahmad Syazwan, M. S. B. Shah Rizam, M. T. Nooritawati," Categorization Of Watermelon Maturity Level Based On Rind Features", 1877–7058 © 2012 Published by Elsevier Ltd. Open access under CC BY-NC-ND license doi: 10.1016/j.proeng.2012.07.327

Note: All the figures and tables in this chapter were made by the Authors

Human Machine Interaction in the Digital Era – Prof. J. Dhilipan et al. (eds)
© 2024 Taylor & Francis Group, London, ISBN 978-1-032-54998-9

Heart Disease Prediction and Analysis Using Ensemble Classifier in Machine Learning Techniques

17

Jasmine Jinitha A.[1]

Research Scholar, Vels Institute of Science Technologies and
Advanced Studies (VISTAS), Chennai

S. Mangayarkarasi[2]

Associate Professor, Department of Computer Science,
Vels Institute of Science Technologies and Advanced Studies (VISTAS), Chennai

Abstract Heart disease is the most dangerous diseases among the people worldwide. It can diagnose earlier using medical history, but this method has been found to be unreliable. Earlier identification of heart disease accurately and promptly will increase the preventing heart failures. Manual methods for diagnosing cardiac disease must be more accurate and subject to inter-examiner variability. Machine learning techniques are more effective and dependable in classifying heart diseases or not. Previous studies had several drawbacks, like computing slow processes, sometimes speedy but inaccurate. To overcome this, we proposed an Ensemble classifier to achieve classification accuracy. In this study, we analyzed numerous machine learning algorithms utilized for predicting and identifying the heart diseases. The proposed mechanism involves data preprocessing, and the resultant data is applied to the proposed system for predicting heart disease. The experimental work determines the enhanced performance of the proposed system in the aspect of ROC, AUC, recall, precision, F-measure, and accuracy. The proposed Ensemble classifier has achieved 98.8% accuracy in predicting heart diseases, whereas existing approaches such XGBoost [Farhat et al.2022] has achieved 90%, and Smote-XGboost [Ishaq et al. 2021] has achieved 85.6% of accuracy.

Keywords Classification, Machine learning, Preprocessing, Feature selection, Heart disease and disease prediction

1. Introduction

Cardiovascular disease is a serious heart disease; worldwide, it has one-third of the annual death rates [1]. Many scientists have created algorithms for diagnosing heart disease to improve the current situation using machine learning to extract usable data from available medical datasets. Utilizing these diagnostic techniques can facilitate clinical evaluations of medical diagnoses, improve the diagnosis procedure, and reveal disease-related information will save more lives. An accurate diagnosis and effective treatment can preserve the lives of patients, several tests are necessary for predicting cardiac disease. Several machine learning-based approaches have been used to forecast the probability of finding heart disease. Most of these techniques use publicly accessible datasets to train and test models. The availability of these datasets has improved the efficiency of machine learning-based predictive analytics and created new opportunities for scholars to create enhanced algorithms for estimating the risk of developing cardiac disorders.

[1]jasminejinitha@gmail.com, [2]smangai.research@gmail.com

DOI: 10.1201/9781003428466-17

Machine learning (ML) algorithms are more effective in dealing with nonlinear and complex features. Several algorithms, including KNN, SVM, and LR, have been effectively used to address a range of illnesses classification and prediction difficulties, such as an advanced indicator for Electro Cardiogram (ECG) detection and prediction the congenital heart disease [Che et al. 2021]. Researchers have proposed various machine learning models for improved cardiac disease prediction, including KNN, LR, DT, NB, FR, SVM, and others. Creating an advanced and economical method of accurately forecasting heart disease is essential. An Ensemble classifier as the machine learning classifier is proposed for predicting heart diseases. The proposed system is executed with the medical dataset containing the records of heart diseases.

Organization of this paper: in section 1 introduction is discussed, related works in section 2, the proposed mechanism in section 3, experimental works in section 4, and the conclusion in section 5.

2. Literature Review

Ali et al. applied a feature selection mechanism to execute the low-dimensional data obtained from the sensor data and medical records. To get the dataset, they employed a feature selection method that considered data flow and feature ranking. By using an ensemble deep learning method, they achieved higher accuracy than the existing models.

Ishaq et al. integrated random forest for feature ranking and SMOTE technique for data balancing. Each model's prediction accuracy on balanced data was much higher than on unbalanced data without treatment. The prediction results of nine popular algorithms were analyzed on data containing both SMOTE-generated samples and untreated outliers.

Joo et al. considered the cardiovascular disease dataset to predict cardiovascular disease risk. In which, they extract 25 variables from dataset by integrating the results of health examinations and questionnaire responses. The information from the physician's medication is significant for performing feature selection.

Umarani et al. developed a machine learning-based heart disease prediction mechanism by implementing XGBoost. A combined architecture using hybrid synthetic minority over-sampling technique-edited nearest neighbor (SMOTE-ENN), density-based spatial clustering of applications with noise (DBSCAN), heart disease prediction model (HDPM), clinical decision support system (CDSS) is applied for outliner detection and distribution of the trained data. Finally, XGBoost is implemented for predicting heart diseases.

Abdul et al. concentrated on utilizing machine learning (ML) techniques to increase the forecast accuracy of heart disorders. The author applied hyper parameter tuning in the ML algorithms in this work. The trained dataset is executed to the nine classifiers with hyperparameter tuning and without hyperparameter tuning.

For precise and prompt detection of heart diseases, Farhat et al. proposed a robust machine learning approach. It is a two-phase work initially analyzing the performance of SVM, SGD, KNN, LR, RF, DT, and NB with the Cleveland heart disease dataset. Next, LASSO, mutual information, MultiSURF, variance threshold and ANOVA are implemented to extract the relevant features.

Yang et al. proposed a Smote-Xgboost Algorithm and Feature Optimization model for predicting heart diseases. Initially, feature optimization is applied to obtain the significant features from the dataset and resolve the overfitting problems. Next Smote-Enn algorithm is applied for data balancing and then Xgboost algorithm is implemented to get the final prediction result.

3. Proposed Methodology

The proposed ensemble classifier, a machine learning approach comprising various processes such as preprocessing, applying ensemble classifier, bagged trees, prediction, feature selection, and performance evaluation. Figure 17.1 illustrates the proposed architecture, and the proposed workflow is described in upcoming sections.

4. Data Preprocessing

Data preprocessing is the mechanism for "cleaning and transforming " data; the data may be in various formats, including structured, semi-structured, or unstructured. These data have been taken directly from the input dataset or extracted in another way. Data processing is computational biology's most complex and time-consuming machinelearning task. This complexity is due to the medical dataset's irregular, redundant, unnecessary, and noisy information. When addressing data preparation, the terms "data creation" and "filtering procedures" are essential. The main advantage of preprocessing is it accelerates

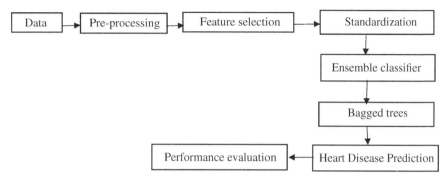

Fig. 17.1 Proposed architecture

total execution. Data preprocessing includes filtering, standardization, instance determination, feature selection, extraction, transformation, and so forth. The resulting training set is the dataset obtained by data preprocessing.

Ensemble classifier with Bagged Trees

The ensemble classifier belongs to the machine learning algorithms that build a group of classifiers and then categorize incoming data points by weighing the predictions/classifications of those classifiers. Compared to the conventional machine learning approaches, which attempt to study one hypothesis from the training dataset. Ensemble learning trains several learners to tackle the same issue. Ensemble methods attempt to create hypotheses sets and combine them for use in prediction.

A bagging tree is an ensemble approach applied for accurately classifying the data. It is also referred to as Bootstrap Aggregation. Initially, the base classifiers c1, c2,..cn are built on the bootstrap samples D1, D2,.., and Dn with a replacement before generating the decision trees. Later, all basic classifiers c1, c2,... cn are combined with the majority votes to create the final decision tree. Thus, this proves that bagging plays a vital role in medical diagnosis and can be used with any classifiers.

5. Feature Selection

Accurate feature selection is significant in improving classification accuracy. In the proposed system, the resultant data from the ensemble classifier is used for feature selection using various attributes. The selection of a few essential aspects results in accurate prediction outcomes at minimum time. The sample distribution in most disease datasets needs to be more balanced, fewer samples fall into the positive category whereas more fall into the negative group. The performance evaluation is discussed in the experimental and result section.

6. Experimental Results

Experimental work is discussed in this section; the dataset taken for this research is covid-19 dataset obtained from the ECDC repository. ECDC repository is the widely utilized repository for research works and contains many dataset sets linked to COVID-19. We have treated both regular and ICU-admitted patients daily and weekly.

MATLAB is used to implement the proposed work, and the result is obtained in Recall, F-measure, and Precision.

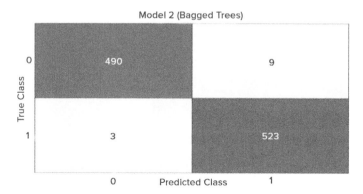

Fig. 17.2 Number of observations (Confusion Matrix)

Figure 17.22 defines the observation of the proposed ensemble classier with the input dataset. The result shows the bagged tree formation in which the x-axis signifies the prediction class, and y-axis signifies the true class.

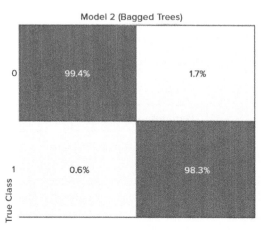

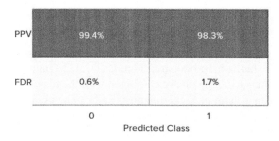

Fig. 17.3 PPV and FDR

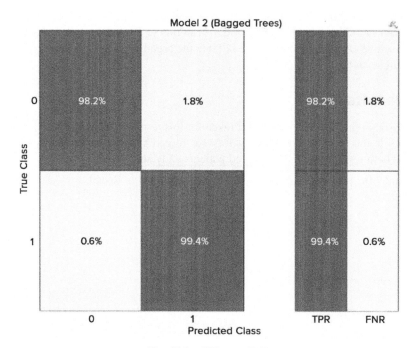

Fig. 17.4 TPR and FNR

Figure 17.3 defines the observation of the proposed ensemble classier with bagged trees from the input dataset. The result shows the Positive Prediction Value (PPV) and the False Discovery Rate (FDR). The x-axis signifies the prediction class, and y-axis signifies the true class.

Figure 17.4 defines the observation of the proposed ensemble classier with bagged trees from the input dataset. The result shows a True Positive Rate (TPR) and a False Negative Rate (FNR). The *x*-axis represents the prediction class, and y-axis represents the true class.

6.1 Scatter Plot

A Scatter plot is a mathematical plot or representation that determines the result of two data collection variables. Scatter plot applies Cartesian coordinates, according to which the horizontal axis determines the one variable value and the vertical axis determines the other variable value. The plotting points visualize the shape, color, or size.

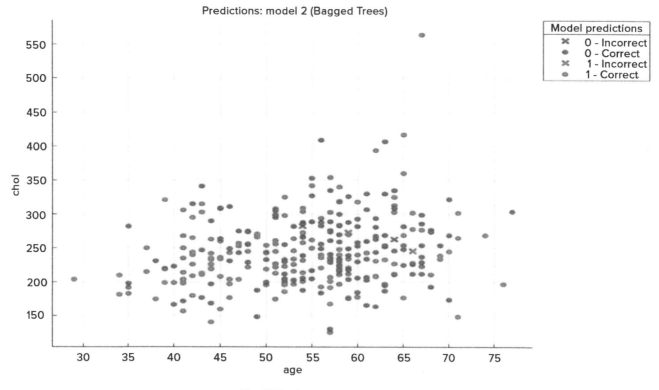

Fig. 17.5 Scatter plot age vs. chol

Figure 17.5 shows the observation of the scatter plot between the ages and chol. The bagged tree-based prediction model determines the correct and incorrect prediction concerning age and chol. The color and symbol uniquely determine each prediction. The x-axis signifies the age, and y-axis signifies the chol.

6.2 Area Under ROC Curve (AUC)

The area under the ROC curve (AUC) is similar to the integral ROC curve, which contrasts true positive rate (TPR) values and false positive rate (FPR) values in the range of 0 to 1. The AUC offers a thorough performance evaluation over all conceivable levels. AUC values vary from 0 to 1, and higher numbers indicate more effective classifiers. In order to achieve a true positive rate of 1 for any threshold, a perfect classifier constantly gives positive class tags to positive class data. An AUC value of 1 is obtained for this ideal classifier by drawing a line connecting the points [0, 0], [0, 1], and [1, 1]. On the other hand, the threshold values, true positive rate and false positive rate will be the same for a random classifier that assigns scores at random. The AUC is 0.5, and the ROC curve is diagonal.

Figure 17.6 shows the Bagged tree prediction result based on ROC with 0 positive class. The y-axis signifies True Positive Rate (TPR) and x-axis signifies false positive rate (FPR).

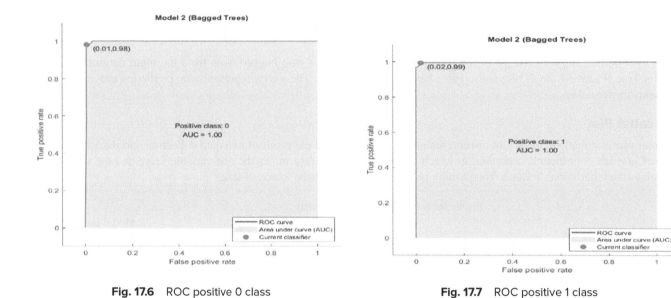

Fig. 17.6 ROC positive 0 class   **Fig. 17.7** ROC positive 1 class

Figure 17.7 shows the Bagged tree prediction result based on ROC with positive 1 class. The x-axis signifies false positive rate (FPR), and y-axis signifies the True Positive Rate (TPR).

Table 17.1 Detailed accuracy value by class

=== Detailed Accuracy by Class ===								
TP Rate	**FP Rate**	**Precision**	**Recall**	**F-Measure**	**MCC**	**ROC Area**	**PRC Area**	**Class**
0.913	0.108	0.913	0.913	0.913	0.805	0.974	0.980	absent
0.892	0.087	0.892	0.892	0.892	0.805	0.974	0.968	present
0.904	0.099	0.904	0.904	0.904	0.805	0.974	0.975	

Table 17.1 describes the detailed accuracy value by class obtained from the dataset. The prediction classes are absent and present. The parameters considered are TP rate, Recall, FP rate, F-Measure, Precision, PRC area, MCC and ROC. The obtained values respective to the parameters are mentioned in the above table.

7. Comparison Results

To evaluate the proposed ensemble classifier efficiency, a comparison work is conducted with the existing XGBoost and Smote-Xgboost approaches. The accuracy is determined by important parameters such as TP rate, Recall, FP rate, F-Measure, Precision, PRC area, MCC and ROC. The obtained results with the proposed system are 98.8%, It proves that the accuracy achieved by the proposed ensemble classifier is far better than the others.

Table 17.2 Accuracy comparison table

Algorithms	Accuracy in %
Ensemble Classifier	98.8%
XGBoost [12]	90%
Smote-Xgboost [15]	85.6%

8. Conclusion

In this work, an Ensemble classifier with the bagged tree is proposed to predict heart disease from the covid-19 dataset from the ECDC repository. This work mainly focuses on overcoming the drawbacks of existing approaches such as lack of accuracy and maximum time consumption. The research on machine learning approaches dealing with the combination of Covid-19 and heart disease is minimum. The major problem with machine learning is that sometimes a dataset can be accurately classified, even though it can be improved if its attributes are successfully extracted. In the proposed system, preprocessing is initially computed with the dataset to remove the noisy information. Next, an ensemble classifier with the bagged tree is applied to get the prediction model on attributes like TP rate, Recall, FP rate, F-Measure, Precision, PRC area, MCC and ROC. Finally, the proposed Ensemble classifier performance is evaluated with MATLAB, which results in an accuracy of 98.8%. The obtained results prove its accuracy level in prediction is more efficient than the existing approaches.

References

1. Cardiovascular Diseases. Available online: https://www.who.int/health-topics/cardiovascular-diseases/(accessed on 10 September 2022).
2. K. Polat and S. G¨unes¸, "Artificial immune recognition system with fuzzy resource allocation mechanism classifier, principal component analysis and FFT method based new hybrid automated identification system for classification of EEG signals," Expert Systems with Applications, vol. 34, no. 3, pp. 2039–2048, 2008.
3. Ishaq, A.; Sadiq, S.; Umer, M.; Ullah, S.; Mirjalili, S.; Rupapara, V.; Nappi, M. Improving the Prediction of Heart Failure Patients' Survival Using SMOTE and Effective Data Mining Techniques. IEEE Access 2021, 9, 39707–39716. [CrossRef]
4. F. M. J. M. Shamrat, M. A. Raihan, A. K. M. S. Rahman, I. Mahmud, and R. Akter, "An analysis on breast disease prediction using machine learning approaches," " Int. J. Sci. Technol. Res.vol. 9, no. 2, pp. 2450–2455, 2020.
5. Joo, G.; Song, Y.; Im, H.; Park, J. Clinical Implication of Machine Learning in Predicting the Occurrence of Cardiovascular Disease Using Big Data (Nationwide Cohort Data in Korea). IEEE Access 2020, 8, 157643–157653. [CrossRef]
6. Che, C.; Zhang, P.; Zhu, M.; Qu, Y.; Jin, B. Constrained transformer network for ECG signal processing and arrhythmia classification. BMC Med. Inform. Decis. Mak. 2021, 21, 184. [CrossRef]
7. Hoodbhoy, Z.; Jiwani, U.; Sattar, S.; Salam, R.; Hasan, B.; Das, J. Diagnostic Accuracy of Machine Learning Models to Identify Congenital Heart Disease: A Meta-Analysis. Front. Artif. Intell. 2021, 4, 197. [CrossRef]
8. Umarani Nagavelli, Debabrata Samanta and Partha Chakraborty, "Machine Learning Technology-Based Heart Disease Detection Models", Hindawi Journal of Healthcare Engineering Volume 2022, Article ID 7351061, 9 pages https://doi.org/10.1155/2022/7351061
9. Abdul Saboor, Muhammad Usman, Sikandar Ali, Ali Samad, Muhmmad Faisal Abrar, and Najeeb Ullah, "A Method for Improving Prediction of Human Heart Disease Using Machine Learning Algorithms", Hindawi Mobile Information Systems Volume 2022, Article ID 1410169, 9 pages https://doi.org/10.1155/2022/1410169
10. Farhat Ullah, Xin Chen, Khairan Rajab, Mana Saleh Al Reshan, Asadullah Shaikh, Muhammad Abul Hassan, Muhammad Rizwan, and Monika Davidekova, "An EfficientMachine LearningModel Based on Improved Features Selections for Early and Accurate Heart Disease Predication", Hindawi Computational Intelligence and Neuroscience Volume 2022, Article ID 1906466, 12 pages https://doi.org/10.1155/2022/1906466
11. Yang, J.; Guan, J. A Heart Disease Prediction Model Based on Feature Optimization and Smote-Xgboost Algorithm. Information 2022, 13, 475. HTTPS:// doi.org/10.3390/info13100475

Note: All the figures and tables in this chapter were made by the Author

Study of Factors associated with Diets and Prevalence of Overweight and Obesity among Adolescent Girls

18

A. Maria Vinitha[1]

Research Scholar, Presidency College, Triplicane,
Chennai, Tamil Nadu, India

T. Pramananda Perumal

Principal (Retired), Presidency College, Triplicane,
Chennai, Tamil Nadu, India

Abstract Overweight and Obesity among adolescent girls have become a growing health concern in urban cities of India. For the past few decades, the adolescent population of urban cities is more inclined to consume more western foods rather than opting for territorial cuisines. The adolescent girls of India in terms of acquiring fashion have got special love and psychological tendency towards modern western culture and cuisines and their alluring technologies. Culture and fashions have emerged among humans based on the geographical location and climate of their living place. The influence of exposure to western food habits and modern processed foods have not been elaborately investigated yet. Moreover, the actual food items and eating lifestyle for the development of overweight and obesity are yet to be carefully analysed because we mix both foreign and local styles in our living. The aim of this investigation is to determine the authentic causes, especially of food items and eating habits for overweight and obesity among adolescent girls of school and college goers in Chennai, India. Being in India, many places are in a nutritional transformation phase, in which overweight and obesity have become common prevalence. Nutritional evaluation needs to be done in the age of adolescence as it is the transitory period between childhood and adulthood. Thus concentrating on the adolescent age group will help to prevent the occurrence of disease and to improve the healthy lifestyle for future in the society.

Keywords Adolescent girls, Diet type, Food types, Non-communicable diseases, Overweight, Obesity, Physical activity, Risk factors

1. Introduction

Nearly 39% of the adult population globally have been grouped into overweight with a Body Mass Index (BMI) of 25.0–29.9 kg per m^2 or obese with BMI of more than 29.9 kg per m^2 in 2014 and it has been increasing by two folds from 1975. While the obesity incidence are 6.4% and 3.2% among men and women respectively during1975, it has risen to 14.9% and 10.8%, correspondingly during 2014 (Di Cesare M. et al., 2016). The growing incidence of overweight and obesity, especially in developing nations like India, have been correlated significantly (Tandon N. et al., 2018, Dandona L. et al., 2017, Prabhakaran D. et al., 2018). The incidence of overweight and obesity has been rising rapidly from 8.4 % in 1998 to 15.5% in 2015 among females. The incidence of obesity has grown from 2.2% to 5.1% during the above period for the same female population

[1]a.vinialex@gmail.com, [2]pramanandaperumal@yahoo.com

DOI: 10.1201/9781003428466-18

(Mumbai IIPS, 2017). Almost 80% of adolescent individuals would continue to be obese and it is a serious threat factor in the development of Non-Communicable Diseases (NCDs), namely Cardio-Vascular Disease (CVD), Non-Insulin-Dependent Diabetes Mellitus (NID-DM) and Cancer. In many countries, Obesity is found to be causative of different psychological issues among adolescents, primarily due to unhealthy dietary and physical inactivity (Catherine M. Levy, 2017). Besides the physical impacts of flab, overweight and obesity, adolescents have commonly been developed with psychological problems in addition (Ebbeling CB, Pawlak DB, and Ludwig DS. 2002).

The investigation has been performed to find and evaluate the risk factors that cause the incidence of overweight and obesity among adolescent girls and assist in planning the strategies that are needed to strengthen adolescent well-being and health. The purpose of this study is to identify the key factors and its relevant variables from numerous risk indicators influencing obesity among adolescent girls based on the hidden factors in anthropometric profile, dietary pattern, food frequencies and physical activity. Further the study includes Chi-Square test, is to find the statistical significance among BMI and risk factors.

2. Details of the Study

2.1 Study Sites and Participants

The data for this study is randomly collected from the girls belonging to ages of 16 and 19 years from seven different schools and five different colleges, located in Chennai, an urban area in Tamil Nadu, India for the period from January through March 2019 as a diverse background. The survey has been performed based on the consent, obtained from the parents of the participants as well as from the participants before data collection.

2.2 Sample Size Estimation

This study has planned with a sample size of more than 1537 (but in our study, we have taken the sample size as 2000) in order to have a confidence level of 95% and a margin of error within ±2% by setting population proportion 0.2. The sample size has been estimated by applying the formula $N = \dfrac{z^2 \hat{p}(1-\hat{p})}{\epsilon^2}$.

2.3 Data Collection

Using a standard questionnaire, data with respect to social, demographic, physiological, anthropometry details have been recorded. Self and other family members like parents, siblings and grand parents' health details about NCDs namely CVD, Hypertension, DM, Osteoporosis and Cancer were too collected. By applying the WHO (World Health Organization) standards. Adult BMI is calculated using the formula BMI = Weight (Kg)/Height^2 (m^2). The adult BMI cutoff values of less than 18.50 are considered Underweight, if between 18.50 and 24.9 Normal, if between 25 and 29 Overweight and if more than 30 as Obese (WHO, 2016).

3. Results and Discussion

3.1 Respondents' Profile

The mean age of participants is 17.63 years with Standard Deviation (SD) of 1.15. The mean weight of participant's have 48.94 kilogram with SD of 9.26. The mean height of participants have 155.6 cm with SD of 6.62. The mean BMI of participants have 20.19 with SD of 3.49. The mean waist circumference of participants have 88.7 cm with SD of 8.8. The median physical activities in a day of the participants have 25 minutes Inter Quartile Range (IQR: 20–60 minutes). The median of income per month of the participant families has Rs. 34,000 with IQR of Rs. 19,000 –1,00,000. 32(1.6%) of 2000 participants have at least one overweight /obese parent. The obesity prevalence has 20(1%) while overweight prevalence has 113(5.65%). Both overweight and obesity have observed to be more in late adolescences with maximum obesity and overweight in 19 years 10(0.5%) and 33(1.65%) respectively.

On assessing the lifestyle associated risk variables for overweight and obesity, it has been noted that there exist high prevalence of overweight and obesity among adolescents who have taken meat based diet 108(5.4%), those who have taken meal frequently 103(5.15%) and those who have skipped meals 56(2.8%). Frequently consuming unhealthy dietary and instant food has got statistical significance with prevalence of overweight and obesity 91(4.55%), 85(4.25%), 96 (4.8%), 103(5.15%), 73(3.65%) and 79(3.95%) respectively. Moreover, the individuals of 117(5.85%) had overweight and obese have not done any form of physical activity like exercise or playing sports.

Table 18.1 The prevalence of overweight and obesity due to the significant factors as follows

Factors	Range	χ^2	df	P
Age (in completed years)	16 to 19	37.7	6	0.000**
Income in Rs.	>10000 to <=100000	69.41	12	0.000**
Diet Type	Vegetarian, Vegan, Lacto Veg, NonVeg	8.71	9	0.044*
Diet Regularity	Regular, Irregular	14.72	3	0.002*
Meal Skipping	Yes, No	7.2	3	0.047*
Meal Skip	Breakfast, Lunch, Dinner	8.71	9	0.046*
Meal Skip Freq.	Daily, Weekly Once, Weekly Thrice, Rarely	15.04	9	0.089*
Physical Activity	Very Poor, Poor, Good, Very Good, Intense	25.26	15	0.046*
Presence of self NCD	No NCD, Obesity, BP, Diabetes, CVD	154.6	12	0.000**
Obesity family members	No, Yes	49.76	3	0.000**
CVD in family members	No, Yes	13.17	3	0.004*
BP in family members	No, Yes	15.88	3	0.001*
DM in family members	No, Yes	21	3	0.000**
Fish	Regular, Fairly, Rarely	26.42	6	0.004**
Mutton and Chicken	Regular, Fairly, Rarely	11.96	6	0.014*
Oily Food	Regular, Fairly, Rarely	15.96	6	0.043*
Cholesterol	Regular, Fairly, Rarely	26.77	6	0.000**
Animal Fat	Regular, Fairly, Rarely	12.18	6	0.044*
Sweet and Chocolate	Regular, Fairly, Rarely	36.32	6	0.000**
Instant Foods	Regular, Fairly, Rarely	14.72	6	0.039*

* - Significant at 5% level (p < 0.05), **- Significant at 1% level (p < 0.01)
Source: Made by the Authors (Based on the execution of the results)

Table 18.2 Inferences, deduced from the values of, df and P of Table 18.1

Sl.No	Inferences from Table 1
1	In Demographic profile, the p value of age and income are less than 0.01 which is highly significant at 1% level.
2	In Dietary habits, the p value of diet type, diet regularity, meal intake frequency and meal skipping frequency are less than 0.05 which is significant at 5% level.
3	There is significant (p value<0.05) association between physical activity and BMI, that is high prevalence of overweight and obese.
4	In Family history, affliction of NCD, family obese and family DM are associated with BMI because the p values are less than 0.01 at 1% level. Similarly, family CVD and family BP have the p values are less than 0.05 which are significant at 5% level.
5	In Food frequency, the p value of cholesterol, sweet and chocolate are less than 0.01 which is highly significant at 1% level.

Source: Made by the Authors (Inferences obtained from Table 18.1)

3.2 Sampling Adequacy Test by KMO and Bartlett's Tests for Sphericity of Dataset

KMO measure of sampling adequacy: 0.694

Bartlett's test Chi-Square value : 11749.79

Bartlett's test p-value : 0.000

In our study, the KMO test value is found to be 0.694 which is more than that of the threshold value 0.6. The p-value for the Bartlett's test is 0.000 which is less than 0.01. Hence it is determined that the dataset is satisfied for conducting factor analysis.

3.3 Principal Component Analysis (PCA)

PCA assists in determining a sequence of linear variable combinations which improves interpretability and minimizing information loss. The procedure of finding principal (statistically significant) components/factors is done using a scree plot.

When the factors in the dataset are having Eigen value greater than 1, they are considered to be the major factors. In our study, the scree plot of Eigen values greater than 1 reveals that dimensionality of entire 41 factors in the dataset has reduced into 9 grouped factors. The loadings of factor analysis have been found to be 38.2% of the total variance.

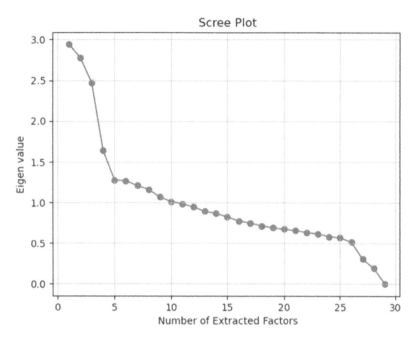

Fig. 18.1 Eigen value while performing sklearn decomposition factor analysis in Python

Source: Made by the Authors (Based on the result of the work)

Table 18.3 Loadings of factor analysis

Factors	Fac1	Fac2	Fac3	Fac4	Fac5	Fac6	Fac7	Fac8	Fac9
WEIGHT	**0.9584**	0.0042	0.0478	-0.0082	0.2431	-0.1076	0.0008	0.0552	-0.0505
HEIGHT	0.1487	0.0444	-0.0119	-0.0150	**0.9767**	-0.0766	-0.0050	-0.0137	-0.1052
BMI	**0.9655**	-0.0179	0.0578	-0.0007	-0.2133	-0.0720	-0.0006	0.0638	0.0044
WAISTCIRCUM	**0.7433**	0.0510	0.0779	0.0233	0.1227	0.0351	0.0133	0.0408	-0.0447
MTIMING	-0.0070	-0.2199	0.0930	-0.0831	0.0077	**0.6375**	-0.0310	-0.0636	-0.0428
SKIPPING	-0.0034	**0.9388**	-0.0964	0.0968	0.0183	-0.1934	-0.0182	-0.1235	0.0141
MSKIP	0.0302	**0.7683**	-0.0666	0.0599	0.0275	-0.1353	-0.0022	-0.1193	0.0192
RCEREALS	-0.0654	0.0058	0.2916	0.0418	-0.0373	-0.0197	0.0620	0.0236	**0.4105**
PULSES	-0.0280	0.0064	0.0791	0.1921	-0.0408	0.0320	0.0208	-0.0149	**0.5428**
YOVEG	-0.0432	0.0231	-0.0110	**0.5345**	-0.0328	-0.0337	0.0222	0.1885	0.1397
GLV	0.0025	-0.0147	0.0415	**0.5165**	0.0565	-0.0306	0.0471	-0.0436	0.1087
FRUITS	-0.0035	0.0182	0.0755	**0.5437**	0.0033	-0.0537	0.0231	-0.0278	-0.0257
FISH	0.0204	0.0313	0.0540	0.1869	0.0082	0.0324	**0.6160**	0.1152	-0.0272
MANDCHICKEN	-0.0143	-0.0493	0.2960	-0.0583	-0.0092	0.0549	**0.6047**	-0.0446	0.0575
OILYFOOD	-0.0188	-0.0502	**0.5506**	-0.0495	0.0359	-0.0160	0.1257	0.1148	0.1139
SWECHO	0.0790	-0.0290	**0.5145**	0.0748	0.0052	-0.0030	-0.0010	-0.0164	-0.0591
INSTANTFOOD	0.0126	0.0031	**0.5415**	0.0521	-0.0369	0.0714	-0.0170	0.0113	-0.1044

Source: Made by the Authors (Based on the execution of the result)

Table 18.4 Factor loadings of variables

Sl. No	Factor	Variables
1	Fac1	Weight, BMI, Waist Circumference
2	Fac2	Meal skipping, Meal skip frequency
3	Fac3	Oily Food, Sweet and Chocolate and Instant Food
4	Fac4	Yellow Vegetables, Green Leafy Vegetables, Fruits
5	Fac5	Height
6	Fac6	Meal timing
7	Fac7	Fish, Mutton and Chicken
8	Fac8	No loadings value more than 0.45
9	Fac9	Rice Cereals, Pulses

Source: Made by the Authors

Additionally, the hormonal variations during puberty, coupled with mentioned diet habits and lifestyle, have resulted in increasing of overweight and obesity in the adolescent age group.

4. Conclusions and Future Study

The prevalence of adolescent overweight and obesity are 5.65% and 1% respectively in the study, performed here. Frequent consumption of cholesterol food, animal fat, sweets and instant foods, lack of physical activity, meal intake and skips are used as the variables which are significantly connected to the development of overweight and obesity. Further, the other major controlling factors for overweight and obesity are age, income and diet habits. Creating awareness of healthy food habits as well as physical activity and lifestyle changes can be readily embraced and adopted as soon as the study is done in the concerned age group. Further, there prevails a space for closer study in this domain where broader investigations may additionally determine a cause-effect relationship.

5. Acknowledgments

The authors thank Professor K.S.Easwarakumar of Anna University CEG campus, Chennai for useful discussions, his interest and encouragement. We are grateful to "**Baby-Perumal Research Institute**" at Chennai for supporting us with computing facilities.

References

1. Di Cesare M. et al, (2016). "Trends in adult body-mass index in 200 countries from 1975 to 2014: A pooled analysis of 1698 population-based measurement studies with 19.2 million participants", Lancet 387, 1377– 1396.
2. Tandon N. et al, (2018)."The increasing burden of diabetes and variations among the states of India: the Global Burden of Disease Study 1990–2016", Lancet Glob. Heal. 6, e1352–62.
3. Dandona L. et al, (2017). "Nations within a nation: variations in epidemiological transition across the states of India, 1990–2016 in the Global Burden of Disease Study", Lancet 390, 2437–60.
4. Prabhakaran D. et al, (2018). "The changing patterns of cardiovascular diseases and their risk factors in the states of India: the Global Burden of Disease Study 1990–2016", Lancet Glob. Heal. 6, e1339–51.
5. Mumbai IIPS, (2017). "National Family Health Survey (NFHS-4) 2015– 16 India", International Institute for Population Sciences (IIPS) and ICF.
6. Catherine M. Levy,(2017). Master Thesis on "The Prevalence of Overweight and Obesity among Adolescents with Chronic Health Conditions", the Aquila Digital Community, University of Southern Mississippi.
7. Ebbeling CB, Pawlak DB, Ludwig DS. (2002). "Childhood obesity: public health crisis, common sense cure", Lancet 360(9331):473– 48.
8. WHO. (2016). Obesity and Overweight: Fact sheet 311, WHO Media Centre.

Human Machine Interaction in the Digital Era – Prof. J. Dhilipan et al. (eds)
© 2024 Taylor & Francis Group, London, ISBN 978-1-032-54998-9

Comparison of Distinct Mel-Frequency Cepstral Co-efficient (MFCC) Selection for Heart Disease Prediction Using Machine Learning Algorithms

19

K. Vetriselvi[1]

Ph.D Scholar, PG and Research Department of Computer Science,
Periyar Government Arts College, Cuddalore, Tamil Nadu, India

G. Karthikeyan[2]

Assistant Professor, PG and Research Department of Computer Science,
Periyar Government Arts College, Cuddalore, Tamil Nadu, India

Abstract The Mel Frequency Cepstral Coefficient is designed to model features of audio signal and is widely utilized in various fields. Most of the researchers are used only 2-13 coefficients or 39 coefficients. MFCC selection count is an essential phase. Here MFCC selection phase is applied on heart disease prediction system. Heart produces different kinds of sounds such as normal and abnormal sounds if people affected by heart disease. First the heartbeat audio sounds are pre-processed. Then the feature extraction MFCC technique is applied on the heartbeat sounds. In this research proposes different kinds of MFCC filters are used with machine learning algorithms such as K-Nearest Neighbour, Support Vector Machine, Decision Tree,Random Forest, Naïve Bayes and Logistic Regression. Precision, recall, f1score and accuracy of the classifier with different count of MFCC features are used as the performance measures. This research finds MFCC count which provides maximum accuracy and better performance of mentioned six algorithms for identifying heart disease. The MFCC feature count depends on problem in different audio frequencies. Here selectively, Random Forest gives better accuracy with the MFCC feature of 40 for heart disease prediction system.

Keywords Decision tree, KNN, Logistic regression, MFCC, Naïve Bayes, Random forest, SVM

1. Introduction

The design of a system for identifying heart illness depends on the extraction and selection of the optimal parameters from heartbeat acoustic signals. It may affect the classification task inferentially. By eliminating information unrelated to illness analysis of the audio data and that primarily contributes to the detection of phonetic beat sound disparities, the goal of representation selection is to compress the audio data. When a substantial amount of reference information is kept, such as distinct cardiac signal types, the need for efficient information storage becomes crucial.

The solid illustration is created using MFCCs. The logarithm of the short-term energy spectrum's final product, the MFCC coefficients, was developed by Pols in 1977. According to Bridle and Brown (1974) and Davis and Mermelstein (1980), the MFCCs are more effective than any other feature. The MFCC cepstral's 0th coefficient is ignored by automatic speech recognition systems because it is unstable (Picone, 1993). For each frequency band in the signal under analysis, the zeros

---
[1]vetriselvik2009@gmail.com, [2]gkarthikeyan2007@gmail.com

DOI: 10.1201/9781003428466-19

coefficient can be viewed as a collection of average energies. Another key element for automatic speech identification is the signal's energy. The frame energy and first- or second-order time derivatives are two often utilized energy-related properties. Numerous studies have shown that adding energy information as another model element in addition to cepstrums can increase the system performance (Huang et al., 1996).

In this paper numerous investigates are designed and concluded to evaluate the effects of distinct implementations and of how the MFCC filters are integrated. If we use different filters the performance of the system also changed. Here different ML algorithms are checked by using different Mel filter numbers. The rest of the paper is explained chapter 2 about related work, chapter 3 explains different machine learning algorithms, chapter 4 give details performance evaluation metrics of machine learning algorithms, chapter 5 explores proposed methodology, chapter 6 explains about experimental results and chapter 7 concluded with future scope.

2. Related Work

A read speech database is used to determine the language from the speech signal. As spectral characteristics, MFCC are used. Using a Gaussian mixture model, fifteen different language identification models were created. The effects of various spectral feature counts, including 6, 8, 13, 19, 21, 29, 35, and 40, are investigated (Koolagudi et al., 2012). Features from the (MFCC) are retrieved for each audio input. An MFCC 13 dimension feature vector is produced for every signal input. These attributes are utilized to develop sentiment models using classifiers from Deep Neural Networks and Gaussian Mixture Models (Abburi et al., 2017).In order to classify EEG signals, the authors give an experimental evaluation of MFCCs. On the CHB-MIT Scalp EEG Database, the MFCC characteristics have been tested. With 15-MFCCs, 25-MFCCs, and 35-MFCCs, three alternative settings are evaluated to see how they affect accuracy. According to Rajesh (2016), the 25-MFCCs arrangement performs better than the competition with 97% accuracy, 98% sensitivity, and 96% specificity.

Using all systole and diastole intervals, 50 Mel-Frequency Cepstral Coefficients based attributes are mined to improve feature quality in comparison to earlier methods. Using a 26-dimensional MFCC vector, seven distinct SVM and K-Nearest Neighbors based classifiers are trained to identify and categorize cardiovascular illnesses. Fivefold cross-validation is utilized to test the classifiers, and a 20% holdout validation strategy was applied. Gaussian SVM classifier achieves 92.6% classification accuracy using chosen features (Ahmad et al., 2019).

3. Heart Disease Prediction by Machine Learning Algorithms

Machine learning techniques are typically employed to forecast cardiac disease. The six types of machine learning algorithms used here are KNN, Naive Bayes, Logistic Regression, Decision Tree, SVM, and Random Forest.

4. Performance Evaluation metrics of Machine Learning Algorithms

True positives – an outcome in which the model precisely calculates the positive class (heart disease patients)

True negatives – an outcome where the model forecasts the unfavourable class (non-heart disease patients)

False positives – an outcome in which the model wrongly predicts the positive (heart disease) class

False negatives – an outcome where the model makes an error in its prediction of the negative class

Precision

$$\text{Precision} = \frac{\text{True Positive}}{\text{True Positive} + \text{False Positive}} \tag{1}$$

Recall

$$\text{Recall} = \frac{\text{True Positive}}{\text{True Positive} + \text{False Negative}} \tag{2}$$

F1 Score

$$\text{F1 Score} = \frac{\text{Precision} \cdot \text{Recall}}{\text{Precision} + \text{Recall}} = \frac{TP}{TP + 1/2(FP + FN)} \tag{3}$$

5. Comparative Analysis of MFCC with Machine Learning Algorithms

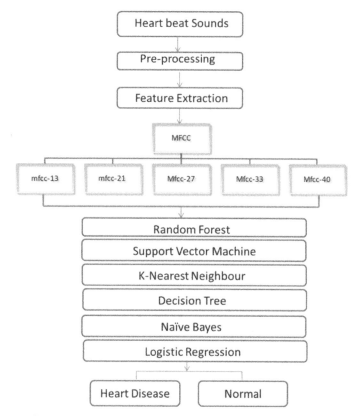

Fig. 19.1   Block diagram of comparative analysis of MFCC feature component with machine learning algorithms

The raw heart beat sounds are given input to the system. First pre-processing applied on the heart beat audio signal for filtering unwanted noise. Data analysis would be applied by using onset detector, onset backtrack and onset strength. The audio signal frequency has been changed into numbers. Furthermore MFCC have been applied on the audio signals. By choosing different filter number of MFCC count the machine learning algorithms working differs. From this experiment proves for this heart disease dataset the machine learning algorithms works better when the MFCC count would be 40 and also Random Forest Classifier gives highest accuracy for predicting heart disease.

5.1 Pre-processing

The PCG signals are pre-processed by using linear minmax scalar. After pre-processing analyse the data by using librosa library. Onset detector found the high peak amplitude and onset backtrack is used to find the preceding local minimum slice points. Onset strength is used to find the onset envelope or formants of the frequency. After analyse the data feature extraction is handled by using MFCC.

5.2 Feature Extraction Technique

After pre-processing the heart beat signals are fed into the further process feature extraction technique as Mel-frequency cepstral coefficient. The computation of MFCC is discussed in further chapters.

5.3 Mel Frequency Cepstrum Coefficients

Numerous studies have shown that the mel-frequency range—which may be characterized as a linear frequency spacing under 1,000 Hz and a logarithmic spacing higher than 1,000 Hz—is used by the ear to hear frequency components in speech (Ahmad et al., 2019). Filters spaced logarithmically at high frequencies and linearly at low regularities were employed to record the phonetically relevant parts of speech (Repaka et al., 2019). In order to determine how the physical frequency and the mel-frequency relate to one another, use the following formulas (Picone, 1993):

$$M(f) = 2595 \log_{10}\left(1 + \frac{f}{700}\right) \qquad (4)$$

where f is the frequency expressed in hertz. The mel-frequency cepstrum coefficient is suggested in (Bridle and Brown, 1974) based on this supposition.

The following steps can be used to calculate the MFCC:

1. The windowed speech segment is transformed into the frequency domain using the Discrete Fourier transform. The short-term power spectrum P is obtained by squaring and adding the short term speech spectrums real, imaginary components (f).
2. The spectrum P(f) is warped into the axis of the Mel-frequency, where M is the Mel-frequency, as P(M).
3. The triangle band pass filter P(M) is then convolved with the resulting warped power spectrum to produce (M). By discretizing with P(M), instances of the critical-band power spectrum are generated (Mk).
4. In equation (4), the MFCC is calculated. The lower dimensions of the MFCC calculation are used to compress the signal components.

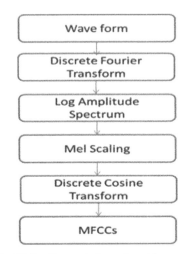

Fig. 19.2 Computation of mel-frequency cepstral coefficients

5.4 Comparisons on MFCC Implementation

The effectiveness of MFCC could be impacted by the following elements: (1) the filters count; (2) the filters shape. so design several comparison experiments with various numbers of filters to determine which MFCC filter is more crucial.

5.5 Different Filter Numbers in MFCC

One element that may have an impact on the performance of the classifier is the quantity of triangular band-pass filters. Results from various numbers of filters are presented in experimental. Here different filter numbers such as 13,21,27,33 and 40 were applied on the heart disease dataset. At filter number K = 40, the classifier performs at its peak level.

6. Experimental Results

6.1 Evaluation Using Random Forest

Table 19.1 Evaluation using random forest

Heart Beat Sounds	Precision					Recall					F1 Score				
	MFCC 13	MFCC 21	MFCC 27	MFCC 33	MFCC 40	MFCC 13	MFCC 21	MFCC 27	MFCC 33	MFCC 40	MFCC 13	MFCC 21	MFCC 27	MFCC 33	MFCC 40
Artifact	1.00	0.86	1.00	1.00	1.00	1.00	1.00	1.00	1.00	1.00	1.00	0.92	1.00	1.00	1.00
Extrahls	0.75	0.64	1.00	0.80	0.87	1.00	1.00	1.00	1.00	1.00	0.86	0.78	1.00	0.89	0.93
Extrastole	1.00	1.00	0.93	0.92	1.00	0.87	0.64	0.76	0.86	1.00	0.93	0.78	0.84	0.89	1.00
Murmur	1.00	0.53	0.78	0.77	0.91	0.6	0.32	0.52	0.47	0.43	0.75	0.40	0.62	0.59	0.59
Normal	0.90	0.67	0.79	0.74	0.83	0.99	0.83	0.93	0.89	0.96	0.94	0.74	0.86	0.81	0.89

6.2 Evaluation Using Support Vector Machine

Table 19.2 Evaluation using support vector machine

Heart Beat Sounds	Precision					Recall					F1 Score				
	MFCC 13	MFCC 21	MFCC 27	MFCC 33	MFCC 40	MFCC 13	MFCC 21	MFCC 27	MFCC 33	MFCC 40	MFCC 13	MFCC 21	MFCC 27	MFCC 33	MFCC 40
Artifact	1.00	0.60	0.99	0.88	0.89	1.00	1.00	1.00	1.00	0.89	1.00	0.75	1.00	0.93	0.89
Extrahls	0.67	0.78	1.00	0.80	0.85	1.00	1.00	1.00	1.00	0.85	0.80	0.88	1.00	0.89	0.85
Extrastole	0.12	0.71	0.28	0.67	0.75	0.13	0.18	0.29	0.29	0.18	0.13	0.29	0.29	0.40	0.29
Murmur	0.53	0.64	0.88	0.77	0.71	0.50	0.28	0.52	0.56	0.43	0.51	0.39	0.65	0.65	0.54
Normal	0.73	0.58	0.70	0.70	0.68	0.71	0.84	0.80	0.86	0.90	0.72	0.69	0.75	0.77	0.77

6.3 Evaluation Using Decision Tree

Table 19.3   Evaluation using decision tree

Heart Beat Sounds	Precision					Recall					F1 Score				
	MFCC 13	MFCC 21	MFCC 27	MFCC 33	MFCC 40	MFCC 13	MFCC 21	MFCC 27	MFCC 33	MFCC 40	MFCC 13	MFCC 21	MFCC 27	MFCC 33	MFCC 40
Artifact	1.00	0.75	1.00	1.00	1.00	1.00	1.00	1.00	0.86	0.78	1.00	0.86	1.00	0.92	0.88
Extrahls	0.67	0.64	1.00	0.86	0.87	1.00	1.00	1.00	0.75	1.00	0.80	0.78	1.00	0.80	0.93
Extrastole	0.34	0.83	0.52	0.57	0.77	0.87	0.71	0.76	0.86	1.00	0.49	0.77	0.62	0.69	0.87
Murmur	0.33	0.33	0.37	0.56	0.43	0.50	0.24	0.37	0.53	0.57	0.40	0.28	0.37	0.54	0.49
Normal	0.84	0.70	0.71	0.71	0.80	0.48	0.75	0.63	0.68	0.66	0.61	0.72	0.67	0.69	0.73

6.4 Evaluation Using K-Nearest Neighbour

Table 19.4   Evaluation using K-Nearest neighbour

Heart Beat Sounds	Precision					Recall					F1 Score				
	MFCC 13	MFCC 21	MFCC 27	MFCC 33	MFCC 40	MFCC 13	MFCC 21	MFCC 27	MFCC 33	MFCC 40	MFCC 13	MFCC 21	MFCC 27	MFCC 33	MFCC 40
Artifact	1.00	0.67	1.00	1.00	1.00	1.00	1.00	0.80	1.00	1.00	1.00	0.80	0.89	1.00	1.00
Extrahls	0.60	0.58	0.91	0.89	0.75	1.00	1.00	0.91	1.00	0.92	0.75	0.74	0.91	0.94	0.83
Extrastole	0.29	0.56	0.47	0.31	0.64	0.47	0.32	0.41	0.36	0.53	0.36	0.41	0.44	0.33	0.58
Murmur	0.37	0.52	0.62	0.61	0.46	0.5	0.48	0.59	0.39	0.48	0.43	0.50	0.60	0.47	0.47
Normal	0.85	0.60	0.76	0.64	0.75	0.65	0.66	0.80	0.74	0.74	0.73	0.63	0.78	0.69	0.74

6.5 Evaluation Using Naïve Bayes

Table 19.5   Evaluation using Naïve Bayes

Heart Beat Sounds	Precision					Recall					F1 Score				
	MFCC 13	MFCC 21	MFCC 27	MFCC 33	MFCC 40	MFCC 13	MFCC 21	MFCC 27	MFCC 33	MFCC 40	MFCC 13	MFCC 21	MFCC 27	MFCC 33	MFCC 40
Artifact	0.80	0.71	0.83	0.78	0.90	1.00	0.83	1.00	1.00	1.00	0.89	0.77	0.91	0.88	0.95
Extrahls	0.67	0.43	0.92	0.73	0.76	1.00	0.86	1.00	1.00	1.00	0.80	0.57	0.96	0.84	0.87
Extrastole	0.19	0.01	0.21	0.2	0.25	0.87	0.01	0.82	0.79	0.76	0.31	0.01	0.33	0.31	0.38
Murmur	0.71	0.50	0.94	0.64	0.75	0.50	0.16	0.56	0.25	0.26	0.59	0.24	0.70	0.36	0.39
Normal	0.84	0.52	0.69	0.6	0.66	0.32	0.83	0.29	0.37	0.43	0.46	0.64	0.40	0.46	0.52

6.6 Evaluation Using Logistic Regression

Table 19.6   Evaluation using logistic regression

Heart Beat Sounds	Precision					Recall					F1 Score				
	MFCC 13	MFCC 21	MFCC 27	MFCC 33	MFCC 40	MFCC 13	MFCC 21	MFCC 27	MFCC 33	MFCC 40	MFCC 13	MFCC 21	MFCC 27	MFCC 33	MFCC 40
Artifact	1.00	0.71	0.80	0.75	1.00	1.00	0.83	0.80	0.86	1.00	1.00	0.77	0.80	0.80	1.00
Extrahls	0.60	0.43	0.90	0.50	0.80	1.00	0.86	0.82	0.62	0.92	0.75	0.57	0.86	0.56	0.86
Extrastole	0.01	0.01	0.01	0.01	0.01	0.23	0.01	0.01	0.12	0.12	0.15	0.01	0.12	0.13	0.01
Murmur	0.92	0.50	1.00	0.67	0.80	0.55	0.16	0.37	0.17	0.17	0.69	0.24	0.54	0.27	0.29
Normal	0.78	0.52	0.66	0.57	0.63	0.95	0.83	0.99	0.91	0.94	0.86	0.64	0.79	0.70	0.76

6.7 Performance Accuracy of Machine Learning Algorithms

Table 19.7 Performance accuracy of machine learning algorithms

Machine Learning Algorithms	Performance Accuracy				
	MFCC-13	**MFCC-21**	**MFCC-27**	**MFCC-33**	**MFCC-40**
Random Forest	84.62	71.12	83.07	78.46	87.69
Support Vector Machine	63.14	61.23	70.01	73.07	71.53
Decision Tree	57.22	67.15	63.84	66.92	73.07
K-Nearest Neighbour	63.13	58.21	71.53	63.07	70.01
Naïve Bayes	46.27	50.02	50.12	45.38	53.84
Logistic Regression	78.23	52.26	68.12	58.19	71.14

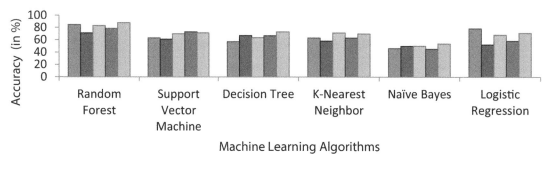

Fig. 19.3 Performance Accuracy of Machine Learning Algorithms.

7. Conclusion and Future Scope

In this research focused heart disease identification using different Mel-frequencies. Here the problem compute different number of filters that produces MFCC features component of 40 with Random Forest will give more accuracy than other count of filters comparatively machine learning algorithms. Here we have compared distinct machine learning algorithms with different filters which provide best accuracy. From this research we have found that MFCC filter count wise 13,21,27,33 and 40. This type of feature extraction sense to the application makes simpler and more straightforward. That helps to decrease the feature size and choosing the best MFCCs. In future, neural network models it may help to increase the accuracy and finding the best parameters.

8. Dataset

The pascal challenge database is the source of both normal and abnormal patient datasets. Data has been gathered from two sources: first from the general public via the iStethoscope Pro iPhone app, provided in Dataset A, and second from a clinic trial in hospitals using the digital stethoscope DigiScope, provided in Dataset B (Bentley et al. 2011). Dataset A contains Normal, Murmur, Extra Heart Sound and Artifact heart sounds and Dataset B contains Normal, Murmur, Extrasystole heart sounds. Here combined dataset A and B used to predict heart disease patients.

References

1. Abburi, H., Prasath, R., Shrivastava, M., &Gangashetty, S. V. (2017).Multimodal sentiment analysis using deep neural networks. In *Mining Intelligence and Knowledge Exploration: 4th International Conference, MIKE 2016, Mexico City, Mexico, November 13-19, 2016, Revised Selected Papers 4* (pp. 58–65). Springer International Publishing.

2. Ahmad, M. S., Mir, J., Ullah, M. O., Shahid, M. L. U. R., & Syed, M. A. (2019).An efficient heart murmur recognition and cardiovascular disorders classification system. *Australasian physical & engineering sciences in medicine*, *42*, 733–743.

3. Bentley, P., Nordehn, G., Coimbra, M., Mannor, S., & Getz, R. (2011). Classifying heart sounds challenge. Retrieved from Classifying Heart Sounds Challenge: http://www. peterjbentley. com/heartchallenge.

4. Bridle, J.S. and Brown, M.D., 1974.An experimental automatic word recognition system. *JSRU report*, *1003*(5), p.33.

5. Davis, S., and Mermelstein, P. (1980).Comparison of parametric representations for monosyllabic word recognition in continuously spoken sentences. *IEEE transactions on acoustics, speech, and signal processing*, *28*(4), 357–366.

6. Gnana Rajesh, D. (2016). Analysis of MFCC features for EEG signal classification. *Int. J. Adv. Sig. Img. Sci*, *2*(2).

7. Huang, X.D., Acero, A., Alleva, F., et al (1996) "From SPHINX-II to WHISPER – making speech recognition usable," pp. 481–508, in book "Automatic speech and speaker recognition: advanced topics". C.H. Lee, F.K. Soong and K.K. Paliwal eds. USA: Kluwer Academic Publishers, 1996.

8. Koolagudi, S. G., Rastogi, D., &Rao, K. S. (2012).Identification of language using mel-frequency cepstral coefficients (MFCC). *Procedia Engineering*, *38*, 3391–3398.

9. Picone, J. W. (1993). Signal modeling techniques in speech recognition. *Proceedings of the IEEE*, *81*(9), 1215–1247.

10. Pols, L. C., 1977.Spectral analysis and identification of Dutch vowels in monosyllabic words.

11. Repaka, A. N., Ravikanti, S. D., & Franklin, R. G. (2019). Design and implementation heart disease prediction using natives Bayesian. In *International conference on trends in electronics and information*.

Note: All the figures and tables in this chapter were made by the Author (Vetriselvi K.)

Human Machine Interaction in the Digital Era – Prof. J. Dhilipan et al. (eds)
© *2024 Taylor & Francis Group, London, ISBN 978-1-032-54998-9*

Artificial Intelligence Based Virtual Health Assistant and Patient Monitoring System

20

D. Kanchana[1]

Assistant Professor, Department of Computer Applications (MCA),
SRM Institute of Science and Technology, Ramapuram Campus, Chennai, Tamil Nadu, India

J. Shobana[2]

Assistant Professor, Data Science and Business Systems, School of Computing,
SRM Institute of Science and Technology, Kattankulathur, Chennai, Tamil Nadu, India

A. V. Kalpana[3]

Assistant Professor, Data Science and Business Systems, School of Computing,
SRM Institute of Science and Technology, Kattankulathur, Chennai, Tamil Nadu, India

Abstract Hospitals plays an important role in our lives as it provides a lot of facilities to everyone who are suffering from different diseases or any problems which occurs due to many things like climate change, work stress etc. This system presents an experimental investigation on the use of conversational interfaces (CIs) to help doctors to provide occupational health consultations. The CI was carried out using a chatbot assistant that send message with actual assistance and a web-based information dashboard. There were two different system designs created, the first employing proactive chatbots and the second using on-demand interactions. Eight healthcare consultations were used in a field trial to examine the efficiency of the recommended CI and the 2 different chatbot designs. Quantitative findings revealed that workplace health doctors were eager to use these technologies in their work, and the CI was well-liked as a reliable tool for use in medical consultations. The qualitative analysis of the data supported our design concept.

Keywords Artificial intelligence, Conversational interfaces, Health assistant bot, Electronic health records, Ciphertext policy attribute-based encryption

1. Introduction

Lack of optimal human and material resource utilisation to provide coordinated medical services to prevent illnesses and treat ailments after they occurs.[1] Insights show that Arab countries suffer from the negative consequences of a high prevalence of medical problems, such as diabetes, liver disease, and parasite disorders including histamine intolerance and intestinal illness.

At the same time, it is necessary to develop a health care web application to assist people in their needy times. [2] In today's traditional method people have to approach hospitals to do basic diagnosis. Thus, this project focuses on developing web application to assist people during emergencies. Where people can upload their body temperature, blood report, pulse rate etc. in the application to know their health condition and to know the problems with their body [3]. It also helps them to know the food habits to be followed. It provides prescriptions to the patient based upon the uploaded tests reports. Thus, it provides a

[1]kanchand@srmist.edu.in, [2]shobanaj1@srmist.edu.in, [3]kalpanaa2@srmist.edu.in

DOI: 10.1201/9781003428466-20

complete medical assistance through the application. Therefore, it is essential to develop such an application in this busy world where people take least care on their health, it helps them to improve their health more efficiently [4].

2. Literature Review

This section of the study aimed to review the literature on virtual medical assistance used in healthcare, identify and describe these cutting-edge technologies and their prospective applications and highlight any associated challenges. [15]. This study discusses a sophisticated AI medical chatbot for users, particularly during the pandemics and emerging situation.

Despite significant efforts in computer science to advance technological aspects [6], it is still unclear how valuable these aspects are for enhancing clinical operations. This is pertinent to the complexity that comes from the challenging nature of tasks involving healthcare decision-making [7]. To improve the efficacy of CDSTs, S Alexander [8] recommended that the systems deployment not be intrusive and intrusive to present clinical settings. Similar criticisms were made by Alford et al. [9] regarding the ease with which technical advantages could not be converted into practical and cost-effective CDST implementations. In a more contemporary argument, Allen et al. [10] claimed that the implementation and architecture of CDSTs should take into account the clinical context's unique qualities, including workflow patterns and healthcare's collaborative nature. Previous research together point to a lack of HCI concerns and the necessity to resketch the work.

Early detection could prevent certain medical problems from arising or lessen their difficulties. This is due to a combination of arranging, operating, and specialised factors. This would result in a significant increase in the scope of social insurance if we had the chance to defeat them. Also, lack of readily available essential consideration data frameworks is a flaw, this is arguably the most extraordinary programming that legitimately supports all regulatory and specialised human services activities, ensuring that the healthcare establishment has full control over all of its activities and resources [11]. The success of these advanced systems doesn't depend on a precise selection of hardware and programming. Maybe their success depends on how reasonable they are with different clients from medical service providers, such as doctors, nurses, professionals, and even executives, since each of these classes has a different perspective on the world, different needs for information, and different advantages from each of these frameworks [12].

3. Proposed Methodology

To assist doctors in providing occupational health consultations, conversational interfaces (CIs) are used and is presented in this system as an experimental study. The CI was carried out using a chatbot assistance that sent texts with real-time assistance and a web-based information dashboard. There were two different system designs created, the first employing proactive chatbots and the second using on-demand interactions. Eight healthcare consultations were used in a field trial to examine the performance of the two distinct chatbot designs and the suggested CI. Quantitative results showed that occupational health specialists were eager to use this technology in their work and that the CI was well-liked as a reliable tool to be used during patient visits. The qualitative data was analysed and demonstrated that our proposed design. In current system people have to approach the nearby hospital or clinic to diagnose their problem. Even for knowing about normal fever, diet food habits, symptoms about diseases people have to go to the hospital. It is more time consuming as people have to wait in a long queue to meet doctor just to do basic diagnosis. It is more expensive too. People have to approach different doctors to get details about different diseases. People have to take appointments to meet the doctors. Doctors will be available at only specified timings this would be a drawback in case of emergency need.

This system aims to develop an application that will provide basic medical assistance during emergency needs. In this application people have to upload their body temperature, blood reports and other test details to know their health condition. People are in return provided by the food habits to be followed based upon their health condition. Even the people are provided by medicine prescription for their diagnosed problem. Thus people need not wait in a long queue to visit doctors, it is cost reducing too, it is available every time and accessible worldwide. People can get assistance for various problem from single source. Therefore, it is an essential application in today's busy world.

In this study, we provide Health Assistant Bot, an artificial intelligence - based assistant that supports patients. HAB enables users in particular to:

(i) Use a Symptom Checker (SC) to determine their condition;

(ii) Select the right doctor for their needs by employing a recommender platform

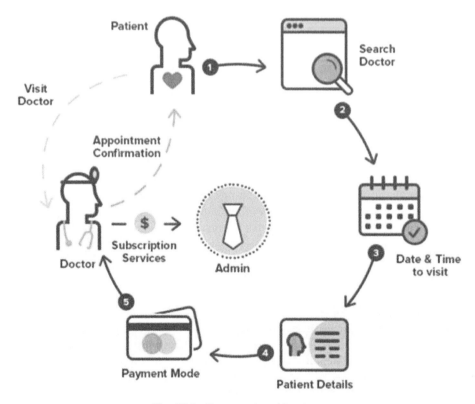

Fig. 20.1 Proposed architecture

(iii) To facilitate recording of medical procedures and wellbeing indicators;

(iv) To increase the user's awareness of related disorders and their symptoms.

We are conscious of the reality that a skilled physician cannot be adequately replaced by an automated system. Health Assistant Bot will assist the user in scheduling a doctor's appointment in the event that the conditions are serious or unknown.

The following contributions are made through this system: We created an IVA that is modular and allows users to featuring customised features like the ability to track treatments and health metrics, as well as doctor recommendations. We release a knowledge base in Italian for mapping symptoms with diseases. We introduce a method to identify the user's condition and clinical area based on the symptoms described through the platform. Finally, we assess the effectiveness of our design choices in an in-vivo research where we asked the user to complete tasks using both our conversational interface and a standard web interface.

We suggested a secured cloud-based EHR system that relies on various levels multi-authority CP-ABE to authorise get to control approaches to ensure the confidentiality and protection of health data stored in the cloud. The suggested system offers social insurance providers, patients, and professionals a higher level of synchronization, interoperability, and exchange of EHRs. The characteristic area authorities in the building deals with a different quality space and is free to work. Additionally, the administration authority does not perform any computational work, and multiple aspects of candidate validation have been acknowledged and sealed. Any legislature that offers treatment administrations to the majority of resident patients and has a distributed computing architecture is able to accept the recommended plan. Future work will involve implementing and evaluating the suggested plot in a real-world setting.

4. Conclusion

In Conclusion, based on legitimate examination and evaluation of the structured framework, it very well may be securely presumed that the framework is a productive, usable and solid records the board framework. It is working appropriately and satisfactorily meets the base desires that were set for it at first. The new framework is relied upon to give benefits expanded by and large profitability, execution and productive records the executives.

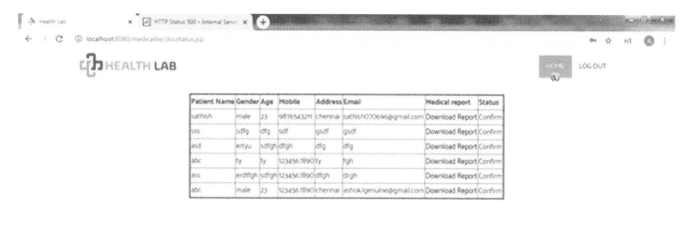

Fig. 20.2 Patient details

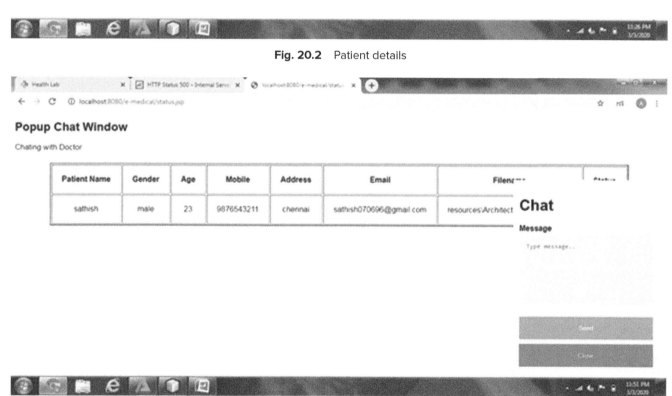

Fig. 20.3 Popup chat window

References

1. Abbasi, B. and S. Z. Hosseinifard. 2014. On the issuing policies for perishable items such as red blood cells and platelets in blood service. Decision Sciences 45(5): 995–1020.
2. Abernethy, M. A. and A. M. Lillis. 2001. Interdependencies in organization design: A test in hospitals. Journal of Management Accounting Research (13): 107–129.
3. M. Almalki and F. Azeez, "Health Chatbots for fighting covid-19: A scoping review," Acta Informatica Medica, vol. 28, no. 4, p. 241, 2020.
4. Abernethy, M. A. and J. U. Stoelwinder. 1991. Budget use, task uncertainty, system goal orientation and subunit performance: A test of the 'fit' hypothesis in not-for-profit hospitals. Accounting, Organizations and Society 16(2): 105–120.

5. Ahadiat, N. and M. Gomaa. 2018. Healthcare fraud and abuse: An investigation of the nature, and most common schemes. Journal of Forensic & Investigative Accounting 10(3): 428–435. (Survey questions pertaining to either Medicare/Medicaid or private insurance companies).

6. M. Almalki and F. Azeez, "Health Chatbots for fighting covid-19: A scoping review," Acta Informatica Medica, vol. 28, no. 4, p. 241, 202.

7. Aidemark, L. and L. Lindkvist. 2004. The vision gives wings: A study of two hospitals run as limited companies. Management Accounting Research (September): 305–318.

8. Aldhizer, G. R. and P. Juras. 2015. Improving the effectiveness and efficiency of healthcare delivery systems. The CPA Journal (January): 66–71.

9. Alexander, J. A., M. L. Fennell and M. T. Halpern. 1993. Leadership instability in hospitals: The influence of board-CEO relations and organizational growth and decline. Administrative Science Quarterly 38(1): 74-99. (JSTOR link).

10. Alford R. R. 1974. Research note: Problems of data and measurement in interorganizational studies of hospitals and clinics. Administrative Science Quarterly 19(4): 485-490. (JSTOR link).

11. Allen. H and S. Sullivan. 2006. Seeing the "health care costs. Harvard Business Review (February): 48–56.

12. M. Polignano ve dierleri, "HealthAssistantBot: A personal health assistant for the Italian language", IEEE Access, vol. 8, pp. 107479–107497, Haziran 2020.

13. Shivani Singh, Manmeet, Tanwar and sharma, "Design and Development of Conversational Chatbot for Covid-19 using NLP: an AI application", 6th International Conference on Computing Methodologies and Communication (ICCMC) I 978-1-6654-1028-1/22/ ©2022 IEEE I DOI: 10.1109/ICCMC53470.2022.975389.

Note: All the figures in this chapter were made by the author.

Human Machine Interaction in the Digital Era – Prof. J. Dhilipan et al. (eds)
© 2024 Taylor & Francis Group, London, ISBN 978-1-032-54998-9

A Novel OLPP Based LVQ-LM Network for Health Care Dataset

21

Poornima V.[1] and Surya Susan Thomas[2]

Department of Computer Science, SRMIST, Ramapuram, Chennai

Abstract A Cardiovascular disease which is additionally referred to as heart diseases that have been a common and steady issue in the field of medical research. Nowadays various methods were applied which is not robust for the prediction of human being expenses and disease risks for patients. This paper proposed an Orthogonal Local Preserving Projection (OLPP) method to reduce the function dimension of the input high-dimensional data. The dimension reduction improves the prediction rate with the help of hybrid classifier i.e. Linear Vector Quantization method combine with the Levenberg-Marquardt (LM) training algorithm in the neural network. The combination of LVQ (Learning Vector Quantization) and LM (Levenberg-Marquardt) algorithms is a hybrid approach to solve optimization problems, particularly in pattern recognition and machine learning.LVQ is a supervised learning algorithm that is used to classify data into pre-defined categories. It uses a set of prototype vectors that represent each class and updates the prototypes based on the training data. LM, on the other hand, is a non-linear optimization algorithm that is used to minimize the sum of squared residuals between the model and the data. To combine LVQ and LM, the LVQ algorithm is used to initialize the model parameters, and then the LM algorithm is used to optimize the parameters and it determines the best network parameters such as weights and bias that minimizes the error. This combination results in a hybrid approach that benefits from the strengths of both algorithms and can produce better results than using either algorithm alone. The combination of LVQ and LM has been applied to various pattern recognition and machine learning tasks, such as image classification, speech recognition, and data clustering, among others. The final output of the optimization technique is combined with the performance metrics as accuracy, sensitivity, and specificity. From the result, it is observed that hybrid optimization techniques increase the accuracy of the heart disease prediction system.

Keywords Data mining, Machine learning algorithms, OLPP, Heart disease prediction, UCI dataset, Optimizaion

1. Introduction

Heart disease has easily identified over the last decade and has become the leading cause of death for people in maximum countries around the world. Doctors are trying to diagnose quickly and accurately [1]. Heart disease can affect the cardiac tissue and cause sudden cardiac death as a result of heart attack [2]. Coronary heart disease is consistently recording the highest fatality rate among non-infectious diseases, and the rate is still increasing [3]. Coronary heart disease is defined as the problems of the heart that happens because its blood circulation is decreased; it leads to the fatty deposits build upon an inner layer of the blood vessels that provide the heart muscles with blood, resulting in contraction [4]. Various heartrelated defects are called as heart disease that mainly spoils the heart. It causes death all over the world. The hazardous disease causes several people in different countries including India. A large number of people died due to cardiovascular diseases that are yearly increasing [5].

[1]poornimv@srmist.edu.in, poornimasudhaagar@gmail.com; [2]suryasut@srmist.edu.in

DOI: 10.1201/9781003428466-21

Despite significant advancements in early diagnosis, screening, and patient care, breast cancer still affects one in eight women globally and accounts for the majority of cancer-related fatalities in females[6]. Both benign and malignant breast tumors do not always develop to cancer. However, a combination of preoperative tests can improve the accuracy of the diagnosis [7].

The combination of LVQ and LM has been applied to various pattern recognition and machine learning tasks, such as image classification, speech recognition, and data clustering, among others. Levenberg-Marquardt (LM) is a numerical optimization algorithm that is commonly used in regression problems. It can be used to fit a model to a set of data by minimizing the sum of squared errors between the model predictions and the actual data. By combining the OLPP-LVQ and LM algorithms, a more flexible model can be obtained that can handle non-linear relationships between the features and output while preserving the local structure of the data. In this combined approach, the OLPP-LVQ is used as a pre-processing step to classify the data into different regions, and then the LM algorithm is used within each region to perform regression. This allows for a more accurate model that can capture the complex relationships between the features and the output, while still preserving the local structure of the data. Projections (OLPP) and Learning Vector Quantization (LVQ). OLPP is a dimensionality reduction technique that preserves the local structure of the data while reducing its dimensionality. LVQ is a classification algorithm that uses a set of prototype vectors to represent the classes in the data.

In the OLPP-LVQ algorithm, the OLPP technique is applied first to reduce the dimensionality of the data, and then the LVQ algorithm is applied to classify the data into different regions based on the reduced features. This results in a division of the feature space into Voronoi regions, where each region is represented by a prototype vector. According to relevant studies, many ML classification algorithms are routinely used to predict heart disease in various investigations. The other portions of this study are organised as follows: segment 3 presented a proposed classifiers for prediction of heart disease, segment 4 summarised the results of the prediction using various classifiers, and segment 5 concluded the report.

2. Literature Survey

A variety of preoperative diagnostics, including physical examination, digital breast tomosynthesis, ultrasound, and magnetic resonance have improved breast cancer screening and detection [8]. KNN, SVM, Random Forest, and Decision Tree classification results were compared by Arpita Joshi and Dr. Ashish Mehta [9] studied the effectiveness of the Support Vector Machine and found that the Artificial Neural Network algorithm was beneficial in today's medical systems. In a comparison study between ANN and SVM, Kalyani Wadkar et al. [10] came to the conclusion that ANN was a superior classifier than SVM since ANN had a greater efficiency rate. By combining machine learning and deep neural network techniques with support value, Anji Reddy Vaka et al. [11] presented a pioneer method to detect BC. The results of the simulation showed that The DNN method offered advantages in terms of potential, effectiveness, and picture quality all factors that are crucial in current medical systems.

Automatic methods for classification of plant diseases also help taking action after detecting the symptoms of leaf diseases. This paper presents a Convolutional Neural Network (CNN) model and Learning Vector Quantization (LVQ) algorithm based method for tomato leaf disease detection and classification. The dataset contains 500 images of tomato leaves with four symptoms of diseases. We have modeled a CNN for automatic feature extraction and classification. Color information is actively used for plant leaf disease researches. In our model, the filters are applied to three channels based on RGB components. The LVQ has been fed with the output feature vector of convolution part for training the network. The experimental results validate that the proposed method effectively recognizes four different types of tomato leaf diseases [12]. Poornima et al [13] stated that hybrid neural network classifier uses a heart disease prediction method. The Group Search Optimization algorithm and the Levenberg-Marquardt algorithm are combined in this novel model GSO-LM network classifier. The feed forward neural network is trained using the dimensionality-reduced output from the OLPP in order to analyse the system's performance. By combining machine learning and deep learning approaches, Monica Tiwari [14] proposed a novel way to identify breast cancer. According to a comparison of ML and deep learning techniques, the accuracy achieved by CNN and ANN models (99.3 percent) was higher than that of machine learning models (97.3 percent). A novel approach to detecting BC by image categorization using machine learning approaches was proposed by Abdullah-Al Nahid and Yinan Kong [15].

3. Proposed Methodology

The initial stage of our system is pre-processing which helps to extract useful data from raw heart disease datasets. Here, the input dataset is high dimensional and it is a great difficulty for prediction. Therefore, feature dimension reduction method is used to reduce the features' space without losing the accuracy of prediction. Here, OLPP is applied to reduce the feature dimension. Once the feature reduction is applied, the prediction is done based on the hybrid classifier. In the hybrid classifier, LVQ Algorithm is used with LM Training Algorithm. The overall architecture of our proposed system is illustrated in Fig. 21.1, providing a visual representation of the different stages and components involved in the process[13].

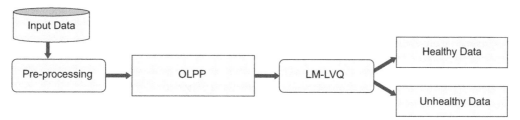

Fig. 21.1 Overall structure of OLPP based LVQ-LM

3.1 Data Preprocessing

The first phase is data preprocessing where redundant and insignificant attributes such as columns having constant are discarded To preserve essential attributes, missing data of a column are replaced with their corresponding mean value [16] [17]. Then normalization process begins. This represents all continuous values in form of binary form within range of 0 to 1. With this approach, we have 14 attributes, and among that one binary class label representing presence or absence of heart disease.

$$\text{Normalization} = \frac{(x_i\text{-meanvalue of coloumn})}{(\text{maximum value of column} - \text{minimum value of coloumn})}$$

$$i = 1, 2, 3, \ldots, n$$

3.2 Orthogonal Locality Preserving Projection for Dimensionality Reduction

Dimensionality reduction techniques reduced the dimensions of databases other words reduce the number of features. The dimensionality procedure expands the accuracy and performance of the proposed model. The pre-processed input data is given to an OLPP [18,19] for dimensionality reduction which results in better classification accuracy. This dimensionality reduction method has a various stage of processing. The initial stage is the PCA projection which reduces the dimensionality of the input data by performing a covariance analysis between two factors. Let the input pre-processed data is of size $m \times n$.

Step 1: Compute the mean μ of the matrix.

Step 2: Calculate the Eigenvector e and Eigenvalues λ of the covariance matrix K. If is a square matrix, a non-zero vector e is an eigenvector of K if there is a scalar such that solve $Ke = \lambda e$

$$k = \left(\frac{1}{n}\right) \sum_{i=1}^{N} (x_i - \mu)(x_i - \mu)^T \tag{1}$$

Step 3: The Eigenvalue and Eigenvectors are arranged and paired. The transformation matrix of PCA is denoted by T_{PCA}. By PCA projection, the extracted features are statistically uncorrelated and the rank of the new data matrix is equal to the number of features (dimensions).

Step 4: Constructing the adjacency graph and choosing the weights: if the node i and j are connected the weight W_{ij} calculated using below equation.

$$W_{ij} = e^{-\frac{E_i - E_j}{t}} \tag{2}$$

Here, $X = (E_1, E_2, \ldots, E_K)$ be a set of e-mails. t is a constant. If the node i and j are not connected means we put $W_{ij} = 0$. The weight matrix W of graph G models having the local structure of varies e-mail.

Step 5: Computing the orthogonal basis functions: After finding the weight matrix W we calculate the diagonal matrix M.

$$M_{ii} = \sum_{j} W_{ji} \tag{3}$$

After that, we calculate the Laplacian matrix L using diagonal matrix M and weight matrix W.

$$L = M - W \tag{4}$$

The orthogonal basis vectors $(o_1, o_2, o_3, ..., o_k)$ can be calculated as follows:

Step 6: Compute o_1 as the eigenvector of associated with the smallest Eigenvalue.

Step 7: Compute o_k as the eigenvector of associated with the smallest Eigenvalue of J_k

$$J_K = \{I - Z^1 A_{K-1} B_{K-1}^T\} Z^{-1} \{XLX^T\} \tag{5}$$

$$A_{K-1} = (o_1, o_2, o_3, ..., o_{k-1}), \ B_{K-1}^T = A_{K-1}^T Z^{-1} A_{K-1}$$

Step 8: OLPP embedding: Let $T_{OLPP} = (o_1, o_2, o_3, ..., o_l)$ embedding is followed,

$$Y \rightarrow XT^T$$

$$T = T_{PCA} \, T_{OLPP} \tag{6}$$

Where; T represents the transformation matrix; Y is the one-dimensional representation of X. This transformation matrix reduces the dimensionality of the feature vectors of the e-mail. This dimensionality reduced features given to the classification process

3.3 Levenberg-Marquardt (LM) Training Algorithm

(i)   Initialize the input weights

(ii) Determine the Jacobian matrix J for the input matrix in LM algorithm.

(iii) Evaluate the error gradient G (fitness) from LM algorithm in neural network.

$$J = \begin{bmatrix} \dfrac{\partial F(a_1, W)}{\partial w_1} & \cdots & \dfrac{\partial F(a_N, W)}{\partial w_W} \\ \dfrac{\partial F(a_N, W)}{\partial w_1} & \cdots & \dfrac{\partial F(a_N, W)}{\partial W_W} \end{bmatrix} \tag{7}$$

(iv) The weight update vector δ of Hessian matrix is calculated as

$$((J_r J + Dl) \, \delta G \tag{8}$$

Here D is the Levenberg's damping factor and δ is the weight update vector used to determine the updated weight. These best weights obtained are again given as input to the LVQ neural network for further processing.

3.4 Learning Vector Quantization (LVQ)

LVQ is a derivative form of the artificial neural network which uses supervised learning and nearest neighbor pattern classifier [20]. It adopted competitive learning and has a similar architecture with the Kohonen Self Organizing Map (SOM) founded by Prof. Teuvo Kohonen in 1982. The basic concept of this method is to get as near as possible to the distribution of input vector in order to minimize the error of classification. This can be done by calculating the Euclidean distance between the input vector and weight vector [21]. The smallest Euclidean distance will be called as the winning vector, where the winning vector will be updated and continued until the termination condition [22]. For more details on the process, the algorithm of LVQ training can be described as follows:

Step 1: Initialized the initial weight vector with learning rate α

Step 2: For each input x, calculate the Euclidean distance and choose the winning vector with the minimum Euclidean distance using Equation (9)

$$D(j) = \sqrt{\sum (\mathbf{x_i} - \mathbf{w_{ij}})} \tag{9}$$

with x as the input vector, w as the weight vector (winner), and n as the number of attributes.

Step 3: Update the weight vector using Equation10 if the class in the neuron j is equal to the target

$$(new) = \mathbf{w_j}(old) + \alpha[x - \mathbf{w_j}(old)] \tag{10}$$

Step 4: Update the weight vector using Equation 11 if the class in the in the neuron j is not equal to the target

$$(new) = \mathbf{wj}(old) - \alpha[x - \mathbf{wj}(old)] \tag{11}$$

Step 5: Perform steps 3 and 4 for each input vector in the training

Step 6: Reduce α

Step 7: Until specified number of epoch is reached, repeat step 2 to 6

Step 8: Test for stopping condition

Finally, the best-updated weights are obtained after n number of iterations through LVQ algorithm and given as input to the neural network for training. Further, two best groups of weights are obtained through the LM algorithm and the LVQ algorithm in a single iteration. Consequently, we compare the obtained two best groups of weights to determine a new best weight. The determined new best weight is again applied to both the LM algorithm and LVQ algorithm to determine the new updated best weights to train the neural network. This process is repeated up to n number of iterations. This classifier classifies the data as healthy and non-healthy with the high optimal rate.

4. Experimental Results

For the implementation of this study, a Heart Disease dataset was utilized, which was sourced from the UCI Machine Learning repository [23]. The dataset was derived by selecting 14 out of the 76 available attributes, representing various tests, and organizing them into the processed Cleveland heart disease database. The dataset comprises 303 instances with 14 attributes, and the class distribution indicates that 54% of instances are labeled as heart disease absent, while 46% are labeled as heart disease present. In order to achieve improved accuracy, the Levenberg-Marquardt (LM) Method was combined with the LVQ network. The objective of this hybrid network was to attain an optimal design that incorporates factors such as productivity, strength, reliability, efficiency, and utilization. The proposed network, based on Orthogonal Locality Preserving Projections (OLPP), LVQ, and LM, was implemented using Python 3.11. The outcomes of the implementation are presented in Table 21.1 and Fig. 21.2, providing visual and tabular representations of the results obtained.

Table 21.1 Comparisons of various classification methods

Classification Methods	Accuracy	Sensitivity	Specificity
GSO-LM[13]	78.4	72.2	75.4
LVQ_LM	98.6	95.41	94

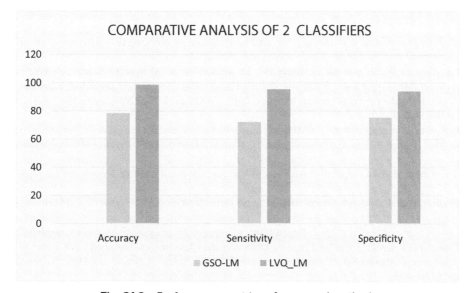

Fig. 21.2 Performance metrics of proposed methods

Figure 21.2 shows that Accuracy, Sensitivity and Specificity of two algorithms such as GSO-LM and LVQ-LM. Based on this table, we can make the following observations: The "LVQ_LM" classification method outperforms "GSO-LM" in terms of accuracy, sensitivity, and specificity. It achieves significantly higher values for all three metrics. "LVQ_LM" has a very high accuracy of 98.6%, indicating that it correctly classifies a large portion of the data. The sensitivity of "LVQ_LM" is also high at 95.41%, suggesting that it has a strong ability to correctly identify positive cases. The specificity of "LVQ_LM" is 94%, which means it performs well in correctly identifying negative cases. On the other hand, "GSO-LM" has lower values for all three metrics compared to "LVQ_LM," indicating lower overall performance. In summary, "LVQ_LM" appears to be the superior classification method in this comparison, achieving higher accuracy, sensitivity, and specificity compared to "GSO-LM." However, it's important to consider other factors such as computational complexity, dataset size, and specific application requirements when choosing a classification method.

5. Conclusion

This research paper presents a novel technique called Orthogonal Local Preserving Projection (OLPP) combined with the hybrid classifier LVQ-LM (Linear Vector Quantization-Levenberg-Marquardt) for predicting human expenses and disease risks in patients. The OLPP-LVQ-LM method effectively reduces the dimensionality of high-dimensional input data, leading to improved prediction accuracy. The OLPP algorithm partitions the feature space using orthogonal projections, while the LVQ-LM hybrid classifier combines the LVQ method with the LM training algorithm in a neural network. This combination offers a powerful tool for handling non-linear regression problems and maintaining the local structure of the data. The experimental results indicate that OLPP-based LVQ-LM is particularly well-suited for use with machine learning algorithms. Performance metrics were compared with GSO-LM and LVQ-LM approaches. By combining OLPP-LVQ and LM algorithms, researchers achieved a more flexible model that can effectively handle healthcare datasets. OLPP-based LVQ-LM proves to be a viable solution for generating an optimal classification model, enhancing the efficiency of health data analysis. The findings demonstrate that the OLPP-based LVQ-LM model significantly improves accuracy, specificity, and sensitivity in predicting heart disease using the UCI dataset.

References

1. Nguyen CL, Phayung M, Herwig U. A highly accurate firefly based algorithm for heart disease prediction. J ExpSys Appl 2015; 1–11.
2. Yosawin K, Chanin N, Tanawut T, Thanakorn N. Datamining of magneto cardiograms for prediction of ischemicheart disease. J EXCLI 2010; 9: 82–95.
3. Jae-Kwon K, Jong-Sik L, Dong-Kyun P, Yong-Soo L,Young-Ho L, Eun-Young J. Adaptive mining predictionmodel for content recommendation to coronary heartdisease patients. J Clust Comp 2013; 1–11.
4. Shamsher BP, Pramod KYS. Predict the diagnosis of heartdisease patients using classification mining techniques. JAgrVeter Sci 2013; 4: 61–64.
5. Srinivas RRG. Rough-fuzzy classifier: a system to predictthe heart disease by blending two different set theories. JSci Eng 2014; 39: 2857–2868.
6. Nolan J, Dunne SS, Mustafa W, Sivananthan L, Kiely PA, Dunne CP. Proposed hypothesis and rationale for association between mastitis and breast cancer. Med Hypotheses. 2020; 144. doi: 10.1016/j.mehy.2020.110057 [PubMed] [CrossRef] [Google Scholar]
7. DeSantis CE, Ma J, Gaudet MM, et al. Breast cancer statistics, 2019. CA Cancer J Clin. 2019; 69(6). doi: 10.3322/caac.21583 [PubMed] [CrossRef] [Google Scholar]
8. Cai D, Lin T, Jiang K, Sun Z. Diagnostic value of MRI combined with ultrasound for lymph node metastasis in breast cancer: Protocol for a meta-analysis. Med (United States). 2019;98(30). doi: 10.1097/MD.0000000000016528 [PMC free article] [PubMed] [CrossRef] [Google Scholar]
9. Arpita Joshi and Dr. Ashish Mehta Comparative Analysis of Various Machine Learning Techniques for Diagnosis of Breast Cancer (2017).
10. Kalyani Wadkar, Prashant Pathak and Nikhil Wagh Breast Cancer Detection Using ANN Network and Performance Analysis with SVM (2019).
11. Anji Reddy Vaka, Badal Soni and Sudheer Reddy Breast Cancer Detection by Leveraging Machine Learning (2020).
12. M. Sardogan, A. Tuncer and Y. Ozen, "Plant Leaf Disease Detection and Classification Based on CNN with LVQ Algorithm," 2018 3rd International Conference on Computer Science and Engineering (UBMK), Sarajevo, Bosnia and Herzegovina, 2018, pp. 382–385, doi: 10.1109/UBMK.2018.8566635.
13. Poornima, V., &Gladis, D. (2018). A novel approach for diagnosing heart disease with hybrid classifier. Biomed Res, 29(11), 2274–2280.

14. Monika Tiwari, RashiBharuka, Praditi Shah and Reena Lokare Breast Cancer Prediction using Deep learning and Machine Learning Techniques.

15. Abdullah-Al Nahid and Yinan Kong Involvement of Machine Learning for Breast Cancer Image Classification: Asurvey (2017).

16. Osman, M. S., Abu-Mahfouz, A. M., & Page, P. R. (2018). A survey on data imputation techniques: Water distribution system as a use case. *IEEE Access*, *6*, 63279–63291.

17. Fazakis, N., Kostopoulos, G., Kotsiantis, S., & Mporas, I. (2020). Iterative Robust Semi-Supervised Missing Data Imputation. *IEEE Access*, *8*, 90555–90569,

18. Rashedi E, Hossein NP, Saeid S. GSA: a gravitational search algorithm. J Info Sci 2009; 179: 2232–2248.

19. Vijay K. A. Face Hallucination using OLPP and kernel ridge regression. Proc IEEE Int Conf Image Proc 2008; 353–356. 27. Gill MW.

20. V. Badbe, V. Londhe, & G. Shirole, Analysis of Heart Disease by LVQ in Neural Network,*International Journal on Recent and Innovation Trends in Computing and Communication,*Vol. 4, pp. 603–607(2016)

21. J. S. Sonawane & D. R. Patil, Prediction of Heart Disease Using Learning Vector Quantization Algorithm, IEEE *Conferenceon IT in Business, Industry, and Government (CSIBIG)* (2014)

22. A. Dongoran, S. Rahmadani, M. Zarlis, & Zakarias, Feature Weighting Using Particle Swarm Optimization for Learning Vector Quantization Classifier, *2nd International Conference on Computing and Applied Informatics,* Series 978(2018)

23. Datasets from (http: //archive.ics.uci.edu/ml/ datasets.html)

Human Machine Interaction in the Digital Era – Prof. J. Dhilipan et al. (eds)
© 2024 Taylor & Francis Group, London, ISBN 978-1-032-54998-9

Effectiveness of Chat Bots to Detect Health Insurance Fraudulent Claims

22

Surya Susan Thomas[1]
Department of Computer Science, College of Science & Humanities,
SRMIST Ramapuram, Chennai, India

Roby Jose[2]
PG Department of Computer Science, Assumption College,
Changanasserry, Kerala, India

Anu Joseph[3]
Department of Computer Applications, Kristu Jyothi College of
Management and Technology, Chethipuzha, Kerala, India

Abstract Chat bots are nothing but artificial intelligence integrated software program which can assist a person in finding a solution for a quest to a great extent at any point of time. Chat bots are enriched with Natural Language Processing (NLP) to support a person without another human intervention or it can be more precisely defined as an 'Automated Support'. The imparted intelligence of chat bot is now widely used in almost all sectors of business life. Insurance sector mainly implements the chat bots in applying for a policy, renewing it, enquiring the premium of a policy and registering complaints of a policy holder. And some newly developed chat bots can even detect fraudulent claims in terms of incorrect submission of data, mismatch of proofs etc and thereby rejecting the payment or escalating further process to a real human to avoid discrepancies. This paper discusses the effectiveness of chat bots developed by various insurance organisations in detecting fraudulent claims and thus provide a benchmark on them.

Keywords Artificial intelligence, Virtual assist, Chat bots, Insurance claims, Fraud detection

1. Introduction

1.1 Digitization of Insurance Claim Process

Insurance sector is getting digitized as the world is becoming more digital day by day by the usage of the innovative technologies. Policy holders need to get access to their policy information in less time. Now taking new policies, enquiry of a police lead, reviews about insurance policy schemes are all available in a click or touch. Smart and seamless handling of customers is the success of any organisation. If a provider did not meet the expectations of the customer, they will shift to a competitor. A more personalized experience is the expectation of any customer. Old fashioned handling of customers and policy documents are subjected to change as competitions among insurance companies tightens up. To strengthen the customer bond, insurers and insurtechs are deepening their ideas in AI -driven insurance chat bots to elevate customer experience.

---
[1]susann.research@gmail.com, [2]roby.research@gmail.com, [3]anujoseph005@gmail.com

DOI: 10.1201/9781003428466-22

The pandemic has expedited the advancement towards a digital world in every aspect. Nearly 70% of customer interactions are now digitized through the soaring technologies such as Artificial Intelligence, Machine Learning, deep learning, chat bots and messaging and it is nearly five times than in 2018 [1].

Artificial intelligence institutes a touch less claim process and avoids unwanted human intervention. AI enhanced chat bot is now available with almost all direct customer dealt website or application to apprehend the interest in them. Chat bots are like artificial customer executives which will assist a customer to clear the queries and doubts of a customer for the website or application they have visited [8].

These AI -driven chat bots in insurance e-commerce have certain effective uses like [1]

- Customer Engagement
- Claims processing and settlement
- Premium Payment collection
- Lead Generation
- Cross selling and Up-selling
- Pers-onalized Customer support
- Live Executive Support

Reasons to execute Chatbots in Insurance [6]

- Automated Claim support
- Interactive bots addressing queries through text/voice
- Enhance customer executive support
- Cost cutting
- Happy Customers
- Helping hands for Insurance agents
- Initial state Anomaly detection

1.2 Chatbots as 'Smart bots 'for Insurance Fraud Detection

Digitized Insurance services have diminished the complexity of claim applying and filling process but it has widened the opportunities of fraudulent activities too. Fraud costs are expected to reach $12 billion by 2026[7]. Insurtechs are trying hard to create and incorporate AI driven smart ideas into the bots to detect and halt the processing of fraudulent claims at the initial stage itself. These smart bots are trained to navigate through the proof of documents through image processing as one way to stop fraudulent activities during policy document submissions. The interaction of these smart chat bots have leveraged the profit in insurance to a new level as frauds can be detected early and loss can be decreased.

Policy holders and customers can either type in or voice message their queries concerning to different policies and the chat bots processes to deliver a personalized experience to the customer. The bots also recommends promotions, policy explanations, lead generation, product generation etc to the new and existing customers [2].Advanced Machine learning integrated with neural network aides the chat bots to achieve the retention of customers. With the advancement of AI and the invention of smart bots, claim verification and authenticity can be smoothly verified without the customer to fill lengthier forms or be in person in an office[3]. Post covid era also paved way for unabridged digitization of any business organisation.

2. Literature Review

An organisation [4] called "Lemonade Insurance", introduced a chat bot called "Jim" which does policy explanation to the customer and interacts with the customer as a real customer executive. Claim initiation process will be kicked off satisfying the needs of the customer and the algorithm used here in the chatbot searches for similar claims in the database. If the details provided for claim submission are genuine, then the customer can make the payment or if doubted for its genuity, the real human "Jim" comes into the scene to wash out the complexity of the scenario.

Oza et al. [9] studied about the Robotic Process Automation (RPA) methodology used in chat bots to enhance business which involves activities similar to human users. This idea was much effective in handling high volume and complicated data with less execution time for processing requests to detect the legitimacy of claims and thus project out the benefits of using chat bots in insurance sector.

Riikkinen et al. [5] came up with a three complimentary theoretical perceptive of artificial intelligence, service logic and reverse usage of customer data to create value in insurance and it will be an innovative thought to apply in the implementation of chat bots in insurance claim processing as well.

Tebenkov et al. [10] made a survey of different types of Artificial Intelligence methodologies employed in chat bot implementation in various arenas. The implementation of chat bots mainly comprised of supervised learning, unsupervised learning and semi-supervised learning methodologies. NLP, Neural Networks, Decision tree algorithms, Bayesian algorithms, Regression algorithms were majorly used in developing effective chat bots to mimic humans.

Wu, Yan et al. [11] overviewed on the possibilities of using deep learning technique in chat bots like open domain conversation modelling including both retrieval-based methods and generation-based methods to process the queries of a user and revert with a best possible answer which clearly serves the need of the user.

Pirila et al. [12] discusses the customer chatbot of a Finnish insurance company for their customer satisfaction and after an online survey from over 200 customers, it was found that the customers were happy with the chat bot but never preferred artificial intelligence over a human.

Pillay et al. [13] explains about a chat bot that uses audio and text to communicate with the human costumer. It easily recognizes the text and responses correctly. The chat bot which was conceived in the pretext of being an entertainment factor for the humans is now able to help them to perform jobs of high quality and also in detecting anomalies in claim processing.

Nuruzzaman et al. [14] proposed an IntelliBot which uses four various strategies to get response in a specialized way in relation to the queries generated by the clients, and could process claims to the next level if the claims are found to be genuine. The bot was benchmarked with other similar bots and was found to be superior in claim processing.

3. Methodology

The major technical concepts used in the development of chatbots to detect fraudulent claims in health insurance [15] are

1. Rule –based concepts
2. Retrieval-based concepts
3. Generative-based concepts

Every method can use any of the following techniques

(a) Parsing

Parsing is a method that gets text as an input and extract meaningful information from it by converting that text into a set of more simple words (lexical parsing) that can be easily stored and operated. The parsing technique was used in an early chatbot, ELIZA to parse the input text for the particular word in the sentence. The word is then matched against the documents in the collection to find the appropriate response to the user.

(b) Pattern-Matching

Pattern matching is the most commonly used technique in chat bots to classify the user input as <pattern> and produce an apt reply stored in <template>. These <pattern> <template> pairs are user defined. Although pattern matching techniques are used in both early and modern chatbots, the complexity of the algorithms used in them differs.[16]

(c) Artificial Intelligence Mark-up language

AIML (Artificial Intelligence Mark-up Language) has been derived from XML (Extensible Mark-up Language). AIML is employed to create conversational flow in chatbots. AIML is made up of data objects which are called AIML objects. These AIML objects, which are also called AIML. A Survey on Conversational Agents/Chatbots Classification and Design Techniques 949 elements, are made up of units which are called topics and categories. The topic is an optional top-level element that has a name attribute and a set of categories related to a specific topic.[17]

(d) Ontologies

Domain ontologies are used in chatbots to replace user defined domain knowledge with ontological domain knowledge. Implementation of ontologies is common as they have been used within specific dialogue system modules, for example in language generation as the basis of the systemic grammar approach[18]

(e) Markov Chain Model

A Markov chain model is a probabilistic model and the rules in the Markov chain process are based on probabilities. When a chatbot deploys this method, it will generate an output that is in par with the state transition. This allows the chatbot to formulate sentences as replies that are more suitable probabilistically but being different every time and are more or less coherent. The conceiving state can be based on the input by the user, giving the reply some relevance. Markov Chains are a popular method to create chatbots for entertainment purposes that mimic simple human conversation. [19]

(f) Artificial Neural Networks

Artificial Neural Networks are the primary source for Machine learning and Deep learning concepts which are basic approaches implemented using supervised, unsupervised and semi-supervised learning methodologies. The use of deep learning neural networks has come up in the field of conversational modelling, especially the recurrent neural network (RNN), sequence to sequence and long short term memory networks (LSTMs) have ruled the field [20].

4. Analysis

The study of certain chatbots of prominent insurance companies which provide health insurance is given below in the following figures.

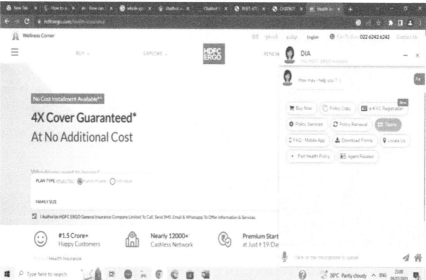

Fig. 22.1   Chatbot 'DIA' [21]

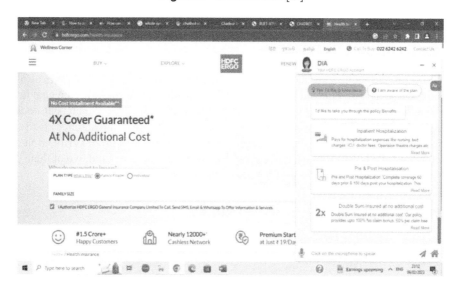

Fig. 22.2   Chat Bot "DIA" in conversation with customer who needs to take a Health insurance policy [21]

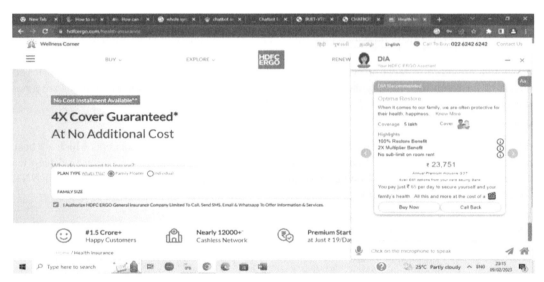

Fig. 22.3 Chat bot " DIA" processing a policy when the details provided by the customer are seemed to be legitimate [21]

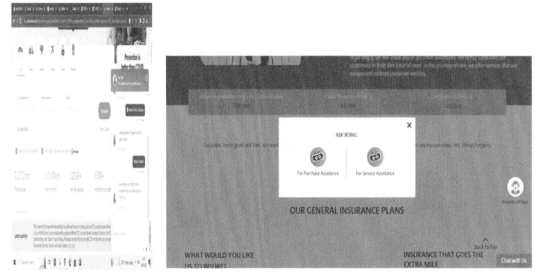

Fig. 22.4 Chatbots of other health insurance companies during claim processing [22]

5. Conclusion

This paper discusses the several methodologies and techniques that are deployed in the development of chat bots used in the insurance claim processing sector. The role played by the human is now intercepted by AI driven chat bots or virtual assistants in a positive perceptive in order to support their human peer in completing a task. The human peer could be a customer or an insurance techie. The paper also deals with the discussion of the effectiveness of health insurance chat bots and their efficiency in stopping a claim from being further processed if encountered with a illegitimacy in proof verification details provided by the client or customer. The chat bots of a few health insurance providers in India was tested and two of them proved to detect health insurance fraudulent claim while verifying the proof of details in the second stage itself.

References

1. https://yellow.ai/chatbots/use-cases-of-chatbots-in-insurance-sector/
2. https://marutitech.com/ai-in-the-insurance-industry/

3. https://suyati.com/blog/how-can-chatbots-help-insurance-providers-win-more-customers/
4. https://www.lemonade.com/blog/the-empathetic-bot/
5. Riikkinen, Mikko, et al. "Using artificial intelligence to create value in insurance." *International Journal of Bank Marketing* (2018).
6. Reis, Thoralf, et al. "An Information System Supporting Insurance Use Cases by Automated Anomaly Detection." *Big Data and Cognitive Computing* 7.1 (2022): 4.Adamopoulou, Eleni, and Lefteris Moussiades. "Chatbots: History, technology, and applications." *Machine Learning with Applications* 2 (2020): 100006.
7. https://kount.com/solutions/insurance-fraud-prevention/
8. Zumstein, Darius, and Sophie Hundertmark. "CHATBOTS--AN INTERACTIVE TECHNOLOGY FOR PERSONALIZED COMMUNICATION, TRANSACTIONS AND SERVICES." *IADIS International Journal on WWW/Internet* 15.1 (2017).
9. Oza, Divyang, et al. "Insurance claim processing using RPA along with chatbot." *Proceedings of the 3rd International Conference on Advances in Science & Technology (ICAST)*. 2020.
10. Tebenkov, Evgeny, and Igor Prokhorov. "Machine learning algorithms for teaching AI chat bots." *Procedia Computer Science* 190 (2021): 735–744.
11. Wu, Wei, and Rui Yan. "Deep chit-chat: Deep learning for chatbots." *Proceedings of the 42nd International ACM SIGIR Conference on Research and Development in Information Retrieval*. 2019.
12. Pillay, Craig Paul, and James Kariuki Njenga. "Opportunities for reducing expenses through digital innovation: The case of an insurance company." *The African Journal of Information Systems* 13.1 (2021): 5.
13. Pirilä, Tommi, et al. "The Role of Technical and Process Quality of Chatbots: A Case Study from the Insurance Industry." *Proceedings of the 55th Hawaii International Conference on System Sciences*. 2022.
14. Nuruzzaman, Mohammad, and Omar Khadeer Hussain. "IntelliBot: A Dialogue-based chatbot for the insurance industry." *Knowledge-Based Systems* 196 (2020): 105810.
15. Hussain, Shafquat, Omid Ameri Sianaki, and Nedal Ababneh. "A survey on conversational agents/chatbots classification and design techniques." *Web, Artificial Intelligence and Network Applications: Proceedings of the Workshops of the 33rd International Conference on Advanced Information Networking and Applications (WAINA-2019) 33*. Springer International Publishing, 2019.
16. Wallace, R.S.: The anatomy of ALICE. In: Parsing the Turing Test, pp. 181–210. Springer, Dordrecht (2009)
17. Shawar, B.A., Atwell, E.: Chatbots: are they really useful? Zeitschrift für Computerlinguistik und Sprachtechnologie, p. 29 (2007)
18. Al-Zubaide, H., Issa, A.A.: Ontbot: Ontology based chatbot. In: 2011 Fourth International Symposium on Innovation in Information & Communication Technology (ISIICT). IEEE (2011)
19. Ramesh, K., et al.: A Survey of Design Techniques for Conversational Agents. In: International Conference on Information, Communication and Computing Technology. Springer, Singapore (2017)
20. Csáky, R.: Deep Learning Based Chatbot Models (2017). https://doi.org/10.13140/rg.2.2. 21857.40801
21. https://www.hdfcergo.com/health-insurance/
22. https://www.bajajallianz.com/health-insurance-plans.html

Human Machine Interaction in the Digital Era – Prof. J. Dhilipan et al. (eds)
© 2024 Taylor & Francis Group, London, ISBN 978-1-032-54998-9

Polycystic Ovary Syndrome Classification Using Machine Learning Algorithm

23

V Sumathy[1], Rexline S. J.[2], Kavya M.[3], and Mary Heaven J.[4]

Loyola College, Chennai

Abstract Polycystic Ovary Syndrome (PCOS) cannot be treated completely but its symptoms can be effectively managed through medication, lifestyle modifications, or a combination of both. However, the existing approaches for detecting and predicting PCOS are inadequate. To address this issue, we propose the implementation of a machine learning system consisting of various algorithms, such as logistic regression, decision tree, random forest, and support vector machine, to facilitate early detection and prediction of PCOS. PCOS cannot be cured completely but its symptoms can be effectively managed through medication, lifestyle modifications, or a combination of both. The existing approach to the detection and prediction of PCOS is inadequate. To address this issue, we propose the implementation of a machine learning system consisting of various algorithms, such as logistic regression, decision tree, random forest, and support vector machine, to facilitate early detection and prediction of PCOS.

Keywords PCOS, Machine learning algorithm, Logical regression, Decision tree, Random forest, Support vector machine

1. Introduction

Polycystic ovary syndrome (PCOS) affects approximately 1 in 10 women of childbearing age, leading to hormonal imbalances, metabolic issues, and various health concerns. PCOS, first identified by Stein and Leventhal, is a widespread and untreatable cause of infertility, impacting around 6-26% of women, according to the World Health Organization (WHO) mentioned by Vedpathak and Thakre(2020). The syndrome encompasses ovulation failure, fertility problems, and mental health complications. Its effects range from acne, obesity, and irregular menstruation to infertility, ultimately affecting the overall quality of life. Furthermore, PCOS is a significant contributor to emotional distress and infertility, with reported rates of depression and anxiety at 28-39% and 11-25% respectively according to Kinjal Raut , Chaitrali Katkar (2022). Nigam et al. (2022) proposed the ccommon characteristics of PCOS include anovulation, elevated androgen levels, and the presence of multiple small ovarian cysts. Symptoms may include irregular or missed menstrual periods, excessive hair growth, acne, infertility, and weight gain. WHO estimates the prevalence of PCOS to be between 6% and 26%, while in India, it ranges from 9.13% to 36%.

1.1 Methods

The main aim of this research paper is to analyse data on Polycystic Ovary Syndrome (PCOS) using different machine learning algorithms, including Logistic Regression, Decision Tree Classifier, Random Forest Classifier, Support Vector Classifier, and Principal Component Analysis (PCA) for feature selection. This analysis can provide valuable insights to patients and physicians, assisting them in making informed decisions regarding treatment options. The illustration in Fig. 23.1 gives the process flow of the procedure followed in the paper.

{sumathy, rexlinesj, 22pds015, 22pds010}@loyolacollege.edu

DOI: 10.1201/9781003428466-23

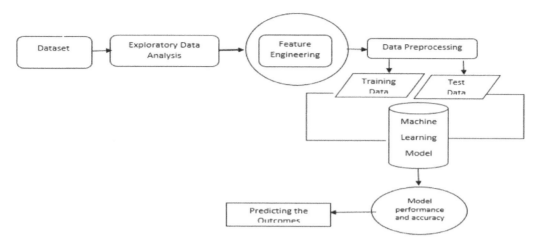

Fig. 23.1   Pipeline of the machine learning procedure

The analysis incorporates a comprehensive list of features, such as Patient File No, Weight (Kg), Height (Cm), BMI, Blood Group, Pulse rate (bpm), RR (breaths/min), Hb (g/dl), Cycle (R/I), Cycle length (days), Marriage Status (Yrs), Pregnant (Y/N), No. of abortions, I beta-HCG (mIU/mL), II beta HCG (mIU/mL), FSH (mIU/mL), LH (mIU/mL), FSH/LH, Hip (inch), Waist (inch), Waist-Hip Ratio, TSH (mIU/L), AMH (ng/mL), Exercise (Y/N), BP_Systolic (mmHg), BP_Diastolic (mmHg), Follicle No. (L), Follicle No. (R), Avg. F size (L) (mm), Avg. F size (R) (mm), Endometrium (mm), PCOS (Y/N), etc.

The models get trained using the dataset, enabling the machine to learn from the various features and predict the appropriate treatment. To identify the most appropriate treatment, Machine learning algorithms are employed. Abhishek Gupta, Sannidhi Shetty (2022) include Logistic Regression, Decision Tree Classifier, Random Forest Classifier, and Support Vector Classifier. Evaluation metrics to study the performance of each model are used.

1.2 Logistic Regression

Logistic regression is a technique used to estimate or forecast the probability of a binary event (yes or no). Its primary objective is to categorize data points into different classes using a linear approach. Unlike linear regression, Logistic Regression doesn't directly fit a straight line to the data. Instead, fits an S-shaped curve known as the Sigmoid curve as per Wang Weiying (2022) . This mathematical function can assign any input value to a range between 0 and 1, making it suitable for binary classification. The sigmoid function restricts its output to only two values: 0 and 1.

$$S(x) = \frac{1}{1 + e^{-x}} \tag{1}$$

1.3 Decision Tree

The problem solved by a decision tree employs a tree structure where each leaf node represents a class label and the attributes on the internal nodes. A Decision tree can express any Boolean function by discrete values. The process begins at the root node and involves dividing the data based on the feature that yields the highest Information Gain (IG).

1.4 Entropy

Entropy quantifies the level of disorder or randomness in the outcomes following each split at every level. The purpose is to decrease the randomness at each new level referred by Hassan and Mirza (2020). The Entropy value always lies between 0 and 1, indicating the degree of impurity or uncertainty.

$$(S) = \sum_{i=1}^{c} -p_i log_2 p_i \tag{2}$$

1.5 Gini Impurity

Gini Impurity shares a similar concept with entropy but utilizes a slightly different formula. In many cases, the Gini impurity has its computational efficiency over other methods. Zigarelli and Hyunsun Lee (2022) mentioned unlike entropy, Gini impurity does not involve a logarithmic function, making it faster to compute. Gini impurity value ranges from 0 to 0.5, representing the degree of impurity or uncertainty.

$$Gini = 1 - \sum_{i=1}^{c}(p_i)^2 \tag{3}$$

Information Gain is a metric used to evaluate the amount of information obtained at each level in forming the decision tree model. Its primary objective is to maximize the information at each step. As the entropy decreases, the information gain increases, indicating a reduction in uncertainty and a better understanding of the data mentioned by Munjal and Gautam (2020). Information Gain is crucial for making informed decisions when building the decision tree.

$$Gain(S, A) = E(S) - \sum_{v \in Values(A)} \frac{|S_v|}{|S|} E(S_v) \tag{4}$$

1.6 Random Forest Classifier

The random forest classifier generates a collection of decision trees using a randomly chosen subset of the training data. According to Rakshitha Kiran (2022), each decision tree is formed from a different randomly selected subset of the training set [10]. For the final prediction, the random forest classifier gathers votes from all the decision trees in the set and determines the majority vote as the final prediction. This ensemble approach helps improve the accuracy and robustness of the classifier by combining the predictions from multiple decision trees referred by Bhat (2021)

Hyperparameter tuning:

Hyperparameters are the parametric tuning technique to improve the performance of the model. It includes the following parameters,

- estimators — the number of decision trees.
- criterion — loss function used to determine the model outcome
- bootstrap — True indicates bootstrapping principles are used else False is assigned
- max_depth — max. depth of the trees
- max_features — max. number of features the model considered for the split

1.7 Support Vector Machine (SVM)

The objective of the SVM algorithm is to establish an ideal boundary or decision line that can categorize the n-dimensional space, making it simple to assign additional data points to the appropriate category in the future. The ideal decision boundary so formed is the hyper-plane.

2. Performance Analysis

In this section the performance of the implemented classifiers is discussed. Five evaluation metrics like Accuracy, Precision, Recall, F1-score and Confusion matrix are found.

Table 23.1 Confusion matrix

Predicted Class	Actual Class	
	Positive (P)	Negative (P)
Positive (P)	True Positive (TP)	False Positive (FP)
Negative (P)	False Negative (FN)	True Negative (TN)

Confusion matrix gives the classification of correct and incorrect predictions.

$$Accuracy = \frac{TP + TN}{TP + TN + FP + FN} \tag{5}$$

The weighted average of precision and recall is given by,

$$F1 - score = \frac{2 + Precision + Recall}{Precision + Recall} \tag{6}$$

The proportion of positive prediction is given by,

$$\text{Precision} = \frac{TP}{TP + FP} \tag{7}$$

True Positive rate is given by,

$$\text{Recall} = \frac{TP}{TP + FN} \tag{8}$$

Table 23.2 Accuracy and F1 score of various classification algorithms

Classification Algorithms	Accuracy Score	F1 Score
Logistic Regression (Before Smote)	82.95%	74.72%
Logistic Regression (After Smote)	82.95%	75.78%
Random Forest	78.51%	64.19%
Decision Tree	71.85%	62.74%
SVM	**84.44%**	**76.40%**

From the above models, it is inferred that the SVM returns the highest accuracy and F1 score. SVM hyper tuning with radial basis function as kernel, 100 as regularization parameter, and 0.001 as kernel coefficient to get the optimal model for the PCOS dataset is done.

3. Conclusion

This research aims to detect PCOS based on clinical and metabolic parameters using various machine learning classification models. The dataset details 541 patients with 43 attributes. Analysis inferred that the most important feature influencing the outcome is FSH. A good accuracy score is obtained when 10 features are taken for modelling. After the implementation of five classifiers: Logistic Regression before and after smote, Random Forest, Decision tree and SVM on 10 features, the SVM model is found to be most accurate with an accuracy score of 84.4% and F1 score of 76.4%.

References

1. Shreyas Vedpathak, Vaidehi Sunil Thakre:PCOcare: PCOS Detection and Prediction using Machine Learning Algorithms. Article in Bioscience Biotechnology Research Communications ·December 2020.
2. Dr. Ashok Munjal, Dr. Rekha Khandia, Brijraj Gautam: a machine learning approach for selection of polycystic ovarian syndrome (pcos) attributes and comparing different classifier performance with the help of wekaand pycaret, Volume - 9 I Issue - 12 I December – 2020.
3. Shakoor Ahmad Bhat: Detection of Polycystic Ovary Syndrome using Machine Learning Algorithms, August 2021.
4. Kinjal Raut 1, Chaitrali Katkar 2, Prof. Dr. Mrs. Suhasini A. Itkar 3. PCOS Detect using Machine Learning Algorithms. Volume: 09 Issue: 01 I Jan 2022.
5. Prashant Richhariya, Madhuri Nigam, Bharti Bhattad, Anita Soni and Pankaj Richhariya: An Efficient Way of Detecting PCOS Using Machine Learning, April 2, 2022.
6. Abhishek Gupta, Sannidhi Shetty, Raunak Joshi, Ronald Melwin Laban:SUCCINCT differentiation of disparate boosting ensemble learning methods for prognostication of polycystic ovary syndrome diagnosis, august 16, 2022.
7. Wang Weiying: Machine Learning Prediction Models for Diagnosing Polycystic Ovary Syndrome Based on Data of Tongue and Pulse, 2022.
8. Malik Mubasher Hassan, Tabasum Mirza: Comparative Analysis of Machine Learnin Algorithms in Diagnosis of Polycystic Ovarian Syndrome International Journal of Computer Applications (0975 – 8887) Volume 175 – No.17, September 2020.
9. Angela Zigarelli, Ziyang Jia, Hyunsun Lee: Machine-Aided Self-Diagnostic Prediction Models for Polycystic Ovary Syndrome: Observational Study, 2022.
10. Rakshitha Kiran P1, Naveen N. C2: Op-RMSprop (Optimized-Root Mean Square Propagation) Classification for Prediction of Polycystic Ovary Syndrome (PCOS) using Hybrid Machine Learning Technique: (IJACSA) International Journal of Advanced Computer Science and Applications, Vol. 13, No. 6, 2022.

Note: All the figures and tables in this chapter were made by the author.

Human Machine Interaction in the Digital Era – Prof. J. Dhilipan et al. (eds)
© 2024 Taylor & Francis Group, London, ISBN 978-1-032-54998-9

Deep Predictive Analytics of Online Learning Materials to Customer Clusters with User Intention and Behavior Patterns

24

J. Dhilipan[1]

Professor & Head, Department of Computer Science and Applications, MCA,
SRM Institute of Science and Technology, Ramapuram Campus, Chennai, India

S. Belina V. J Sara[2]

Assistant Professor, Department of Computer Science and Applications, B.Sc(CS)
SRM Institute of Science and Technology, Ramapuram Campus, Chennai, India

Haarindra Prasad[3] and Raman Raguraman[4]

Faculty of Engineering and Computer Technology, AIMST University, Malaysia

Abstract The system of inline tutoring can be considered as an expanded platform for education, serves the purpose of learning assignment being modelled and assigned for different tribal students by assuring development of Deep Predictive Analytical Student Network with respect to student performance. Usually, large scale distributed online tutoring infrastructure holds tutorials that are abundant that's needs to be assessed to students regarding the cause of execution, As a cause of large number of student enrolling, there is always a presence of complication in the process of identification students based on the accurate syllabus and tutorial on basis of the student performances. In order to design a suitable tutorial allocation, customized knowledge structure has become necessary to various segment of student clusters. On the other hand, previous recommendation systems applying the model of machine learning severely overlooked the dynamics within the student's behavior and intensions on various context, i.e., as fresh technologies constantly developed and progression of the student's behavior and intention in terms of learning and experience is experienced and it drives to a fatigued start and data sparsity problems. In terms of mitigating those complications mentioned in the upper section, a narrative in depth predictive student analytics structural design is employed in this works as deep neural network to outcome knowledge-enabled technology learning recommendation on basis of evolution of student intention and behavior formulation. Initially student dynamics to the learning model and clustering of the student to the online tutoring system has been extracted using various features consisting numerous factors of latent of the student implementing the technique of extraction. Student Latent factor obtained on basis of the learning features of the student's preference according to their experience and knowledge behaviors instead of profile information and geographical information. Moreover, networks based on deep neural methods have been indulged regarding the development of student performance-enabled learning recommendation system to customer clusters on utilization of latent feature extracted. Deep prediction of suitable knowledge structure to student clusters is capable of achieving higher amount of accuracy, latency and reliability. In-depth predictive e-Learning analytics actively schedule the learning to students based on implementing features of latent regarding the evaluation of the function of objective to develop learning proposal with a minimal error resulting in important enhancement based on the performance prediction regarding the information that are discriminative of the current technical information with student. A wide-ranging development is generated based on real-time data to make a comparison on the evaluated model with other conventional practices based on the presentation on possibility of the learning material recommendation to various class of student cluster. The experimental outcomes prove that proposed In-depth architecture of learning can get scalability and effectiveness based on any measurement of student information to the e tutoring system.

[1]hod.mca.rmp@srmist.edu.in, [2]belinavs@srmist.edu.in [3]haarindra@aimst.edu.my, [4]raman_raguraman@aimst.edu.my

DOI: 10.1201/9781003428466-24

Keywords Deep learning, E tutoring system, Learning recommendation, Interconnected data model, Student clustering, Behavioral analysis, Performance analysis

1. Introduction

Online e-tutoring systemsis becoming more effective after covid-19 pandemic on all parts of the world especially to tribal peoples. E tutoring system is a aiding solution which provides distributed learning system to increase the knowledge of the student's to handle various applications based on real world. Especially, these implementations expose the answer as integrated model among artificial intelligence and distributed student scheduling capabilities to cluster and manage the student and application data [1]. Based on the general views, the system regarding e-tutoring relies on education that depends on the principle of selection in an automatic nature. In this principle, the educationalist mainly presents their model of learning to students present in personalized online platforms and networks, and afterward, students evaluate the possible determined technology that can be suitable. However, the principle regarding the selection in an automatic nature allows the student to determine technology relied on their personal preferences. As a result, the system based on particular methods makes the determination process complex regarding the evaluation of perfect technology for the students. In order to mitigate those challenge, huge no of machine learning architecture [3] generally implemented using unverified mechanism with heuristics that are greedy. As a presence of propagation in a large amount of the students and educationalist, this becomes complex to effectively forecast the suitable knowledge-based technology learning proposal ideas to students along complexity of computation in a less amount that leads to issues in the sparsity and cold start.[4].

Specifically, novel deep predictive student analytics model is projected as learning recommendation architecture to solve the complication of sparsitry and chances of cold start based on the implementation of models based on in-depth learning methods. This particular article exposes a network based on deep neural method that has been implemented regarding the evaluation of feature that are latent related to dynamic of student intention and the dynamic of behavior to produce an e- tutoring system. Those multiple latent factors is obtained on employing the feature extraction technique and these feature is projected to deep learning model. Proposed model considered as technology enabled e-learning system based on recommendation for students higher amount of learning latency, accuracy, and reliability. In-depth architecture of learning manages the technology enabled learning knowledge structures to students in the process of embedding features of latent on further utilization of the function of objective to generate the effective suggestion with minimal flaws. It outcomes enhance the performance based on prediction for the information's that are discriminative of the technology enabled learning recommendation module to various class of students.

The enduring study is prepared as review of the relevant work are illustrated in the section number 2, the proposed architecture based on network of deep neural regarding technology enabled student learning recommendation to e-tutoring struture is detailed in the section number 3 and investigational outcomes and efficiency of system that is evaluated is demonstrated in the section number 4 through implementing data based on real time scenario with performance evaluation adjacent to approach state of the art on numerous performance metric. Moreover, article is summarized in the section number 5.

2. Relevant Works

This section exposes, e tutoring system implementing architectures based on machine learning is analyzed relied on details depending upon the architectures about representation of knowledge structures and measure of similarity regarding the information that is processed by the student. Every individual architecture based on machine learning that shows good results regarding the evaluation effectiveness of the process is illustrated vastly and few methods which shows a performance with similarities compared with the evaluated model is described in the following section

2.1 Content based Filtering

This literature section shows filtering based on content usually implemented regarding user prediction for item on basis of the user preferences. Matrix factorization [5] is to determine the association among the user on the e- commerce application based on the non availability of the information of training. Heuristics that is initially greedy is incorporated regarding the determination of the explicitly and implicitly with more features of discrimination to model the space of feature based on user-item recommendation [6]. User space of feature is regenerated based on the feature set of hierarchical random method.

2.2 Collaborative Filtering Learning

Based on this method, user-based item Recommendation system is analyzed regarding the prediction of the item to implement user learning based on preference and similarity of the user. Similarity based on correlation [7] has been employed to calculate the correlation among the item. Measures of Pearson similarity is an extended form responsible for two items linearly relating based on the preference of the user. It is responsible for providing modified list of item based on the user regarding utilizing the model of explicit and implicit feedback.

3. Evaluated Model

This particular section aims to expose requirement of the exhaustive design regarding the evaluated architecture termed as a from of deep predictive analytic network implemented for architecture of learning recommendation on insertion of tuning of parametric of the layers based on deep learning to acquire the prediction of students regarding technology based on the latent factor in the technology and student intention and behavior examination with respect to the student experience clustering.

3.1 E tutoring System

E -tutoring system which composed of students profile, geographic information, Technology information and scheduler of the assignment is implemented for effective changes in the recommendation.

Technology Pool

Pool of technology can be considered as a set of m technology generated by different educationalists based on the different characteristics. Technology belongs to different categories, and it's described by by t_i

$$\mathbf{T} = \{t_1, t_2, t_3 \dots t_n\}$$

Students Pool

The student's pool keeps track of statistics for each student intention and behavior within evaluated the system with the geographic and personnel and entire performance regarding doing and learning in individual subject. These evaluataed statistics are maintained dynamically with the chances of being implemented to direct the scheduling process of the student related to technology. Apart from that, pool of the student holds the data regarding currently available offline and online students available for subject learning with specified number of constraints.

$$\mathbf{D} = \{d_1, d_2, d_3 \dots d_n\}$$

3.2 Estimation of Student Profile

Interms of eliminating the intrinsic rate of error, approaches based on voting such as weighted voting [8] and Bayesian voting [9] has been implemented with respect to student experience based on aggregation regarding the student based on the learning group. Moreover, it helps to compute large number of classes fro students on their feature of latent based on their behavior and experience. In this work, analysis based on latent discriminator is implemented to compute the factor of latent of various students towards the learning technology. The profile of the dynamic student is illustrated based on the form of matrix as a matrix of full projection. The matrix based on full projection acquiring the column describes the geographic and personnel information position of the student relying on the characteristics of intention and the rows denotes the characteristics of experience or behavior. This is further implemented to evade the ratio of fitting overly based on the computation of the matrix adjusted by mean. [6].

Student attribute Correlated Vector for each behavior $\lambda_i = \dfrac{1}{n} \int U \left(\dfrac{dy}{dx}\right)^{-2} \sum\limits_{x \varepsilon C}^{n} \mathrm{I}, (\mathrm{p} - \mathrm{b}_i)$

Combined Student Intention Vector $\lambda_i = \int CU \left(\dfrac{dy}{dx}\right)^{-2} \sum\limits_{x \varepsilon C}^{1} \mathrm{I}$

Vectors with a linear nature upon the student intension, behaviors and experiences are premeditated as a form of subspace on dimensions on multiple formats. The feature of optimal latent for the structure of knowledge preparation is calculated based on integrating the matrix of scatter by

Variation analysis of student Matrix behavior $\lambda_i = \int U(\frac{dy}{dx})^{-2} I_i \frac{ni}{L}$

Projection matrix of student representing profile of the student is processed employing conversion matrix by implementing normalization of matrix on the determined pool technology along equivalent distinctiveness [9]. The aggregation of linear based on parameters of learning-adapted for each student the learning of feature is processed by similarities found by pair-wise of the intention and aspects of behaviour. The vector of latent feature has been determined for pool of technology.

In-depth predictive analytics is architecture of deep learning implemented for student cluster prediction to various class of the learning structures student for the technology learning. This part exposes tuning of parametric of the function of activation for the proposed model for prediction of student cluster on functions of objective relied on implicit behavior and experience of user as a form of latent factors is incorporated. The model of training utilizes max amount of pooling to maps the extracted features of latent within features of pair wise illustration and these representation based on pair-wise is processed in the function of convolution as preference of students towards features of reconstruct for learning recommendation [10].

Layer of max pooling

Based on this particular layer, feature of latent as a form of Subset will be iterated based on features of pair-wise on various pairs of preference of students upon biased intention coefficients and behaviors. The matrix of preference developed using the matrix of factorization. This is represented as student model of the latent factor regarding Predicting the materials of learning for the specified students cluster. The layer of max pooling of the network of neural to produces student feature of latent as layer of embedding. It is as follows

$$\mathbf{x} = \sum_{i=1}^{N} x_i v_i = x_1 v_1 + x_2 v_2 + \dots + x_N v_N$$

The student feature representations to the model which improve the discrimination among students during the computation of the Pearson similarity of correlation based on the preference of the student that have been generally given as an outcome. The components of scheduling of the neural networks that are fully connected for the Prediction based tuning are illustrated in the Table 24.1.

Table 24.1 Component of scheduling for the completely linked neural network

Parameter of hyper of Component of scheduling	Values of parameter
Student size of batch to behaviours matrix	148
Knowledge enabled model Learning Rate	0.05
No of preference and behavioural dimensions of student	15
Epoch value to Embedding layer	45
Max No of features of latent in the subset	1000
Maximum length of student of cluster	1000
Function of error	Cross entropy

Layer of embedding

The layers of embedding collects the feature of preference subset by using their mechanism of inherent hierarchically as a form of features of abstract and discover the features of discrimination within few parameters that are hyper in nature of entirely connected network.

Result regarding the space of feature is dimension of latent connected to the evolving student learning distinctiveness. This particular layer holds various developing distinctiveness of the variable of student intention have been embedded to the activation of function regarding the determination of the student cluster based on the learning syllabus. This particular layer also shows that the features of students of high dimentional value are transformed linearly into the vectors that are low-dimensionally embedded by learning.

Function of activation

The evaluated architecture implies the units of rectified leinaer system (ReLU) function of activation[11] for the recommendation of learning syllabus to developing student clusters holding the vector of preference and vector based on pair-wise values of

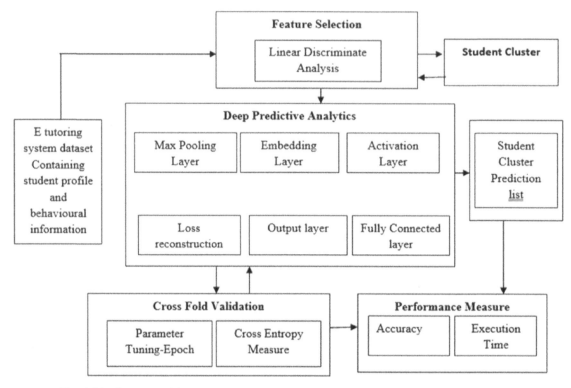

Fig. 24.1 Diagram of the architecture of the recommendation of technology for students

student clusters in the layer that is embedded. The function of activation provides the suitable student based on individual element in matrix of embedding. Vectors that are embedded is processed with values that are parameterized to obtain the proper determination regarding the determined vector of learning syllabus and to individual parameter of epoch updating.

Layer of output

The layer of output of the in-depth predictive analytic contains the prediction result of the student suggestion to particular learning syllabus. Further parametric based feature annihilation of the prediction of student cluster based on the measures of similarity is produced within this particular layer for recommendation of student cluster or recommendation of technology tutorial. Optimization of the soft max [12] is achieved to get the student cluster cross and set mechanism of entropy has implemented to determine the efficiency of the developed prediction of technology for the evolution of student. Tuning of hyper parameter is enabled in the layer of output to take the Predictive decisions in the tutorial of technology list regarding the reason of being close by computing the resemblance of the derived technology on novel representation of students based on various evolution.

Layer of loss

This layer of loss ensures the accuracy of the prediction based on tuning in a fine nature against parameter of refine of various layers regarding in-depth predictive analytics model to ensure the minimum wrror of reconstruction among the layer of feature max pooling and the layer of ReLu activation. Furthermore, the loss function of the cross entropy has been incorporated to handle the outcome of prediction regarding the determination of the student based on the environment of e-tutoring [13].

Algorithm 1: Technology based on In-depth Evolution Learning Recommendation

Input: knowledge-based discriminative learning tutorials and student set

Output: StudentCluster Prediction for Discriminativeknowledge-based learning tutorialrecommendation Process

Linear Discriminant Analysis ()

Determine latent feature Set of the user on intention and behaviour

$F_s = \{u_1(I,B), u2(I,B)\ldots..u_3(I,B)\}$

Process Deep Predictive Analytics Learning ()

Convolution layer ()

    Extraction feature based on kernel and stride

        Feature Map

        Max Pooling ()

Calculate Optimal student Behavior and intention reduce the feature map

    Embedding Layer ()

        Estimate optimal feature for student clusters

            Activation Layer ()

                Parameterized Tuning of ReLu Function for map student cluster to the dynamic syllabus

                    Layer of output ()

SoftMax () --- student cluster recommendation to e learning

This particular function provide enthusiasm to the feature of embedded latent points on map of representative to form learning list of syllabus tutorials to student based on the experience and behavior of the student in e tutoring system as solution of recommendation.

4. Results of Experimental Approaches

In this section, results regarding the experimental tests of the evaluated e tutoring tutorial model of recommendation are analyzed upon the conventional collaborative filtering model for approaches of recommendation based on the evolution of the characteristics of the students in the system of e tutoring. Illustration of the further performance on the architecture that has been proposed is depicted based on outperforming the previous approaches in various components of F1 score, Precision, execution Time and recall.

4.1 Description of the Dataset

Experiments with extensive nature have been named by datasets Edu week [14] that is generally considered as a data set based on real time. The real time dataset holds 100,000 instances of tutorial of technology and student acquiring different scale of ratings of 1–5 based on different variations of experience in the execution of the project. The dataset that has been turned into the form of matrix is implemented for continuation process of the analysis.

4.2 Evaluation of Framework

The framework that has been proposed for recommendation of technology syllabus is calculated upon the measures that are following Completeness of Pair wise (PC), Recall(R), Precision (P); computation time, and F1-scorecs (F).

Precision of the framework

The value of predictive positive of the characteristics of student cluster can be considered as the ratio of details of relevant learning compared to the retrieved learning instance within the characteristics of technology. The Precision can be also counted as a determined number of students that are correct divided by the value of students that are returned of from the list of the student. The true positive can be considered as a number of occurrences based on real positive in the space of data and the false negative can be considered as a number of occurrence for real negative in the space of student. The entire evaluation is calculated upon the dataset depicted in the Fig. 24.2.

$$Precision = \frac{\text{True positive}}{\text{Truepositive} + \text{Falsepositive}} = \frac{TP}{TP + FP}$$

Recall

It can be considered as the ratio of characteristics that are relevant of the student based on characteristics that are retrieved over the total number of students that are relevant. The recall section is a active part of the occurrence with relevancy of student that can be successfully determined within the tutorial of the technologies. The value of true positive is a number of occurrences

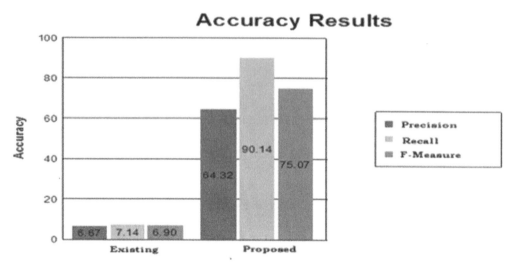

Fig. 24.2 Performance computation of prediction accuracy

based on real positive in the student space and false negative is number of real negative occurrences within the space of the student for tutorials of the technology.

$$\text{Recall} = \frac{\text{True positive value}}{\text{Truepositive} + \text{Falsenegative}} = \frac{TP}{TP+FN}$$

In the analysis of recall, the relationship based on time-dependence syllabus of learning tutorials and student in the application process can reflect a student intension on choosing the latent pattern and preference of the technology tutorials on basis of student performance.

Deep predictive analytics can enable the chances of categorizing the group of student relied on their behavior and performance regarding tutorials. It efficiently addresses the problem of scalability in explorations based on large scale data by obtaining student for recommendation of tutorial within severely similar and smaller tutorials list compared to extracting the entire set of data.

Algorithms based on traditional recommendation relies on machine learning that has been implemented depending on measures of accuracy in that the outcomes of the recommendation holds the lists that is created through containing similar model items. Higher amount of correlation similarity represents diversity in a lower form of the results based on tutorials assignment. This process shows that the prediction regarding individual student will contain the aim of the student that needs to be enhanced further to boost the recommendation's quality.

F Measure

It can be considered as a metric implement to estimate the accuracy of a test regarding the outcome of the recommendation of tutorials towards students and is mentioned as the mean of weighted harmonic of the recall and precision [15] of the data of test or training to recommendation of tutorials on the determination of characteristics of an individual student. As a result, it enhances the recommendation size set that will lead the enhancement proportionally in the value of recall. However, at the similar time it drives to a noticeable decrease in the values of precision.

Execution time

It is mentioned as no of time taken to establish the student instance matching for the various behavioral characteristics evolution among the two heterogeneous tutorials in the e tutoring systems.

The entire performance regarding the time of execution of the architecture that is proposed is generally estimated upon conventional approaches is depicted in the Fig. 24.4 and values of performance of the models has been exposed in the Table 24.2 on the system of recommendation.

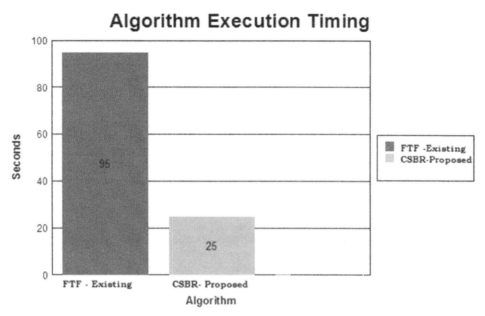

Fig. 24.3 Performance computation

Table 24.2 Performance comparison of methodologies

Dataset	Method	Precision (%)	Recall (%)	F measure (%)	Time of Computation (s)
Edu week Dataset	Proposed- Existing	97.37 94 .61	86.23 84.23	97.23 95.26	10 20

According to the Table 24.2, planned methods evolve the entire complete progression of recommendations on the dataset that is projected will effectively run about 2-3 orders of degree faster compared to the reference of state of art approaches based on source regarding evolution of student. For evaluating the proposed model of the reference student choice by the help of heuristics along with the system of prediction method that has been analyzed as it increase hugely.

5. Conclusion

We targeted along with utilized depth predictive analysis in favor of examining student due to suggestion of tutorials completing to create high accuracy along with scalability within higher education stage. The architecture utilize in depth predictive analysis along parameter related to hyper tuning to remove the reconstruction fault along with loss function in favor of excellent efficiency forecast for future deep absorbing building implements the characteristics within highest pooling coating to support for feature of sparse prophecy by layer of activation. Soft Mix part with layer associated with loss has been confirmed for launching the prediction of discriminative outcome to guider's suggestion. Besides, performance of prediction measured by f measure expects that this is strong for student's comparison through good parameters initialization due to tutorials advice.

References

1. J.A. Iglesias, A. Ledezma, and A. Sanchis, "Creating User Profiles from a Command-Line Interface: A Statistical Approach," Proc. Int'l Conf. User Modeling, Adaptation, and Personalization (UMAP), pp. 90–101, 2009.
2. A. Cufoglu, M. Lohi, and K. Madani, "A Comparative Study of Selected Classifiers with Classification Accuracy in User Profiling," Proc. WRI World Congress on Computer Science and InformationEng. (CSIE), pp. 708–712, 2009.
3. F. J. Ferrer-Troyano, J.S. Aguilar-Ruiz, and J.C.R. Santos, "Data Streams Classification by Incremental Rule Learning with Parameterized Generalization," Proc. ACM Symp. Applied Computing (SAC), pp. 657–661, 2006.
4. R. K. Agrawal and R. Bala, "Incremental Bayesian Classification for Multivariate Normal Distribution Data," Pattern Recognition Letters, vol. 29, no. 13, pp. 1873-1876, http://dx.doi.org/10.1016/ j.patrec.2008.06.010, 2008.

5. J.A. Iglesias, A. Ledezma, and A. Sanchis, "Sequence Classification Using Statistical Pattern Recognition," Proc. Int'l Conf. Intelligent Data Analysis (IDA), pp. 207–218, 2007.

6. Bo Tang, Steven Kay, Haibo He "Toward Optimal Feature Selection in Naive Bayes for Text Categorization "in IEEE Transactions on Knowledge and Data Engineering in Volume: 28, Issue: 9, Sept. 2016

7. H. Ma, D. Zhou, C. Liu, M. Lyu, and I. King, "Recommender systems with social regularization," in Proceedings of the 4th ACM International Conference on Web Search and Data Mining (WSDM), 2011, pp. 287–296.

8. Y. Koren, "Factor in the neighbors: Scalable and accurate collaborative filtering," ACM Transactions on Knowledge Discovery from Data (TKDD), vol. 4, no. 1, pp. 1: 1–1: 24, 2010.

9. P. Wang, H. Wang, X. Wu, W. Wang, and B. Shi, "A Low- Granularity Classifier for Data Streams with Concept Drifts and Biased Class Distribution," IEEE Trans. Knowledge and Data Eng., vol. 19, no. 9, pp. 1202–1213, Sept. 2007.

10. I. Katakis, G. Tsoumakas, and I. Vlahavas, "Dynamic Feature Space and Incremental Feature Selection for the Classification of Textual Data Streams," Proc. Int'l Workshop Knowledge Discovery from Data Streams (ECML/PKDD), pp. 102–116, 2006.

11. [11] Y. Li, Z. Niu, W. Chen, and W. Zhang, "Combining Collaborative Filtering and Sequential Pattern Mining for Recommendation in E-Learning Environment," Proc. 10th Int'l Conf. Advances in Web- Based Learning, pp. 305–313, 2011.

12. P. Lops, M. de Gemmis, and G. Semeraro, "Content-Based Recommender Systems: State of the Art and Trends," Recommender Systems Handbook, pp. 73–105, Springer, 2011.

13. J. Yi, R. Jin, S. Jain, and A. Jain, "Inferring users preferences from crowdsourced pairwise comparisons: A matrix completion approach," in Proc. 1st AAAI Conf. Human Comput. Crowdsourcing, 2013, pp. 207–215.

14. W. Lu, S. Ioannidis, S. Bhagat, and L. V. Lakshmanan, "Optimal recommendations under attraction, aversion, and social influence," in Proc. 20th ACM SIGKDD Int. Conf. Knowl. Discovery Data Mining, 2014, pp. 811–820.

15. J. Wang, P. Zhao, S. C. Hoi, and R. Jin, "Online feature selection and its applications," IEEE Trans. Knowl. Data Eng., vol. 26, no. 3, pp. 698–710, Mar. 2014.

Note: All the figures and tables in this chapter were made by the authors.

Human Machine Interaction in the Digital Era – Prof. J. Dhilipan et al. (eds)
© 2024 Taylor & Francis Group, London, ISBN 978-1-032-54998-9

Intelligent Home Automation with Multiple Languages Voice Control

25

P. Vinayagam[1], R. Ajith[2], R. Arvindkumar[3], N. Kishwanth[4]

Saveetha Engineering College. Chennai

Abstract As IoT devices proliferate, home automation systems must keep up. This technology is widespread in major organizations. Automation improves many elements of life. The "Internet of Things" connects devices remotely. The proposed approach enables mobile appliance remote control. This new Endeavour lets us operate household appliances with our voices in several languages. Bluetooth speaker, light bulb, ceiling fan, and power outlet may be controlled remotely. The Blynk app connects a Raspberry Pi to the appliances remotely. IFTTT ("If This Then That") connects Blynk to Google Assistant. Google Assistant lets users provide commands in their native language. This work emphasizes the linguistic barrier. By allowing users to communicate with the system in several languages, its dependability and efficiency will increase.

Keywords Internet of things, Home automation, Raspberry pi, Blynk, Human-computer interaction system

1. Introduction

Life is getting simpler and less difficult in a variety of industries as a consequence of the advancement of Automation technology. Today's cutting-edge technologies enable us to programme our house to execute a variety of tasks automatically. It's hardly surprise that "home automation" has become a buzzword in India; as the number of second-generation homeowners grows, people want more from their houses than simply basic utilities. As home automation becomes more popular, there will be no shortage of applications for Internet of Things devices designed for the smart home.

The proliferation of Internet of Things (IoT) devices has increased demand for residential and business automation solutions significantly. These innovative automation solutions seamlessly connect to the Internet of Things. Automation enabled by the Internet of Things will be critical for closing the gap between what people are capable of and what technology is capable of. Modern appliances may be monitored and controlled remotely over the internet. Users' internet-based orders will be collected through Wi-Fi modems. We chose Raspberry Pi over the modules and microcontrollers recommended by other authors since it is 50 times faster and superior than other controllers. It is the most precise of all realised concepts since it can move between many activities at the same time, eliminate the barrier without a hitch, and wirelessly link to other devices through Wi-Fi and Bluetooth. A voice-controlled smart home system was constructed using the Raspberry Pi 3B+, Google Assistant, the Blynk app, and If This Then That. The Blynk programme enables Pi interoperability and system security.

It is connected to the Raspberry Pi, allowing us to operate the devices manually using the Blynk app's virtual switches. If This Then That (IFTTT) enables the creation of a connection between the Google Assistant and the Blynk app, allowing appliances to be controlled by voice commands using server-stored applets. IFTTT stipulates that the Blynk app's action will be carried

---

[1]vinayagamap@gmail.com, [2]obethraghu11@gmail.com, [3]arvind123cva@gmail.com, [4]kishwanth1023@gmail.com

DOI: 10.1201/9781003428466-25

out only if and only if Google Assistant provides the necessary trigger (voice commands). Each controllable device can handle as many IFTTT languages as are available.

2. Literature Review

Eva Inaiyah Agustin et al, [3] demonstrated a home automation system that was built with NodeMCU and ESP8266. They used Blynk, which they downloaded, to operate the house electronics. The Blynk app allows for home automation. The Blynk Internet of Things app for Android may be used to turn on and off appliances.

Mustafa Omran's et al., [4] approach enables speech-activated home automation in medical contexts. They used the V3 voice recognition technology. This is a simple medical ward. We utilised an Arduino Uno to create these concepts. Speech recognition has an accuracy rate of 75%. Abiodun E. Amoran et al., [5] devised a system for remotely monitoring and controlling household devices. Users may customise the system using the Blynk software and an Arduino Mega 2560 and a Raspberry Pi 3B+. In order to combine sensors with the raspberry pi and Blynk, Arduino was deployed.

Rohit Jaykar et al., [6] recommended using one's voice to control several components of home equipment. Consumer electronics of several types can be used. Homemade is always better. Create a digital assistant to help us with our daily duties. Salihu Aliyu et al., [7] presented a voice-controlled home automation system powered by Android and Bluetooth. This system makes use of atmega328 microprocessor technology and the HC-06 Bluetooth module.

Reddy Govinda Thotli et al., [8] proposed connecting the Internet of Things with voice-activated digital assistants such as Google Household or Siri to better operate connected household devices. The Raspberry PI controls the Relay, which is in charge of controlling the electrical appliances. There are a few gadgets that can do this, but we prefer to create our own. The Google assistant requires voice commands. IFTTT, a web service that produces if-then conditional statements, is connected to Adafruit, a free cloud-based IoT web server that may be used to build virtual switches.

3. Existing Methodologies

Over the course of the last ten years, a significant amount of research has been conducted to investigate the potential of utilising Google's virtual assistant for voice-activated home automation. For the purpose of this particular project, Google Assistant will only operate in response to voice instructions. It is possible to create if else statements by using the website known as "If This Than That" (IFTTT), which is linked to the Adafruit account.

The Adafruit account is a free cloud-based IoT web server that is utilised in the construction of virtual switches. The voice commands offered by Google Assistant are now supported by the IFTTT service. In this particular configuration of a smart house, the Google Assistant may be instructed to manage several aspects of the home with only a few simple voice commands. These aspects include light switches, ceiling fans, and even motors.

The microcontroller is used to ensure that the relays carry out the actions that are specified by the Google Assistant. This is achieved through the use of the relays. In response to user requests, Google Assistant is able to alter the state of the relay associated with the paired device. Wi-Fi allows for the transfer of information between an application and a NodeMCU (ESP8266) microcontroller (Internet).

4. Proposed Methodology

Numerous firms incorporate Internet of Things (IoT) components in a vast array of consumer items, so tying industry to this initiative. This category includes devices such as televisions, refrigerators, smart speakers, and others. Students should thus take the time to study Internet of Things (IoT) devices and their construction. Current and future engineering designs will incorporate IoT devices often. Currently, only a few of languages are supported by commercially available smart home devices, with English being the most frequent.

Language hurdles may make it difficult for non-English speakers to use the speech recognition functions of their smart home. For those who are unable to communicate in English, this project will create a multilingual internet of things (IoT) home automation system, such as the senior population in India, in doing daily chores such as turning on and off lights using simple voice instructions. It may be remotely controlled by voice commands or smartphone applications. Due of its reliance on the Raspberry Pi, this system will be inexpensive, dependable, and easy to install and operate. Figure 25.1 and Fig. 25.2 describes the working of voice recognition system.

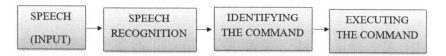

Fig. 25.1 Working of speech recognition

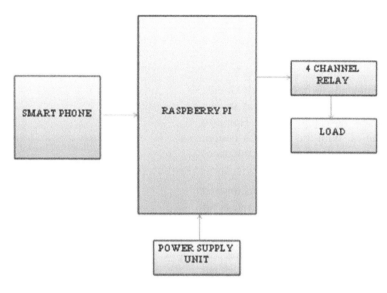

Fig. 25.2 Block diagram of proposed model

4.1 Setting up Google Assistant

1. We can change Google Assistant's language by going to settings > language > and choose our preferred language.
2. Second, access the "Routines" section of Google Assistant's settings.
3. Click on the + floating icon in the menu of Routines to set the voice command.

4.2 App making

1. Start with a new project in Android Studio and add a button navigation bar
2. Enable internet connectivity in the application so that the Raspberry Pi's GPIO may be controlled wirelessly.
3. In other words, establish a menu bar.

4.3 Testing

1. Use 5V DC to power the Raspberry PI.
2. Connecting the Raspberry PI and the mobile device to a wireless network is the second step.
3. Use "hello google" followed by the language-specific voice command.
4. It will turn on the room's lights.
5. The app is manageable inside the app itself.
6. To activate or deactivate the lights, launch the app and utilize the on screen controls.

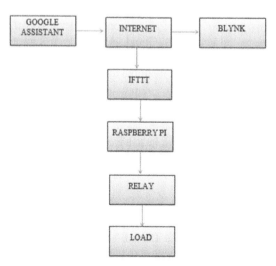

Fig. 25.3 Voice assistant architecture diagram

Table 25.1 Connection between relay board and Raspberry Pi

Raspberry Pi	Relay Board
IO 13	IN 1
IO 15	IN 2
5V	5V
GND	GND

5. Experimental Results

It includes utilizing voice commands to interact with Google Assistant. We've upgraded Google Assistant so that us can operate it with our voice. Users have told Google Home to perform certain household duties. With the push of a button, light bulbs, ceiling fans, motors, and other household appliances may be switched on and off.

The Google assistant transmits commands that are deciphered by the Raspberry pi, which then activates the relays in accordance with the instructions. In response to user voice instructions, Google Assistant toggled power to each relay's linked device. Figure 25.4 and Fig. 25.5 shows that the real time output of the voice controlled home appliance model and android mobile app.

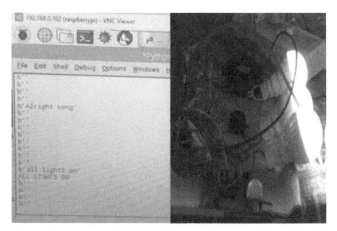

Fig. 25.4 Voice assistance home appliance control model

6. Conclusion

The voice-controlled home automation system based on Raspberry Pi assists the elderly and others with limited mobility in using electrical appliances. This article presents a user-friendly home automation system. With an Android app, consumers can make decisions and remotely operate appliances, which simplify their lives. We will use Google's Assistant for this, and we will be able to operate it via speech recognition in several languages. Google Assistant was chosen because of its reliability, durability, and ease of implementation. Android smart phones make it simple to access and incorporate Google Assistant into home automation projects. Google Assistant may be set up on an iPhone.

The proposed system should integrate Google Assistant. Multiple research projects have studied Raspberry Pi and open source software as home electronics management systems. With such platforms, users may personalize applications.

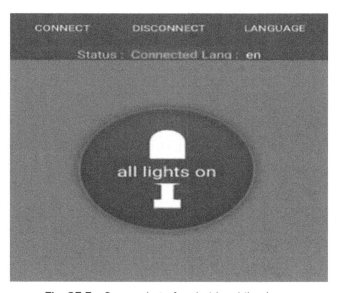

Fig. 25.5 Screenshot of android mobile phone

References

1. Noh Nor, Jaafar Haryati, Mustafa Wan, Syed Idrus Syed Zulkarnain and MazelanHaidiel, "Smart Home with Biometric System Recognition", Journal of Physics: Conference Series, vol. 1529, no. 4, 2020.
2. B Samitha, E Kumar, C Kumar, B Srinath, R Devi and T Venkatesh, "Intelligent Appliance Control System Using IoT", Journal of Physics: Conference Series, vol. 1916, no. 1, 2021.
3. Eva Inaiyah Agustin, RikyYunardi and Aji Akbar Firdaus, "Voice recognition system for controlling electrical appliances in smart hospital room", TELKOMNIKA (Telecommunication Computing Electronics and Control), vol. 17, no. 2, pp. 965, 2019.

4. Mustafa Omran, Wasan Saad, Bashar Hamza and Ahmed Al-Baghdadi, "Designing and Manufacturing of Home Automation Monitoring System Using Internet of Things Technology", Journal of Physics: Conference Series, vol. 1973, no. 1, pp. 01208, 2021.

5. Abiodun E. Amoran, Ayodele S. Oluwole, Enitan O. Fagorola and R.S. Diarah, "Home automated system using Bluetooth and an android application", Scientific African, vol. 11, pp. e00711, 2021, ISBN 2468-2276.

6. Rohit Jaykar, Shraddha Chobe, TejshreeKamegaonkar, Varsha Surwase, "voice control home automation system using raspberry pi", International Research Journal of Modernization in Engineering Technology and Science, Volume:04/Issue:05/May-2022.

7. Salihu Aliyu, Abdulazeez Yusuf, Umar Abdullahi, Mustapha Bola and Ajao Lukman, "Development of a Low-Cost GSM-Bluetooth Home Automation System", International Journal of Intelligent Systems and Applications, vol. 8, pp. 41–50, 2017.

8. Gopal Krishna Reddy Thotli, J. Maheshwar Reddy C. Manoj Kumar Reddy, "IOT Home Automation Using Raspberry Pi with Google Voice Assistant"Compliance Engineering Journal, ISSN NO: 0898–3577.

9. K.Y.Durga Prasadet al., "Voice Recognition Based Home Automation using Raspberry Pi", International Journal of Innovative Science and Research Technology,Volume 3, Issue 7, July – 2018

10. E. Shirisha et al., "IOT Based Home Security And Automation Using Google Assistant", Turkish Journal of Computer and Mathematics Education (TURCOMAT), vol. 12, no. 6, pp. 117–122, 2021.

11. P. Vinayagam, Dr. P. Anandan, Dr. N. Kumaradharan, "Image denoising using a nonlinear pixel likeness weighted frame technique", Intelligent Automation & Soft Computing, Vol.30 No.3 (2021), [Pages: 869–879], [DOI:10.32604/iasc.2021. 016761]

Note: All the figures and tables in this chapter were made by the Authors

Human Machine Interaction in the Digital Era – Prof. J. Dhilipan et al. (eds)
© 2024 Taylor & Francis Group, London, ISBN 978-1-032-54998-9

Commercial Power Hip Kinfolk Organized Commerce Plus the Basic Aimed at Scientific Office

26

A. Jayabal[1]
Associate Professor, Department of Commerce,
SRM IST, Ramapuram Campus

I. Enock[2]
Research Scholar, Asst. Professor, Department of Commerce,
SRM IST, Ramapuram Campus

Abstract In India, the KINFOLK company has a very significant impact on the expansion of the country's gross domestic product (GDP), the creation of new jobs, the growth of the nation, and the accumulation of wealth. The expansion of enterprises like KINFOLK over the course of the past ten years has been both commendable and quite significant. KINFOLK companies have a profound comprehension of the political climate in our nation. As a result, they are able to readily capitalize on the empty niches in the market in which there are unfilled chances for business, and as a result, they have been extremely successful over the course of many years. The majority of KINFOLK firms were initially established by men's fathers and grandfathers, and these families have now grown into new enterprises across a variety of industries and beyond geographic boundaries. While many of these companies have gone on to become pillars of the Indian economy, there are also examples of situations in which such COMMERCIAL houses engaged in manipulative techniques and unethical business activities in order to satisfy their own gains and maximize their own profits without having any altruistic concern for the society as a whole. The Scientific Office is a tool that applies investigation and analytical procedures to identify frauds in order to prevent such activities from occurring. In light of this, the purpose of this article is to investigate whether or not KINFOLK firms in India are operating in a responsible and moral way by meeting their social responsibilities and adhering to the principles of COMMERCIAL POWER. Several case studies have been conducted to investigate the instances of fraud that occurred within KINFOLK-Organized businesses. The findings of these case studies underscore the need for a Scientific Office because conventional methods were unable to resolve the issues.

Keywords Ethics, Scientific office, Moral and stakeholders

1. Introduction

We are already extremely accustomed to the idea that the shareholders are the owners of firms and that it is the role of the board of directors to run the companies. This is a concept that has been around for a very long time. The Board of Directors is also accountable for ensuring that their companies have adequate POWER. The term "COMMERCIAL POWER" refers to the idea that a company's board of directors should run the business in such a way that is profitable, open to public scrutiny, and run for the benefit of the company's stakeholders. The board of directors should also be accountable to the company's stakeholders.

---
[1]reachjayabal@gmail.com, [2]enocki@srmist.edu.in

DOI: 10.1201/9781003428466-26

Stakeholders are people who have an interest in the way a company performs (for example, employees, trade unions, suppliers, the government, and competitors, etc.).

There is always a strong connection between the idea of COMMERCIAL POWER and the phrases Ethics and Moral. Therefore, prior to obtaining COMMERCIAL POWER, it is essential to have a solid understanding of the distinction between ethics and morals. Although they seem similar, ethics and moral are not the same thing at all, with ethics covering a considerably broader and deeper range of topics. It is possible to define ethics as a way of being human, and it is also believed to be the norm of moral conduct that all commercial enterprises should adhere to in order to ensure the longevity of any business. Both "ethics" and "moral" originate from the Greek word "eths," which means "character," while "moral" comes from the Latin word "MOs," which means "custom." Because of this, the two terms are related yet distinct from one another. The word "ethics" comes from the Greek word "ethics," which means "character."

The concept of "COMMERCIAL Social Responsibility" was first articulated in India in the early 1960s. At that time, activists were the ones who began to question the profit-maximizing goals of commercial enterprises. The famed German philosopher Karl Marx once commented on commercial objectives saying—commercial is all green, only philosophy is grey. The comment's deeper meaning was that companies only try to maximise their profits by exploiting society, and as a result, they always remain green (meaning happy and wealthy), but society always remains grey (meaning exploited and sad). This assertion made by Karl Marx does not, however, hold water in the modern world. This is due to the fact that if a firm attempts to exploit society for the purpose of maximising profits, then society will work to ensure that such a business is not permitted to exist by employing the principles of commercial power.

2. Importance of COMMERCIAL POWER and Need for Scientific Office

A good system of COMMERCIAL POWER is important on account of the following:

(a) Investors and other stakeholders in a firm have a need for protection and safety against unfair practises in terms of COMMERCIAL reporting and accountability. It has been found that many companies do not correctly estimate, value, or forecast the future performance of the company in order to generate a picture that is profitable for the organisation. They do this in order to raise the funds necessary for such inaccurate estimates. The interests of investors will be better protected thanks to COMMERCIAL POWER, and as a result, investors will have more confidence in the market.

(b) An appropriate level of COMMERCIAL POWER leads to the prompt and satisfactory resolution of investor complaints. RBI and SEBI have taken a number of steps in this direction, but in the years to come, there will likely be a need for even more.

(c) It will contribute to the establishment of a standard for COMMERCIAL POWER, which will assist investors acquire confidence when they are making decisions regarding their investments. Investors would rather put their money into businesses that have strong internal controls in place and a proven track record of having excellent COMMERCIAL POWER.

(d) Countries that have strong principles of COMMERCIAL POWER will have a greater Global Value in terms of their ability to attract international investments. It will establish a direct connection between COMMERCIAL POWER and flows of foreign investment, which, in the long term, will contribute to the growth and development of the economy in the country.

(e) Nations with robust COMMERCIAL POWER, such as the United States of America and the countries of Europe, make use of Scientific Office, which insures a robust stock market. In order to guarantee honest and open dealings on the market, certain activities, such as trading on inside information, are prohibited. It will boost the confidence of the shareholders while also shielding them from any unfair business practises. It is necessary to implement harsher sanctions and fines in order to make headway against the issue of insider trading. A more robust stock market will lead to general growth and wealth for the country as a whole, which will improve the country's position in the rankings and rankings of countries throughout the world.

(f) A fraud takes place when management deliberately makes a false statement with the goal to deceive others. This practise can be carried out by a single director or by a small number of directors, but when it does occur, the entire COMMERCIAL POWER structure is held accountable. At times like these, the auditors get the blame because they are unable to identify the perpetrators and they do not make segregation. As a result of this, the Scientific Office is required as a tool to identify the really culpable parties inside the management of the corporation and establish a system of justice there.

3. What is KINFOLK-Organized Business?

In most cases, a company is referred to as having a KINFOLK-Organized structure if it is owned and run by one or more families. In the case of a joint stock company, this term refers to a business in which the members of a certain KINFOLK own the majority of the firm's shares, or in which the Board of Directors is comprised of the majority of other members of the same KINFOLK. As a result, members of KINFOLK hold both the ownership and the control of the organisation. Typically, these companies are established by a small number of individuals or KINFOLK members, and over the course of time or several generations, they grow to become one of the most influential corporations in the nation. Such enterprises are prevalent over the world like:

(a) LG, which has been in the KINFOLK family for four generations,

(b) NIKE, a well-known brand of athletic shoes Organised by Knight KINFOLK,

(c) Wal-Mart, which began operations in the 1960s and was founded and is run by the Walton KINFOLK

(d) The information technology company DELL, which Michael Dell founded in 1984, etc.

(e) In the context of India, if we count them all, the list will continue to grow; it includes companies such as Ambani's Reliance Group, the Tata Group, the Birla Group, and many others. KINFOLK organisations in India and elsewhere throughout the world have always been successful because they adhere to a set of core principles and values when conducting their work. These principles and values take into account what is better for everyone and do not assess situations based on their members' own points of view. By putting more emphasis on "WE" rather than "ME," they ensure that the ways and dynamics of KINFOLK play an essential part in shaping the ways in which the firm operates.

4. Cases of COMMERCIAL Scandals and Frauds in KINFOLK Organized Business

(a) Health South Corporation Scandal (2003)

On February 22, 1984, the American company HealthSouth was operating in the commercial sector in Birmingham, Alabama, in the United States of America. Richard M. Scrutiny was the one who initially established it as a KINFOLK private firm under the name AMCARE, INC. The new business began its operations with an initial capital of between $50,000 and $70,00, and very quickly changed its name to Health South Corporation before eventually going public. Before the controversy, it was one of the most successful healthcare companies in the United States that was open to the public.

Summary of the Case:

• In order to satisfy the expectations of the firm's stockholders, the chief executive officer of the company, Richard M. Scrutiny, exaggerated the company's earnings by $1.4 billion. According to the allegations, he instructed accountants to report fictitious transactions in order to boost earnings. The United States Securities and Exchange Commission (SEC) became aware of the fraud or scandal when the company's CEO sold $ 75 million in stock in a single day right before the company disclosed substantial losses. This prompted the involvement of the US Securities and Exchange Commission (SEC).

• The Chief Executive Officer was found guilty on all 36 counts of Office Fraud and received a sentence of 7 years in prison for his actions.

(b) Satyam Scandal (2009)

Satyanarayan Raju, the patriarch of Ramalinga Raju, the central figure in the scandal, started a grape growing company in the Hyderabad area. In addition to establishing KINFOLK, Satyanarayan Raju established Satyam Computer Services Ltd. In the 1970s, the KINFOLK's financial stability allowed the younger Ramalinga Raju to study for a Master of Business Administration in the United States. Upon his return to India, Ramalinga Raju not only established cotton mills and a construction company, but also took over control of the KINFOLK enterprise. Ramalinga Raju founded Satyam Computer Services, an IT firm, the next year, in 1987. The company was a publicly traded organization in India that offered IT services and acted as a back office. The corporation abused its COMMERCIAL POWER by employing unethical auditing practices, which led to a misunderstanding of the company's financial records and, ultimately, to the fraud or scandal.

Summary of the Case:

- The firm inflated the revenue of the COMMERCIAL by $1.5 billion by deceptive means, and as a result, it lied about its revenues, revenue and profit margins, and cash balances to the tune of 50 billion rupees.

- The fraud was uncovered after the company's founder and current chairman, Ramalinga Raju, acknowledged to it in a letter to the company's board of directors. In the letter, he described in detail how the scam was carried out. This led to the discovery of the fraud.

- Due to the fact that Ramalinga Raju's company is owned by KINFOLK, he and his brother were both charged by the legal system with breaking trust, conspiring or creating conspiracy plans, deceiving and manipulating the books of accounts along with the auditors of the company. Despite this, after an initial period of detention, the CBI was unable to file charges against them in a timely manner, which resulted in their release.

(c) The Scandal at Lehman Brothers Holdings Inc. (2008)

Lehman Brothers was a global leader in the provision of financial services and investment banking. In 1844, a German immigrant called Henry Lehman began the company and a small general store in Montgomery, Alabama. Later in 1850, Henry Lehman and his two closest brothers, Emanuel and Mayer, founded the financial firm now known as Lehman Brothers. The Great Depression in the 1930s, two world wars, the failure of Long Term Capital Management in 1998, and the default on Russian debt in 1998 are just some of the adversities that the company has overcome throughout the course of its existence. From its founding in 1850 until the financial crisis of 2008, Lehman Brothers was a thriving business.2017 was the year that Fortune magazine named Lehman Brothers the "Most Admired Securities Firm." The collapse of Lahman sent global stock markets plunging and Wall Street into a frenzy of panic. This bankruptcy filing was considered to be the largest in history due to the fact that it dwarfed those of previous bankrupt titans like WorldCom and Enron.

Summary of the Case:

- There were claims that the company had disguised $50 billion in sales as loans. The corporation allegedly hid loans as well. Lehman Brothers' alleged sale of troubled loans to Cayman Islands banks, with the proper understanding being that the banks would buy back the debts at a later date, is what sparked the investigation into the matter. Lehman Brothers's actions brought the issue to the forefront. It created the appearance, whether intentional or not, that Lehman Brothers had $50 billion more in liquid cash and $ 50 billion fewer in problematic assets than it actually did. When it came to liquid assets, Lehman Brothers reported $50 billion more than it actually possessed. The executives and board members, as well as the company's auditors, were allegedly aware of the suppression, as detailed in the reports.

- As soon as the news broke, the Dow suffered its largest one-day point loss in history; but, after considerable debate, Congress did accept Henry Paulson, the 74th US Secretary of the Treasury,'s bailout assistance plan totaling 700 billion dollars. This series of measures was officially known as the "Troubled Asset Relief Programme." On November 20, 2008, the Dow closed at a new six-year low of 7,552.29, and it continued to fall, reaching a new low of 6626 in March of 2009. Everything started in the fall of 2008 after the announcement.

(d) The Scandal Regarding Kingfisher Airlines Limited (2012)

Because United Breweries Ltd., Vijay Mallya (the chairman), Kingfisher Finest India Ltd., and other companies that are all part of Vijay Mallya and his group of Companies promoted Kingfisher Airlines Limited, it is possible to classify Kingfisher Airlines Limited as a KINFOLK Organised. The patriarch of the Mallya family, Vijay Mallya Sr. It is possible to see Vital Malaya as a pioneering entrepreneur in the annals of Indian history. In 1947 and 1948, Vital Malaya began to buy shares of United Breweries Ltd., and he quickly rose up the ranks to become the company's first director in 1948. The years 1947 and 1948 fall into this time period. His son eventually took over the reins of the family business after it had grown substantially over the years in a variety of industries, including beverages, aviation, investing, and others. The year 2006 marked the beginning of Kingfisher Airlines Limited's operations in India. The company went bankrupt in February of 2012 after suffering enormous losses as a result of a variety of management issues and poor POWER. The following is a rundown of the promoter's holdings in the company in terms of their shareholdings:

Table 26.1 Population data

Sl. No.	Name of the Shareholder	Shares held in numbers	Shares held as % of total
1.	United Breweries (Holdings) Ltd.	500,000	0.05
2.	UB International Trading Ltd.	50,000,000	5.95
3.	Vijay Malaya	16,117,321	1.89
4.	UB Overseas Ltd.	14,563,180	1.69
5.	Kingfisher Finest India Ltd.	3,231	0
Total		81,183,732	9.58

Source: Author's Compilation (Annual Reports, www.moneycontrol.com)

Indian Airlines domestic market share in August, 2012 before bankruptcy of Kingfisher Airlines was as follows:

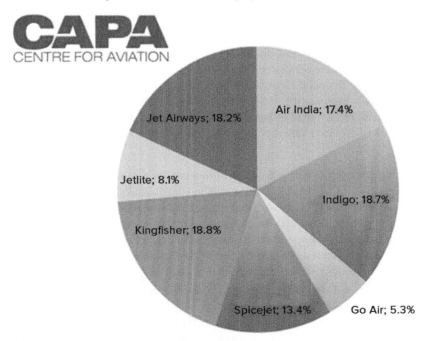

NB: Jet Airways and JetLite combined share is 26.3%

Fig. 26.1 Indian Airlines domestic market share

Source: CAPA – Centre for Aviation and Indian Ministry of Civil Aviations (www.centreforaviation.com)

Summary of the Case:

- The ineffective use of power, the absence of competent management, and the reliance on individual judgement, most notably on the part of Chairman Vijay Malaya, are some of the factors that contributed to the failure of the organisation.
- After the merger with Air Deccan, Kingfisher's annual losses surpassed 2,000 Crore for each of the following three years. This was the result of the transaction. People in India tend to be quite price conscious, both of the goods they purchase and the services they pay for.
- The owner of Kingfisher Airlines, Vijay Malaya, kept the airline's rates at extremely high levels, and the company's advertising costs skyrocketed. This marked the beginning of the company's decline and eventual demise.
- In February of 2013, the Airports Authority of India (AAI) issued a formal notice to Kingfisher that issued a warning and advised them that total debts of Rs. 2,552 million (US$35 million) had accumulated. It had reported a loss of 445 Crores Indian Rupees for the three months that ended in December of 2012. The decrease in revenue was significant, amounting to 1,342 billion rupees (based on the previous revenue level of 1,780 billion rupees). This represents a decrease of around 25 percent. The massive amount of debt that the corporation owed, which totaled 5,000 Crore, resulted in enormous interest

costs for the company, which increased to 350 Crore from 340 Crore the previous year. Therefore, by the end of December, turning things around in 2012 was no longer an option.

- As of the month of December in 2012, it was estimated that the total amount of outstanding debt and bank loans was more than 290 Crores. In addition to this, it owed a significant amount of back taxes, and the chairman of the Central Board of Direct Taxes (CBDT), MC Joshi, stated in a broadcast that the CBDT is considering taking legal action against Kingfisher for unpaid tax dues of approximately Rs. 500 Million.

- The outstanding salaries of employees were large, and there was little possibility of reimbursement. As a result, the company's creditors came to the conclusion that they should file a petition in the Bangalore High Court to have the business shut down. The value of Kingfisher's shares dropped by almost 68 percent in 2012, and the company's market capitalization hit a record low on December 12, 2012, dropping to approximately 212 million dollars.

- Beginning in 2013, Vijay Mallya and his firms have been embroiled in a series of scandals and financial frauds. In 2017, Mallya made the decision to leave India and relocate to the United Kingdom in response to mounting pressure from the Indian government to settle his companies' debts. The government of India is exerting maximum effort to get his extradition from the United Kingdom on charges of financial crimes. The Economic and Financial Crimes Unit (ED) has reportedly claimed that Vijay Mallya faces three to seven years in prison and up to Rs five lakh in fines for defrauding over Rs 9,980 crore and laundering the earnings of crime. This is according to the news organization Business Today.

5. Measures by Indian Government to in Commercial Power in Kinfolk Organized Business and Scientific Office:

The Companies Act of 2014 in India included the following measures in order to ensure that KINFOLK Organised Companies continue to engage in ethical business practises.

(a) Independent Directors and Women Directors: "The Act now necessitates that at least one-fourth of the total directors of a listed company be Independent Directors and have no substantive or pecuniary relationship with the company or related persons." This requirement applies to both male and female directors. It is a legal requirement for publicly traded corporations that have a paid-up share capital that is greater than Rs. 10 Crores or a turnover that is greater than Rs. 100 Crores to have at least three directors who are "Independent Directors."

(b) Commercial Social Responsibility: "Every company having net worth of Rs. 600 Crores or more, turnover exceeding Rs. 1000 Crores, or net profit of more than Rs. 5 Crores is required to constitute a COMMERCIAL Social Responsibility Committee under Section 136 of the Companies Act, 2014 constituting 3 or more directors with at least 1 Independent Director. "The committee will be in charge of CSR initiatives connected to the promotion of education, gender equality, health, the eradication of poverty, environmental protection, and employment opportunities, etc.

(c) Audit Committee: The Act includes a provision for the formation of an Audit Committee. Members of such a committee should have the requisite literacy levels to analyze financial reports. The committee's role is crucial since it will suggest auditor compensation, oversee auditor engagement, and evaluate auditor independence. It is important to have both an Audit Committee and a Nomination and Remuneration Committee.

(d) Serious Fraud Investigation Office: In order to monitor the operations of the business and conduct investigations into any suspected cases of fraud, a Serious Fraud Investigation Office (SFIO) must be set up in accordance with Section 211 of the Act. This department investigates allegations of fraud and keeps tabs on the company's operations. From the State or Federal Registrar or Inspector, or any other relevant government agency. According to (Bob Tracker, 2018), the new laws established by the Companies Act have led to the exposure of several businesses as having engaged in malpractices or fraud. Therefore, the court has started issuing orders for Scientific Audit investigations of these companies.

The Companies Act of 2014 includes some provisions, and in addition to those, the following further measures need to be considered:

(a) SEBI Guidelines: The Securities and Exchange Board of India (SEBI) also put up specific requirements on COMMERCIAL POWER, includes a focus to undertake special audits or Scientific Audits. These standards are binding on all listed businesses, and KINFOLK Organised Business must comply to them.

(b) Separation of Business and KINFOLK Practises: In order to ensure that the business is able to function without any hiccups, it is necessary to define exactly what each member of KINFOLK is responsible for doing within the company. It should not be the purpose of the company to engage in internecine conflict among its members but rather to compete on a global scale and increase wealth to its full potential.

(c) Leadership: There should be no ambiguity in mentioning the KINFOLK company's leadership and protecting them from trivial challenges. Furthermore, all parties involved should have a crystal clear understanding of the succession plan for the firm and its leadership.

(d) Board of Directors: In a KINFOLK-Organized business, the Board of Directors ought to be elected according to democratic principles, on the basis of qualifications, and favouritism ought not to predominate in such appointment.

6. Conclusion

The business and the POWER of KINFOLK business need to develop in order to adapt to the arrival of new generations who will carry the business forward. Each succeeding generation is tasked with gaining an understanding of the culture and society that surrounds them and acting accordingly in order to advance not only the company but also society as a whole. The key to success in the future will be to work intelligently and strike a healthy balance between the forward momentum of the globe and the innovative ideas that will make it a better place. The KINFOLK company ought to strike the appropriate balance in order to conduct its business in a way that will make use of innovation and technology in a sustainable manner for the improvement of the world and, as a result, satisfy the requirements of the society. Therefore, the options are limitless if they can just strike the perfect balance and apply the appropriate POWER.

References

Books:

1. *Dutta Sudipt, August 1998, KINFOLK Business in India, sage Publications, ISBN – 10*: 0803993277 ISBN- 13: 978-0803993273
2. Sukla Hitesh, 2015, KINFOLK Business: Roots to Routes, Himalaya Publishing House, ISBN: 978-93-5142-536-6
3. CA Munish Bhandari, Paperback 2017, A Handbook on COMMERCIAL and Allied Laws, 19th Edition, Bestword Publications Pvt. Ltd., ISBN-10: 9385075195 ISBN-13: 978-9385075193
4. SEBI Manual by Taxmann, 32nd Edition July 2019 edition (2019), ISBN- 10: 9387957411 ISBN-13: 978-9387957411
5. Bob Tracker, Paperback 2018, COMMERCIAL POWER: Principals, Policies and Practices, Oxford University Press, ISBN: 9780198747468, 0198747462

Websites:

1. https://www.oecd.org/daf/ca/COMMERCIALPOWERprinciples/43654301.pdf
2. http://www.icaiknowledgegateway.org/littledms/folder1/module-4-chapter-7.pdf
3. www.moneycontrol.com
4. www.businesstoday.in/markets

Human Machine Interaction in the Digital Era – Prof. J. Dhilipan et al. (eds)
© 2024 Taylor & Francis Group, London, ISBN 978-1-032-54998-9

Review on Student Performance Evaluation System for E-Learning

27

Nisha Raveendran[1]
Research Scholar, SRMIST Ramapuram,
and Christ College (Autonomous), Kerala

N. Vijayalakshmi[2]
Assistant professor (Sr.G),
SRMIST Ramapuram, Chennai

Abstract E-learning is one of the most needed research field in educational mining. It offers many interactive features to make the learning process easier. Educational data analyzes techniques helps the teachers to make the learning environment effective by understanding student's learning difficulties. Generally, learning difficulties of students can be identified in many ways such as, analyzing students learning behaviors, academic performance and assessment evaluation, etc. So, the main objective of the review article is to analyzes various education mining techniques and analyzes recent trends on students' performance analysis. Moreover, this review helps new researches to identify the learning difficulties in e-learning platform, identifies recent student's performance monitoring techniques. It also helps the new researchers to identify suitable learning algorithms to develop student evaluation systems.

Keywords E-learning, Student performance analysis, Machine learning, Deep learning techniques, Educational mining, Recommendation system

1. Introduction

Educational mining [1] is one of the most emerging research filed. The current outbreak of covid-19, increases the importance of this study among the researchers. As, the increased availability of the e-learning mode facilitates students to learn [2] from anywhere in a safe place. It offers teaching and learning in live broadcast and on-demand modes based services. However, the traditional e-learning approaches [3][4][5] are failed to satisfies the current academic need based learning. The main problem with this approach is obtaining their learning experience, understanding level of students, evaluating the quality of the study materials and correlating academic performance. Since, the root cause of the traditional approach [6] is dealing with these massive data [7]. In data analysis, data becomes meaningful, when it is used in right time and right place. For example, the student's academic data and information about the study material helps the teachers to identify where they are excelling and failing off track. Therefore, it is necessary evaluate the student's academic [8] and e-leaning mode based data to develop strong decision making system [9]. This system needs to improve the student's learning experience by predicting the performance of students' [10] [11].

[1]nr5300@srmist.edu.in, nishasudheesh0409@gmail.com; [2]vijinatarajan23@gmail.com, vijayaln@srmist.edu.in

DOI: 10.1201/9781003428466-27

Student's Emotion recognition in e-learning

Emotional recognition [12][13][14] is one of the emerging application field in the e-learning platform. It analyzes student's emotions during the online teaching sessions. It helps to identify the students understanding level. This application uses various artificial intelligent techniques to perform this task.

Automatic student's Feedback analysis in e-learning

Feedback analysis [15] of e-leaning is one of the challenging research application in education mining. In this, the feedback of students during online teaching sessions has been collected in the form of text data. It analyzes [16] various textual feedbacks of students to evaluate the student learning ability and teaching quality.

Teaching quality evaluation

In educational mining, the teaching quality evaluation [17][18] is one of the complicated application research in e-learning. Since, the teaching quality is analyzed with many criteria such as, learning outcomes, quality of study material provided by the teachers, and so on. These type of datasets contains non-leaner and multi variant data.

Student's academic performance prediction

Currently, the student's performance analyzes [19][20] in terms of behavioral, feedback, and emotion is popular in e-learning platform. But accurate student's academic performance prediction is the most challenging research field in the current covid-19 outbreak. Because, this type of application has to analyze student's academic data are challenging task.

Implementing Intelligent ideas in innovative education using big data

The e-leaning based applications[21][22] generates educational data continuously. The present and future e-learning data analyzes has to deal with high volume of data. So, it needs to utilizes advanced AI techniques to perform any type of analysis.

The current educational mining based researches are focusses on student's behavioral analysis, emotion detection, feedback and performance analysis. It doesn't provide any learning difficulties based feedback systems to improve students learning experience. Many e-learning based evaluation approaches are designed by many researchers to evaluate quality of study materials, teachers, student and learning environment. However, the main focus of the study is on student performance prediction and their learning ability (predicted performance) based study material suggestion(video) systems. The main objective of the studies is analyzing the performance of various machine learning algorithms and deep learning algorithms to identify suitable techniques for students' academic performance prediction, predicted performance and course based study material suggestion. The subsequent section reviews on some of the studies on student performance analysis system, feedback analysis, and study material quality evaluation system based studies.

The rest of the study has been organized in the following order, section II describes the review of literature on recent student's performance prediction approaches. Section III and IV discusses the limitations of the machine learning (ML) and deep learning (DL) based educational mining approaches. Section V discuss the various suggestion system. Section VI describes the various benchmark datasets for student's performance prediction. Section VII discusses the conclusion of the analysis.

2. Review of Literature

This section analyzes various study on education mining and big data analysis on educational data. It helps to identify the recent evaluation of the current online education system.

This research [23] has been Introduced a collaborative model to improve the student's and teacher's learning experience. It combines collaborative filtering and deep model to form an effective model to identify students learning experience. It analyzes the course evaluation data and feedback comments of each student to identify their behaviors while reading the study material or learning through the online sessions. The analysis results of the research show that the collaborative deep model improves the optimal evaluation strategy and the students learning quality.

This study [24] has been developed a machine learning (ML) technique based educational mining application. The main motive of this research is predicting the future of current education system and improving the learning experience of students. This ML model identifies the student's attention level to take appropriate corrective actions on time. This model helps to identify the students learning experience.

This research [25] is used an augmented reality based learning model evaluation environment to make the education standard effective. It enhances the quality of education using innovative approach to assess the suitability, acceptance. It develops a statistical and probabilistic index based decision making model to evaluate a specific student's suitable and acceptable learning environment using the AR based user centric application. It analysis age, gender, voluntariness, and experiences of specific students as the main criteria to identify the behavior of student. This model is proposed to help both teachers and students those who are utilizing virtual platform as teaching or learning platform.

This comparison analysis [26] study analyzes the recitals of various machine learning algorithms to identify the student's academic performance. It uses various student data such as, behavioral, personal, and academic data to accomplish this task. It uses mining algorithms to perform preprocessing and regression analysis to identifies performance of each student. The comparison analysis shows that the support vector regression linear (SVRL) method obtains superior performance than other comparison algorithms.

This analysis [27], performs comparison analysis of various video recommendation system used by popular medias such as amazon, Netflix, YouTube, and Hulu. It analysis limitations and advantages of these media platforms. The comparative analysis shows identifies that the most of the video recommendations algorithms using the collaborative filtering to get user's meta data and suggest videos based on user's behaviors (meta data). This study analyzes the performance of the existing video recommendation systems.

This study [28] utilizes the deep learning(DL) model for video recommendation system. It uses various video features such as candidate generation (user history, and other watched contents), video features, and other candidate resources to rank video files to recommend appropriate video to the users. The deep models analysis these features and perform the ranking to suggest videos.

This analysis [29] performance of vibrational auto encoder(VAE) based DL model and ML model to predict the student performance prediction to reduces the prediction error. The DL models used the GRU trained LSTM model and the machine learning models developed using the Extra tree regression(ETR) model. The evaluation results show that the ETR based models predicts the students' performance with less - score rate of 0.720 then the DL model.

In this [30], a hybrid feature selection method is developed to select significant features from the students' performance dataset. This approach combines feature ranking and heuristic method to develop hybrid approach. It assigns feature score to each features and perform ranking to select candidate features, then the heuristic methods find the most correlated features to predict students' performance.

In this [31], a feature selection method is designed to improve the students performces prediction of a machine learning algorithm. It analyzes the prediction performances of two criteria such as students to student's mutual learning (performed between students) based academic performance data analysis and teacher's to students learning based academic performances data analysis. It uses the swarm intelligent techniques to improve the prediction accuracy by selecting significant features.

In this [32], a comparison analysis is performed to identify the performance of two feature extraction techniques. It uses the twitter comments and posts as dataset to perform this comparison analysis. It considers two NLP feature extractors methods such as best match (BM25) and term frequency inverse document frequency (TF-IDF) for the performance analysis. The analysis results show that the TF-DT based approach obtained higher f-measure rate 89.77 % than BM25.

This research [33] has been developed a deep learning model named (GritNet) to predict the students' performance. This model utilizes the bidirectional long short term memory(Bi-LSTM) model to design the model. It contains embedded layers followed by the input layers and the embedded student's data are given to Bi-LSTM layers. The Bi-LSTM and GMP layers are trained the models using ingest past student's events data and predict the log likelihood of future one. The evaluation result proves that the GritNet obtained 91% to 96% for both datasets (like Udacity Nanodegree (ND-A, ND-B)) after a week.

This study [34], utilized a collaborative filtering methods to predict students' performance based on their respective courses. This filtering is applied in three phases such as user based, item based, and extensions based approach. In this, the user based approach, it identifies the location similarity of the users and performs recommendation. In item based approach, it locates the similar items and computes recommendation. In extension based approach, it identifies the outliers and remove them by it disallow the high GPA similarity. The evaluation result proves that the similarity filtering approach improves the prediction error up to 15%.

This section analyzed some of the studies on performance analysis of various machine learning algorithms and deep learning algorithms to identify suitable techniques for students' academic performance prediction. It also analyzes some of the techniques to provide study material (in video format) which are appropriate to their course and predicted performance. The analysis identifies that the deep learning based model perform well than machine learning based approaches. The subsequent section analyzes the performance of various machine learning approaches on students' performance prediction.

3. Limitations of Machine Learning Based Education Mining Techniques

This section discusses the some of the major shortcomings of the recent educational mining and feedback analysis based studies. Some of the major limitations of the present educational mining techniques are given as follows:

1. Learning level (understanding level) of students [35] are not considered to perform analyzes.
2. Teaching effects [36] are not evaluated by evaluating the performances of students.
3. Appropriate feedback [37] outcomes of study materials are not evaluated.
4. The present and future educational data [38] becomes high in volume (big data). So, the traditional prediction models perform poor while handling these data.
5. The traditional machine learning(ML) [39] based students' performance prediction models requires lengthy offline training.
6. The ML models requires unbiased and good quality data to train the model.
7. Suppose, the training the approaches deals with less amount or biased data set, the prediction outcomes can be biased. (For example, in video material suggestion system, it displays irrelevant contents to the users).
8. The ML models take lots of time to train the model. Meanwhile, it uses lot of resources provide accuracy and relevancy.
9. In appropriate choices of mining techniques leads inaccurate interpretation of prediction results. The blunder choices produce chain of errors and it may remain unpredicted for long period of time.
10. It takes massive amount of data and time to train the models. This availability of data leads under fitting or overfitting during the prediction.
11. Analyzing the unstructured data is complicated in ML approaches.
12. The ML techniques performs well with labelled data. However, labelling each students record is time consuming process in ML.

Table 27.1 Limitations of the recent ML model's performance analysis

S. No	Ref	Year	Contribution of study	Limitations of the adopted methodologies
1	[24]	2020	Predicts the future of current education system and improving the learning experience of students	The current and future educational data becomes high in volume (big data) so the ML models performs poor while handing these data.
2	[25]	2020	Developed an augmented reality based learning model evaluation environment to make the education standard effective.	It performs the evaluation based on student's personality analysis, it doesn't consider academic assessment data to perform the analysis
3	[27]	2016	Analyzes the recitals of various machine learning algorithms to identify the student's academic performance.	The current and future educational data becomes high in volume (big data), so the regression models performs poor while handing these data.
4	[35]	2018	Developed a Digital electronics education and design suit(DEEDS) using machine learning techniques Artificial neural network (ANN), Support Vector machine(SVM), logistic regression (LR), Naive Bayes (NB), and Decision tree(DT)	Obtained 0.87 precision rete (ANN). The ML models take lots of time to train the model. Meanwhile, it uses lot of resources provide accuracy and relevancy.
5	[36]	2019	Comparison analysis of SVM, back propagation(BP), Long short term memory (LSTM), gradient boost classifier (GBC) on students' academic performance data set	Obtained maximum accuracy of -80.91% (BP). It takes massive amount of data and time to train the models. This availability of data leads under fitting or overfitting during the prediction.
6	[37]	2020	Developed Ensemble of DT, NB, K-nearest neighbor (KNN), and extra tree(ET) to predict students' performance	Obtained maximum accuracy 91.76%. The traditional machine learning(ML) based students' performance prediction models requires lengthy offline training.
7	[38]	2021	Developed a feature learning algorithms using genetic algorithm and the students' performance has been predicted using the LR, and DT.	Obtained accuracy 96.64%. The ML models requires unbiased and good quality data to train the model.
8	[39]	2021	Analyzed the performance of LR, DT, and RF on student academic performance data set	Obtained maximum of RF-61.08% as accuracy. The traditional prediction models perform poor while handling these data.

The Table 27.1 analyzes the performances of ML models to predict students' academic performances using their academic data. It analyzes efficiency of various ML models for students' performance prediction. This shows that the ML models performs poor while handling the student's academic data. However, these models not provide reliable accuracy rate; due to certain limitations, which are discusses in this section.

4. Benefits and Challenges of Deep Learning Based Educational Mining Techniques

This section discusses the various benefits and challenges faced by the researchers during the DL based educational mining(EM) model designing, training and prediction phase.

1. The DL model [40] based techniques are best choice for state art of performance. It is surprising and continuing to do so in the near future.
2. Unlike traditional ML models, the DL based approach [41] doesn't use feature engineering techniques to improve the performance.
3. It reads the data to correlate and combine independently to enable faster leaning without being explicitly instructed to do so.
4. The DL models [42] performs well with unstructured and different format of data (For examples, data in different formats such as, text, pictures, pdf, and etc.)
5. Labeling student record is not required while training data sets in DL based model [43].
6. Efficient at delivering high quality results.

Challenges

1. The DL models [45] uses high volume of parameters to train the models. However, if large number of parameters with high noisy data leads overfitting.
2. Lack of flexibility on datasets (For example, if same DL model [46] is used to train and predict various problems, then the model produces biased results).
3. The selection of appropriate model for the problem is essential task in DL model. It requires proper knowledge to use proper training methods, other parameters and topology selection.

Table 27.2 Analysis of DL based students' performance prediction models

S. No	Ref	Year	Title of the study	Adopted methodologies for prediction	Acc(%)
1	[33]	2018	GritNet: Student Performance Prediction with deep learning	Vanilla Bi-LSTM	ROC date - 96%
2	[40]	2021	Performance prediction for higher education students using deep learning	Deep models like Deep LSTM and 1D CNN are deployed	3.3% Error rate improved
3	[41]	2020	Student academic performance prediction using deep multi-source behavior sequential network	Developed Sequential prediction based deep network (SPDN) model. Which combine multi-source fusion CNN and Bi-LSTM	79.67% -ROC rate
4	[42]	2020	Online at-risk student identification using RNN-GRU joint neural network	The statistical and time series sequential data of student's are analyzed by integrating Simple RNN, GRU, and LSTM model	80%- accuracy
5	[43]	2021	Student academic performance prediction using hybrid deep neural network	In this, initially input layer, Embedded layer, and Bi-LSTM layers are designed train the data and the Attention layer is added before to output layer to predict student grade	90.16%- accuracy
6	[44]	2020	Transfer learning from deep neural network for predicting student performance	A deep model is constructed using one input layer, two hidden dense layer, and one output layer.	0.7768 or 77.68% accuracy
8	[45]	2017	A neural network approach for students' performance prediction	Utilized RNN to predict student grade using academic data	99%
9	[46]	2018	An approach to predict student's academic performance using recurrent Neural Network(RNN)	RNN is utilized to predict student's final result using academic data.	87%

The Table 27.2 analyzes the performances of various deep learning models in predicting students' academic performances for time series and multi variant academic data. This section analyzes some of the recent studies on DL based students' performance prediction approaches. This section clearly shows that the DL models performs well while handling the student's academic data then ML models. However, these models not provide reliable accuracy rate; due to certain limitations, which are discusses in this section.

5. Recommendation System

This section analyzes the performances and challenges of various recent studies on methodologies for various recommendation system.

Table 27.3 Limitations of the recent DL based students' performance analysis systems

S. No	Ref	Year	Contribution of study	Adapted methodology	Limitations
1	[23]	2021	It combines collaborative filtering and deep model to form an effective model to identify students learning experience.	collaborative filtering and deep model	It doesn't focus to identify the comfort level of students while using the study materials
2	[28]	2016	Developed a deep learning (DL) model for video recommendation system.	deep learning (DL) model	The prediction algorithm suffers with under fitting issues. Since, it suggests videos based on the user's profile data, search and watch history. It leads to suggest irrelevant videos to the users.
3	[47]	2016	Mutual reinforcement of academic performance prediction and library book recommendation	Developed a context aware matrix factorization for Mutual reinforcement method.	Perform well in predicting students' performance prediction, but personalized recommendation are not utilized
4	[48]	2018	A recommendation model based on deep neural network	In this, ranking layer and Feature importance representation layers are added with Deep neural network(DNN) model to perform this tasks.	It doesn't use any complex deep model to train the data.
5	[49]	2019	A novel deep learning based collaborative filtering model for recommendation system	Developed a novel deep model(feed forward neural network-FFNN) to simulate interactions between user and items and to perform recommendation	Relatively complex to train model with video data.
6	[50]	2018	DKN: deep knowledge aware network for news recommendation	Knowledge aware CNN model is designed to recommend news by handling semantic and knowledge level data. It context based model performs the recommendation by the click through rate prediction. It applies content fusion for word, entity data.	To use this model domain knowledge is required.
7	[51]	2021	Artificial intelligent in recommendation system	Performed analysis for various methodologies of recommendation system such as context aware, knowledge based, collaborative filtering based, deep learning based, fuzzy based, transfer learning (TL) based model and so on.	-
8	[52]	2020	Personalized Self Directing learning recommendation system	This analysis suggesting a recommendation model, named as PSDLR.	It used various aspects and choice of e-learning methods are adopted to perform the personalized recommendation.

S. No	Ref	Year	Contribution of study	Adapted methodology	Limitations
9	[53]	2017	A Personalized web content recommendation system in e-learning environment	It utilizes web contents such as navigation behavior, web content, learner's performances, and profile data are utilized to perform the recommendation task.	It utilizes collaborative filters to perform group learners. It cannot handle multi-channel data.
10	[54]	2018	Personalized recommendation system for e-learning environment based student performance	Developed three phases to perform this task. It contains domain module, learner's module, and finally recommendation modules. The recommendation module uses context based and collaborative filters. To improve the performance of this model various mining techniques are adopted to perform this task.	Student learning style is not incorporated with this model.

The Table 27.3 analysis the performance and their limitation of various recommendation system. This analysis helps to identify the suitable recommendation model for the student performance based study material suggestion phase.

6. Bench Mark Dataset Analysis

This section analysis the various benchmark dataset. It contains student's academic and personal data.

Students performance in exam data set

These data [55] [56] are randomly collected form selected students. It contains 1000 instances and 8 attributes such as writing score, reading score, maths score, test preparation completed, lunch time, parents level of education, race/ethnicity, and gender. This data set contains multi-variant data (both nominal, and numeric data).

Student academic performance dataset

These data [57] are randomly collected form selected students. It contains 480 instances and 16 attributes. This features are categorized into three groups such as (1) Demographic features such as gender and nationality. (2) Academic background features such as educational stage, grade Level and section. (3) Behavioral features such as raised hand on class, opening resources, answering survey by parents, and school satisfaction.

The dataset [58] consists of 305 males and 175 females. The students come from different origins such as 179 students are from Kuwait, 172 students are from Jordan, 28 students from Palestine, 22 students are from Iraq, 17 students from Lebanon, 12 students from Tunis, 11 students from Saudi Arabia, 9 students from Egypt, 7 students from Syria, 6 students from USA, Iran and Libya, 4 students from Morocco and one student from Venezuela.

The dataset is collected through two educational semesters: 245 student records are collected during the first semester and 235 student records are collected during the second semester.

Student performance dataset

It contains secondary education (school students) data. This data set contains multi-variant data (both nominal, and numeric data). It is suitable to perform both regression and classification task. This dataset contains totally 649 instances and 33 attributes. This file contains two different subject learning student's information such as mathematics and Portuguese language.

This overall analysis of this study is discusses various applications on education mining, it helps to identify the most emerging research filed in the current e-learning platform. Section II analyzes various related studies on education mining based methodologies. It helps to identify the best methodology and application to perform the analysis. Section III and IV discusses various limitations of the ML and DL based current student's performance prediction approaches. It helps to identify suitable model to develop a student's performance prediction tasks. Section V discusses some of the methodologies of recommendation system. It helps to identify the suitable methodology to design a study material suggestion system. Section VI discusses three different publicly available bench mark datasets to perform the evaluation system. It helps the new researchers to identify the suitable benchmark dataset to evaluate their model's efficiency. The subsequent section discusses the conclusion of the analysis.

7. Conclusion

Thus, the research performs detailed analyzes of various studies on educational mining based applications. The current covid-19 pandemic situation increase the need of e-learning based applications. These e-learning based EM approaches facilitates the teachers and students to use the virtual platform efficiently by improving the learning and teaching quality in virtual mode. The previous sections analysis importance of various e-learning based EM approaches. The analysis shows that there are many e-learning based evaluation approaches are performed to evaluate quality of study materials, teachers, and learning environment. Moreover, it shows that most of the evaluations tasks are accomplished with the help of ML and DL based techniques. So, this study has been analyzed the role of ML and DL techniques in E-learning evaluation systems. It also analyzed various benefits and limitations of these two techniques in the evaluation approaches. The overall analysis of the study has been identified that the student performance prediction and their learning ability (predicted performance) based study material suggestion is the most needed e-learning based evaluation approach in the current situation.

The dataset analysis identifies that the ML based approaches are performs poor while handling academic educational data. Since, these type of educational datasets contains high volume of instances and multi variant data. Therefore, the DL based educational mining approaches has a lot of potentials than ML models. However, it requires to overcome a few challenges in the existing DL models.

According to the section III and IV various model's analysis (ML and DL analysis), the prediction models with more than one hidden layer are more beneficial to train complicated data using nonlinear function efficiently. Moreover, the deep models flexible to incorporate any improvement functions in the middle of the hidden layers during model training or back propagation. The DL models performs well with scalable data (contains complex data) than ML model. Hence, this analysis has been concluded that the hybrid DL model based approach is the most suitable prediction approach for education data analysis and study material suggestion. Collaborative filtering is the simple and cost effective approach to design the study material (video) suggestion system, which are corresponds to the student's predicted performance (understanding level).

References

1. Zhang Y, Yun Y, An R, Cui J, Dai H and Shang X (2021) Educational Data Mining Techniques for Student Performance Prediction: Method Review and Comparison Analysis. Front. Psychol. 12:698490. doi: 10.3389/fpsyg.2021.698490
2. Zhang, Y., Dai, H., Yun, Y., Liu, S., Lan, A., and Shang, X. (2020a). Meta-knowledge dictionary learning on 1-bit response data for student knowledge diagnosis. Knowl. Based Syst. 205:106290. doi: 10.1145/3448139.3448184
3. Zacharis, N. Z.. (2016). Predicting student academic performance in blended learning using artificial neural networks. Int. J. Artif. Intell. Appl. 7, 17–29. doi: 10.5121/ijaia.2016.7502
4. Yang, T.-Y., Brinton, C. G., Joe-Wong, C., and Chiang, M. (2017). Behavior-based grade prediction for moocs via time series neural networks. IEEE J. Sel. Top. Signal Process. 11,716–728. doi: 10.1109/JSTSP.2017.2700227
5. Wang, F., Liu, Q., Chen, E., Huang, Z., Chen, Y., Yin, Y., et al. (2020). "Neural cognitive diagnosis for intelligent education systems," in Proceedings of the AAAI Conference on Artificial Intelligence, Vol. 34, 6153–6161.
6. Su, Y., Liu, Q., Liu, Q., Huang, Z., Yin, Y., Chen, E., Ding, C., Wei, S., and Hu, G. (2018). "Exercise-enhanced sequential modeling for student performance prediction," in Thirty-Second AAAI Conference on Artificial Intelligence (New Orleans, LA).
7. Kushwaha, R. C., Singhal, A., and Swain, S. (2019). "Learning pattern analysis: a case study of moodle learning management system," in Recent Trends in Communication, Computing, and Electronics (Langkawi: Springer), 471–479.
8. Hu, Q., Polyzou, A., Karypis, G., & Rangwala, H. (2017). Enriching Course-Specific Regression Models with Content Features for Grade Prediction. 2017 IEEE International Conference on Data Science and Advanced Analytics (DSAA). doi:10.1109/dsaa.2017.74
9. Hu, Q., and Huang, Y. (2018). A Framework for Analysis Learning Pattern Toward Online Forum in Programming Course[M]. New Media for Educational Change. (Singapore: Springer), 71–80.
10. Cakmak, A.. (2017). Predicting student success in courses via collaborative filtering. Int. J. Intell. Syst. Appl. Eng. 5, 10–17. doi: 10.18201/ijisae.2017526690
11. O. El Hammoumi, F. Benmarrakchi, N. Ouherrou, J. El Kafi and A. El Hore, "Emotion Recognition in E-learning Systems," 2018 6th International Conference on Multimedia Computing and Systems (ICMCS), Rabat, 2018, pp. 1-6, doi: 10.1109/ICMCS.2018.8525872.
12. Clarizia, F., Colace, F., De Santo, M., Lombardi, M., Pascale, F., & Pietrosanto, A. (2018). E-learning and sentiment analysis. Proceedings of the 6th International Conference on Information and Education Technology - ICIET '18. doi:10.1145/3178158.3178181
13. Z. Nasim, Q. Rajput and S. Haider, "Sentiment analysis of student feedback using machine learning and lexicon based approaches," 2017 International Conference on Research and Innovation in Information Systems (ICRIIS), Langkawi, 2017, pp. 1-6, doi: 10.1109/ICRIIS.2017.8002475.

14. Güray Tonguç, Betul Ozaydın Ozkara, Automatic recognition of student emotions from facial expressions during a lecture, Computers & Education, Volume 148, 2020, 103797, ISSN 0360-1315, https://doi.org/10.1016/j.compedu.2019.103797.

15. Joyner D. (2018) Intelligent Evaluation and Feedback in Support of a Credit-Bearing MOOC. In: Penstein Rosé C. et al. (eds) Artificial Intelligence in Education. AIED 2018. Lecture Notes in Computer Science, vol 10948. Springer, Cham. https://doi.org/10.1007/978-3-319-93846-2_30

16. Guannan Li, Lin Xiang, Zhengxing Yu, Hui Li, "Intelligent evaluation of teaching based on multi-networks integration", International Journal of Cognitive Computing in Engineering, Volume 1, 2020, Pages 9-17, ISSN 2666-3074, https://doi.org/10.1016/j.ijcce.2020.07.001.

17. Popenici, S.A.D., Kerr, S. Exploring the impact of artificial intelligence on teaching and learning in higher education. RPTEL 12, 22 (2017). https://doi.org/10.1186/s41039-017-0062-8

18. Zhang Y, Yun Y, An R, Cui J, Dai H and Shang X (2021) Educational Data Mining Techniques for Student Performance Prediction: Method Review and Comparison Analysis. Front. Psychol. 12:698490. doi: 10.3389/fpsyg.2021.698490

19. Adejo, O., and Connolly, T. (2017). An integrated system framework for predicting students academic performance in higher educational institutions. Int. J. Comput. Sci. Inform. Technol. 9, 149–157. doi: 10.5121/ijcsit.2017.93013

20. Xiaogang Wu, Research on the Innovation of Ideological and Political Education in Universities in the Era of Big Data, CIPAE 2020: Proceedings of the 2020 International Conference on Computers, Information Processing and Advanced Education, p.p 336–340, https://doi.org/10.1145/3419635.3419652

21. L. Kuang, W. Wang, B. Qin and L. Gong, "Study on Learning Effect Prediction Based on MOOC Big Data," 2020 IEEE 9th Joint International Information Technology and Artificial Intelligence Conference (ITAIC), Chongqing, China, 2020, pp. 258-262, doi: 10.1109/ITAIC49862.2020.9338830.

22. Ziqiao Wang, Ningning Yu, "Education Data-Driven Online Course Optimization Mechanism for College Student", Mobile Information Systems, vol. 2021, Article ID 5545621, 8 pages, 2021. https://doi.org/10.1155/2021/5545621

23. Asthana P., Hazela B. (2020) Applications of Machine Learning in Improving Learning Environment. In: Tanwar S., Tyagi S., Kumar N. (eds) Multimedia Big Data Computing for IoT Applications. Intelligent Systems Reference Library, vol 163. Springer, Singapore. https://doi.org/10.1007/978-981-13-8759-3_16

24. Kurilovas, Eugenijus. "On data-driven decision-making for quality education." Comput. Hum. Behav. 107 (2020): 105774. DOI:10.1016/J.CHB.2018.11.003

25. Dabhade, P., Agarwal, R., Alameen, K. P., Fathima, A. T., Sridharan, R., & Gopakumar, G. (2021). Educational data mining for predicting students' academic performance using machine learning algorithms. Materials Today: Proceedings. doi:10.1016/j.matpr.2021.05.646

26. Seong-Eun Hong and Hwa-Jong Kim, "A comparative study of video recommender systems in big data era," 2016 Eighth International Conference on Ubiquitous and Future Networks (ICUFN), 2016, pp. 125-127, doi: 10.1109/ICUFN.2016.7536999.

27. Covington, P., Adams, J., & Sargin, E. (2016). Deep Neural Networks for YouTube Recommendations. Proceedings of the 10th ACM Conference on Recommender Systems - RecSys '16. doi:10.1145/2959100.2959190

28. Bansal, V., Buckchash, H. & Raman, B. Computational Intelligence Enabled Student Performance Estimation in the Age of COVID-19. SN COMPUT. SCI. 3, 41 (2022). https://doi.org/10.1007/s42979-021-00944-7

29. Wen Xiao, Ping Ji, Juan Hu, "RnkHEU: A Hybrid Feature Selection Method for Predicting Students' Performance", Scientific Programming, vol. 2021, Article ID 1670593, 16 pages, 2021. https://doi.org/10.1155/2021/1670593

30. Bingsheng Chen, Huijie Chen, Mengshan Li, "Improvement and Optimization of Feature Selection Algorithm in Swarm Intelligence Algorithm Based on Complexity", Complexity, vol. 2021, Article ID 9985185, 10 pages, 2021. https://doi.org/10.1155/2021/9985185

31. A. I. Kadhim, "Term Weighting for Feature Extraction on Twitter: A Comparison Between BM25 and TF-IDF," 2019 International Conference on Advanced Science and Engineering (ICOASE), 2019, pp. 124-128, doi: 10.1109/ICOASE.2019.8723825.

32. Kim, B.-H., Vizitei, E., and Ganapathi, V. (2018). Gritnet: Student performance prediction with deep learning. arXiv preprint arXiv:1804.07405

33. Ahuja R. et al. (2020) Machine Learning and Student Performance in Teams. In: Bittencourt I., Cukurova M., Muldner K., Luckin R., Millán E. (eds) Artificial Intelligence in Education. AIED 2020. Lecture Notes in Computer Science, vol 12164. Springer, Cham. https://doi.org/10.1007/978-3-030-52240-7_55

34. Hussain, M., Zhu, W., Zhang, W., Abidi, S. M. R., & Ali, S. (2018). Using machine learning to predict student difficulties from learning session data. Artificial Intelligence Review. doi:10.1007/s10462-018-9620-8

35. Sekeroglu, B., Dimililer, K., & Tuncal, K. (2019). Student Performance Prediction and Classification Using Machine Learning Algorithms. Proceedings of the 2019 8th International Conference on Educational and Information Technology - ICEIT 2019. doi:10.1145/3318396.3318419

36. R. Singh, S. Pal, Machine learning algorithms and ensemble technique to improve prediction of student's performance, International Journal of Advanced Trends in Computer Science and Engineering, 2020, 9 (3), pp. 3970-3976.

37. Hussain, S., Khan, M.Q. Student-Performulator: Predicting Students' Academic Performance at Secondary and Intermediate Level Using Machine Learning. Ann. Data. Sci. (2021). https://doi.org/10.1007/s40745-021-00341-0

38. Tarik, A., Aissa, H., & Yousef, F. (2021). Artificial Intelligence and Machine Learning to Predict Student Performance during the COVID-19. Procedia Computer Science, 184, 835–840. doi:10.1016/j.procs.2021.03.104

39. Shuping Li, Taotang Liu, "Performance Prediction for Higher Education Students Using Deep Learning", Complexity, vol. 2021, Article ID 9958203, 10 pages, 2021. https://doi.org/10.1155/2021/9958203

40. Li X., Zhu X., Zhu X., Ji Y., Tang X. (2020) Student Academic Performance Prediction Using Deep Multi-source Behavior Sequential Network. In: Lauw H., Wong RW., Ntoulas A., Lim EP., Ng SK., Pan S. (eds) Advances in Knowledge Discovery and Data Mining. PAKDD 2020. Lecture Notes in Computer Science, vol 12084. Springer, Cham. https://doi.org/10.1007/978-3-030-47426-3_44

41. He, Yanbai, Rui Chen, Xinya Li, Chuanyan Hao, Sijiang Liu, Gangyao Zhang, and Bo Jiang. 2020. "Online At-Risk Student Identification using RNN-GRU Joint Neural Networks" Information 11, no. 10: 474. https://doi.org/10.3390/info11100474

42. Yousafzai, B.K.; Khan, S.A. Rahman, T.; Khan, I.; Ullah, I. Ur Rehman, A.; Baz, M. Hamam, H.; Cheikhrouhou, O. Student-Performulator: Student Academic Performance Using Hybrid Deep Neural Network. Sustainability 2021, 13, 9775. https://doi.org/10.3390/su13179775

43. Tsiakmaki M, Kostopoulos G, Kotsiantis S, Ragos O. Transfer Learning from Deep Neural Networks for Predicting Student Performance. Applied Sciences. 2020; 10(6):2145. https://doi.org/10.3390/app10062145

44. Okubo, F., Yamashita, T., Shimada, A., & Ogata, H. (2017). A neural network approach for students' performance prediction. Proceedings of the Seventh International Learning Analytics & Knowledge Conference on - LAK '17. doi:10.1145/3027385.3029479

45. Mondal, Arindam and Joydeep Mukherjee. "An Approach to Predict a Student's Academic Performance using Recurrent Neural Network (RNN)." International Journal of Computer Applications (2018): n. pag.

46. Lian, D., Ye, Y., Zhu, W., Liu, Q., Xie, X., and Xiong, H. (2016). "Mutual reinforcement of academic performance prediction and library book recommendation," in 2016 IEEE 16th International Conference on Data Mining (ICDM) (Barcelona: IEEE), 1023–1028.

47. L. Zhang, T. Luo, F. Zhang and Y. Wu, "A Recommendation Model Based on Deep Neural Network," in IEEE Access, vol. 6, pp. 9454–9463, 2018, doi: 10.1109/ACCESS.2018.2789866.

48. M. Fu, H. Qu, Z. Yi, L. Lu and Y. Liu, "A Novel Deep Learning-Based Collaborative Filtering Model for Recommendation System," in IEEE Transactions on Cybernetics, vol. 49, no. 3, pp. 1084–1096, March 2019, doi: 10.1109/TCYB.2018.2795041.

49. Wang, H., Zhang, F., Xie, X., et al.: DKN: deep knowledge-aware network for news recommendation (2018)

50. Zhang, Q., Lu, J. & Jin, Y. 9Artificial intelligence in recommender systems. Complex Intell. Syst. 7, 439–457 (2021). https://doi.org/10.1007/s40747-020-00212-w

51. Lalitha, T. B., & Sreeja, P. S. (2020). Personalised Self-Directed Learning Recommendation System. Procedia Computer Science, 171, 583–592. doi: 10.1016/j.procs.2020.04.063

52. D. Herath and L. Jayaratne, "A personalized web content recommendation system for E-learners in E-learning environment," 2017 National Information Technology Conference (NITC), 2017, pp. 89–95, doi: 10.1109/NITC.2017.8285650.

53. Fazazi, H.E., Qbadou, M., Salhi, I., & Mansouri, K. (2018). Personalized recommender system for e-Learning environment based on student's preferences.

54. Data source: https://www.kaggle.com/roshansharma/student-performance-analysis/data

55. https://www.kaggle.com/spscientist/student-performance-in-exams/data

56. https://www.kaggle.com/aljarah/xAPI-Edu-Data

57. https://archive.ics.uci.edu/ml/datasets/student+performance

Human Machine Interaction in the Digital Era – Prof. J. Dhilipan et al. (eds)
© *2024 Taylor & Francis Group, London, ISBN 978-1-032-54998-9*

Modified LSTM for Prediction of Pollutants Concentration

28

K. Shyamala[1]

Associate Professor,PG& Research Department of Computer Science,
Dr. Ambedkar Government Arts College (Autonomous), (Affiliated to the University of Madras),
Chennai, Tamil Nadu India

R. Sujatha[2]

Assistant Professor, Department of Computer Applications, J. H. A. Agarsen College,
Madhavaram, (Affiliated to the University of Madras),
Chennai, Tamil Nadu India

Abstract "Air pollution" is the term employed for those situations where airborne emissions are dangerous to the health of people or other living beings or when they negatively affect the environment or materials. A few typical causes of air pollution include cars, factories, homes with combustion gadgets, and forest fires. Carbon monoxide, particulates, nitrogen oxides, and sulphur dioxide are the contaminants that are most harmful to human health. Environmental and indoor air pollution, which causes respiratory illnesses and other disorders, has a major negative impact on morbidity and mortality. In this study, a unique prediction technique called Modified Long Short Term Memory (MLSTM) was created to prognosticate the daily measurements of pollutants such as SO_2, PM_{10}, $PM_{2.5}$, and NO_2 in Salem, Chennai, Thoothukudi, Madurai, and Coimbatore. In terms of predictability and execution time, the Modified LSTM approach's performance was compared to the LSTM method. The Modified LSTM method performs better with higher prediction accuracy and lesser execution time.

Keywords LSTM, Particulate matter, Pollutants, Neural network

1. Introduction

Air pollution is a collective term for substances or other air pollutants that are detrimental to people, animals, or plants. Structures are also harmed. Air contaminants come in a variety of forms. They could be liquid droplets, grains, or gases. Pollutants can be discharged into the atmosphere in a variety of ways, includes through emissions from buildings, machines, planes, and aerosol cans. Most air pollution is generated by people. Secondhand smoke from cigarettes is another source of air pollution. They are known as anthropogenic sources since they were created by people. The ash from fires and the dust from volcanoes are two examples of naturally occurring air pollution. "Natural sources" are what these are called. Large cities have higher air pollution levels because of the buildup of pollutants from many sources. Mountains or significant construction projects can occasionally resist air pollution. This air pollution typically manifests as a haze that obscures visibility [1]. It's known as smog in our language. Smog is the result of the terms "smoke" and "fog" coming together.

[1]shyamalakannan2000@gmail.com, [2]kamalihars2000@gmail.com

DOI: 10.1201/9781003428466-28

Most people think of huge factories or car exhaust when they think about air pollution. Indoor air pollution, however, comes in a wide variety. When fuels such as coal or firewood are burned to heat a dwelling, the air within can get contaminated. Ash and smoke can cause respiratory problems by sticking to foods, clothing, and walls [2]. People who are exposed to air pollution might experience a wide range of negative health effects. Long-term implications and short-term effect are the two sections. Short-term, transient impacts consist of illnesses resembling pneumonia or bronchitis. Infection of the nose, eyes, skin, or throat might occur as consequences in addition to discomfort. Headaches, lightheadedness, and nausea are some of the signs of air pollution.

Air pollution also includes these in addition to the unpleasant smells that come from factories, landfills, or sewage systems. Some scents are unpleasant despite being less dangerous. Air pollution's long-term impacts might last a person's entire life or for quite a while. Even death could happen as a result of them. A few of the long-term physical condition effects of air pollution comprise cardiovascular disease, lung conditions, and respiratory ailments like emphysema. Long-term revelation to environmental pollution can have negative effect on the brain, nerves, liver, kidneys, and other organs. Air pollution is cited by some scientists as the main contributor to birth anomalies. Approximately 2.5 million fatalities per year are attributed to either indoor or outdoor pollution of the air.

1.1 Air Quality Index

A metric used to evaluate the quality of the atmosphere is the Air Quality Index (AQI). The AQI will display change in the amount of atmospheric air pollution. Clean air is essential for preserving the ecosystem and general health. The National Air Quality Index was formed in 2014 to classify the six levels of air quality. The following levels are available: excellent, acceptable, reasonably polluted, poor, extremely severe and poor. This Air Quality Index was created by the Central Pollution Control Board (CPCB) in collaboration with experts and professionals in air quality and IIT-Kanpur. To measure the air quality in the states and cities, a scale from 0 to 500 is used. Some of the instruments used to measure the air quality are PCE-HFX 100, PCE-RCM 05, and PCE-RCM 8.

2. Literature Review

Y. Bai, et. al. [3] proposes The E-LSTM model, which develops multiple LSTMs in various modes and is inspired by ensemble learning, outperformed feed-forward neural networks and single LSTMs in expressions of mean absolute error (19.604% as well as 16.929%, respectively), root-mean-square (12.077 gm and 13.983 gm), as well as correlation metrics (0.994 and 0.991, respectively).

J. Zhao, et. al. [4] proposes artificial neural network (ANN) as well as long short-term memory (LSTM) algorithms lying on the similar database, which contains recordings from 36 stations that monitor air quality in Beijing between 2014/05/01 and 2015/04/30. The outcomes demonstrate the superior prediction performance of our LSTM-FC neural network algorithm.

A. Masih [5] proposes the outcomes with M5P and Support Vector Machines (SVM) for regression, two well-known classification techniques, the model's performance was confirmed. The models were subsequently evaluated using several filtering methods, including the Relief Attribute Evaluator and Principle Component Analysis. The results collected indicate that Random Forest performs better than the two other data mining algorithms in predicting the concentration of NO2 in the atmosphere.

Q. Wu, et. al. [6] proposes the Least Squares Support Vector Machine (LSSVM) parameters that the Bat algorithm (BA) optimizes are appropriate for prognosticate, WD (A) while maintaining the original data of the AQI sequence. These air pollution parameters consist of CO, PM2.5, PM10, NO2, SO2, and O3. The ultimate AQI forecast outcome will be calculated as the sum of each subseries' prediction values. Notably, the offered solution not only fully utilizes the benefits of traditional SD but also addresses the difficulty that the standard time sequence forecasting system, which is focused mostly on decomposition technique, cannot take the influencing factors into consideration.

K. Xiao, et. al. [7] proposes Chengdu's air pollution characteristics indicated both high PM2.5 and O3 concentrations. The centre of Chengdu saw the highest PM concentrations, which may have been caused by the combined impacts of industrial and automobile emissions. In the downtown, automobile emissions were the main cause of ozone pollution, but industry in the northern region, where there were fewer vehicles, had a more significant impact on O3. Winter had the highest concentrations of PM10, NO2, PM2.5, and CO, while summer had the lowest. Winter also saw the highest SO2 concentrations and autumn saw the lowest, while summer saw the highest O3 concentrations.

V. Reddy, et. al. [8] proposes based on the employ of the LSTM and recurrent neural network (RNN) as a predictive model for potential predicting using time-series information on air pollution as well as meteorological data in Beijing. RNNs, and especially LSTM algorithms, are good for time series datasets with a lot of data and a lot of dependencies between the data points. Our findings demonstrate that, for a single time step, the LSTM framework and basic support vector regression both predict future time steps with an accuracy that is comparable.

U. Pak, et. al. [9] proposes the best network model for the suggested ozone predictor was built using the CNN-LSTM model with various architectures. The recommended ozone predictor's metrics (RMSE, MAE, and MAPE) were assessed and compare using multi-layer perceptrons (MLP) and LSTM methods, and the results showed that they were decreased to 83% linked to the MLP technique with 35% comparable to the LSTM method. In terms of accuracy of predictions and seasonal stability, it was established that the recommended CNN-LSTM hybrid form perform well than MLP and LSTM models.

H. Liu, et. al. [10] proposes the long-short-term memory (LSTM) neural network, a deep learning automated process technology, was used to estimate the low-frequency wind speed rate sub-layers. The Elman neural network, a well-known recurrent neural network, was developed to foresee the high-frequency sub-layers. Eleven distinct prognosticating models are used in the actual forecasting tests to verify the proposed model's actual prediction capabilities. According to the experimental findings, the suggested model performs satisfactorily in high-precision wind speed forecast.

C. J. Huang, et. al. [11] suggests using the mean absolute error (MAE), the root mean square error (RMSE), the pearson correlation coefficient parameter, and the index of agreement (IA) as four measurement metrics. Comparing the projected CNN-LSTM method (APNet) to added machine learning techniques, the testing outcome shows that it had the highest prognosticating precision. In this study, the viability and usefulness of the CNN-LSTM method for estimating PM2.5 values are also demonstrated. The major contribution of this learn is the expansion of a deep neural network algorithm employing historical information on the total number of total wind speeds, hours of rain, and overall PM2.5 concentration.

M. Rezamand, et. al. [12] proposes the impact of external factors on the dynamics of bearing failure and is defined using SCADA data, including wind speed and ambient temperature. The vibration signal then determines the failure dynamics for each environmental circumstance. Finally, utilizing the failure dynamics and an adaptive RUL of the damaged bearings is predicted using a Bayesian technique based on the VOC. Using experimental data, the method's effectiveness is tested, and the results show that it is more reliable than the Bayesian method in terms of RUL.

3. Research Area and Data

The study's geographic scope is depicted as including the Tamil Nadu cities of Chennai, Salem, Coimbatore, Madurai and Thoothukudi. In order to monitor the city's five surrounding cities environmental air quality, a network has been constructed [13]. The database contains four pollutants with 648 dataset collected from public database in TNPCB [21]. Four air contaminants, includes PM10, PM2.5, SO2, and NO2, can be regularly monitored at each site. Using the environmental air quality GIS platform, information on the environmental air quality at all of Tamilnadu monitoring stations has been made available to the public. Daily standard information for NO2, SO2, PM2.5, and PM10 were gathered for Chennai, Coimbatore, Salem, Madurai and Thoothukudi from 25th November 2020, to 13th September 2022, and the most recent data was used to fill in the gaps for each type of air pollutant [14]. Additionally, weather information gathered from the Tamil Nadu Pollution Control Board (TNPCB) and the map of monitoring stations were exposed in Fig. 28.1 and the statistical parameters are specified in Table 28.1.

Table 28.1 Statistical parameters of pollutant

Parameters	PM2.5	PM10	SO2	NO2
Unit	$\mu g/m^3$	$\mu g/m^3$	$\mu g/m^3$	$\mu g/m^3$
Mean	20.24	44.35	11.75	19.47
Standard deviation	2.7	2.503	2.458	2.494
Minimum	15	40	7	14
Maximum	29	57	16	26

Source: Made by Author

4. Methodology

We examine machine learning techniques that have been implemented to air quality prognosticating in the past. For predicting air quality, machine learning techniques have shown great success in several fields. In spite of their advantages over conventional

Fig. 28.1 (a) Map of Tamilnadu (b) Map of Chennai (c) Map of Coimbatore (d) Map of Thoothukudi (e) Map of Salem (f) Map of Madurai and the site of monitoring station

Source: Adapted from https://www.researchgate.net/figure/Tamil-Nadu-map-showing-the-selected-districts-for-the-study_fig3_343418281

statistical methods for predicting air quality, neural networks still have potential for development because of problems like high processing costs, poor convergence, over-fitting, and noisy data [15]. A new tool or technique for more accurate modeling of a particular issue often appears in the research-active field of machine learning. The data-driven machine learning approaches offer the chance to evaluate the effects of several air contaminants on various aspects of health at once [16]. The dataset's best prediction model was created using the Long Short Term Memory approach (LSTM) and the framework of prognostication model is exposed in Fig. 28.2.

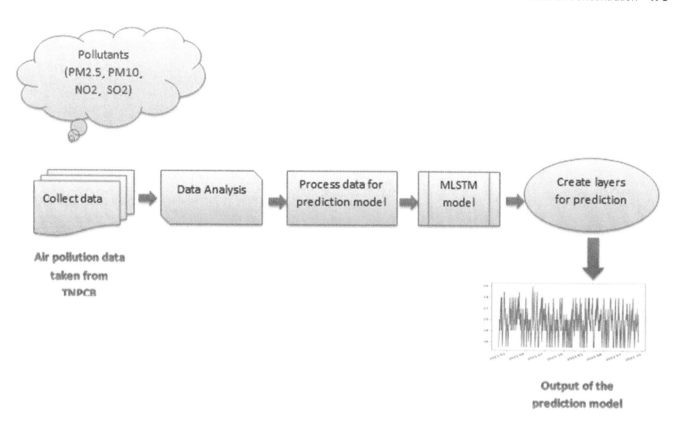

Fig. 28.2   Flow work of modified LSTM method

Source: Made by Author

4.1 LSTM Networks

The term "LSTM" refers to a specific class of RNN called Long Short-Term Memory (LSM) that can learn long-term dependencies. As a result of Hochreiter & Schmidthuber's initial introduction of them in 1997, other writers went on to expand and make them widely used in subsequent research [17]. They provide outstanding outcomes for a range of issues and are being used widely. In particular, the problem of long-term reliance is eliminated with LSTMs. The majority of recurrent neural networks were initially composed of a number of neural network modules that repeated sequentially. Figure 28.3 illustrates a typical instance of this chain-like structure, which in conventional RNNs only consists of a single tanh layer.

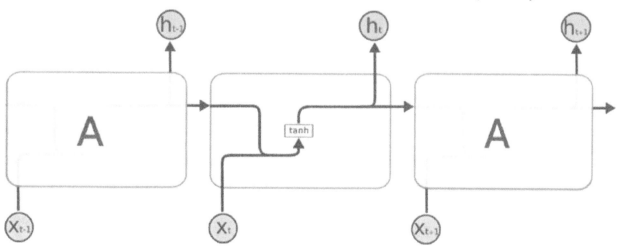

Fig. 28.3   Chain-like structure of LSTM model

Source: Adapted from https://colah.github.io/posts/2015-08-Understanding-LSTMs/

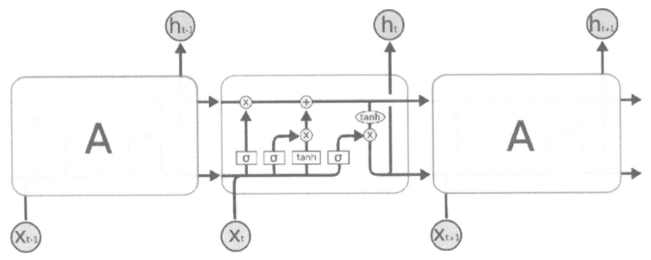

Fig. 28.4 Four network layers of LSTM model

Source: Adapted from https://colah.github.io/posts/2015-08-Understanding-LSTMs/

Although it has a topology similar to a chain, the repetitive module of LSTMs is constructed differently [18]. Figure 28.4 shows how the four layers of a neural network, instead of just one, work together.

In the diagram above, a complete vector is sent from single node's output to other's input along each line. The tiny rectangle boxes indicate trained neural network layer, and the pink circles indicate point-wise processes like vector addition [19]. Merging lines demonstrate concatenation, whereas forking lines demonstrate data capture and transmission to much location.

Understanding LSTMs require knowledge of the horizontal axis that crosses the top of the image and stands for the cell state. The cell's current condition has aspects in common with a conveyor belt. It immediately descends the entire chain with only a few simple linear trades. Unaltered information can spread quickly and easily.

The LSTM can change the state of the cell by adding or withdrawing information, which is meticulously regulated by gates. Gates allow for the voluntary transmission of information. Figure 28.5 illustrates the layers of sigmoid neural network models that they are made up of. They involve a point-wise multiplication mechanism.

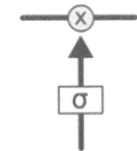

Fig. 28.5 Sigmoid layer of gate

Source: Adapted from https://colah.github.io/posts/2015-08-Understanding-LSTMs/

The outcome of the sigmoid layer is a range of integers among 0 and 1, indicating the amount of each element that should be allowed through. If the value is zero, then "let nothing through," and if it is one, then "let everything through." In categorize to secure and manage the cell condition, an LSTM contains three of these gates.

Stepwise process of LSTM model:

Selecting which data from the cell condition to remove is the primary step in LSTM. It is decided by the sigmoid layer known as the "forget gate layer." Each result in the cell state Ct1 receives a range of 0 to 1 after it analyses ht1 and xt. A1 denotes "complete retention," whereas a 0 denotes "complete removal." Figure 28.6 illustrates the stepwise process of LSTM model including four steps of architecture.

The subsequently step is to make a decision which novel data force be retained in the cell position. The "input gate layer," or sigmoid layer with two parts, is responsible for selecting the values that will be updated first. A tanh layer is use to construct a vector of new potential standards, C_t, which is subsequently used to update the state [20]. An updated state will be created by combining these two in the following phase. It is time to switch from the preceding cell state, C_{t1}, to the present cell state, C_t. Implementing the decisions made in the preceding processes is necessary. While leaving off the earlier items in the list of items to forget, the preceding condition is achieved by walking on both feet at once. The $i_t * c_t$ is then added. Following modifications made to each state value, new candidate entries are created.

As a final step, it is important to choose the output that was intended. Whatever the filtering method, the output will depend on the cell's condition. The part of the cell state that will be yields are first determined using a sigmoid layer. Next, after passing

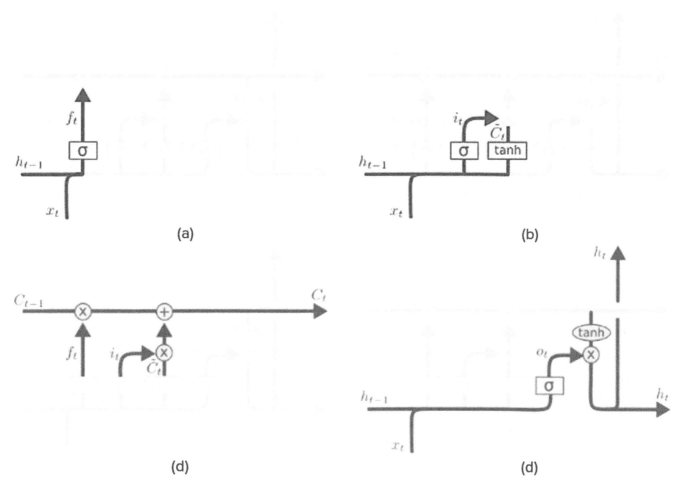

Fig. 28.6   Stepwise process of LSTM model, (a) First step of LSTM model, (b) Second step of LSTM model, (c) Third step of LSTM
model and (d) Final step of LSTM model

Source: Adapted from https://colah.github.io/posts/2015-08-Understanding-LSTMs/

the cell state through the sigmoid gate through tanh, in order to produce only the portions we already decided upon, the cell
position must be multiplied by the results of the gate (push the range between -1 and 1).

Modified Long Short Term Memory (MLSTM)

In Long Short Term Method (LSTM) method, new layer will be created for every iterations irrespective of its presence. Creation
of new layer during every iteration will amount to increased training time of the method.

In this effort, a novel technique called Modified Long Short Term Memory (LSTM) is projected in which layer will not be
created during every iteration. Layer will be created, only if it is not present already. If the layer is already present, its reference
alone only will be created. So, training time will be reduced automatically.

H5 model is implemented in the proposed technique. The LSTM model is not portable, which cannot be run in cloud, whereas
the Modified LSTM can run in local machine and as well as in cloud JCB, AWS Assure etc. It is platform independent.

5. Results and Discussion

The performance of the projected Modified LSTM representation is experienced and trained by data collected from monitor
stations in five cities of Tamil Nadu, from Tamilnadu Pollution Control Board [21]. When compare to the standard LSTM
technique, the Modified LSTM technique produces better results for predicting air pollution, with high accuracy and a low error
percentage using python software packages.

Algorithm 1: Modified Long Short Term Memory (MLSTM)

Input : **Pollutant dataset**
Output : **Prediction of input dataset**

---

1. Initialize the parameters of pollutants for prediction
2. Choose the data to discard from the cell state.

$$\mathfrak{f} = \sigma(w_{\mathfrak{f}}(h_{t-1}, x_t)tb_{\mathfrak{f}})$$

3. Create two layers:
    (i) Input layer for update the cell values.
    (ii) tanh layer for create the new values.

$$\mathfrak{i}_t = (w_{\mathfrak{i}}(h_{t-1}, x_t) + b_{\mathfrak{i}}), C = tanh(w_c(h_{t-1}, x_t) + b_c)$$

4. Combine the old and new cell state.

$$C_t = \mathfrak{f}_t * c_{t-1} + \mathfrak{i}_t * \tilde{c}_t$$

5. Compute the final combined layer using sigmoid function.

$$O_t = ((h_{t-1}, x_t) + b_o)\ h_t =$$
$$o_t * \tanh(C_t)$$

6. for n epochs do
        Train the network using with H5 model
    endfor

Figure 28.7 shows actual and predicted values for the pollutant PM2.5 at Chennai area for the using LSTM and Modified LSTM with accuracy and error percentage. The Fig. 28.7 shows improved accuracy and reduced error percentage. Figure 28.8 shows the prediction charts for the pollutant PM2.5 for five citys.

	Date	Actual	Predicted	Difference	ErrorPercentage
0	2022-08-29	22	19.282020	2.717979	12.354452
1	2022-08-30	21	17.323109	3.676891	17.509006
2	2022-08-31	20	17.194635	2.805365	14.026823
3	2022-09-01	20	17.274542	2.725458	13.627291
4	2022-09-02	19	19.876879	0.876879	4.411551
5	2022-09-03	20	20.482666	0.482666	2.356461
6	2022-09-04	21	21.973896	0.973896	4.432059
7	2022-09-05	20	22.298111	2.298111	10.306303
8	2022-09-06	21	19.922277	1.077723	5.132012
9	2022-09-07	21	21.660357	0.660357	3.048687
10	2022-09-08	20	22.381058	2.381058	10.638719
11	2022-09-09	20	22.896149	2.896149	12.649065
12	2022-09-10	21	21.887690	0.887690	4.055657
13	2022-09-11	18	18.081650	0.081650	0.451562
14	2022-09-12	16	17.877950	1.877951	10.504284

```
Total Error Percentage :  7.814519436000086
Model Accuracy :  92.18548056399992
```

	Date	Actual	Predicted	Difference	ErrorPercentage
0	2022-08-29	22	21.817528	0.182472	0.829419
1	2022-08-30	21	22.262568	1.262568	5.671257
2	2022-08-31	20	21.897802	1.897802	8.666634
3	2022-09-01	20	20.918655	0.918655	4.391560
4	2022-09-02	19	20.277124	1.277124	6.298351
5	2022-09-03	20	20.483902	0.483902	2.362353
6	2022-09-04	21	20.620400	0.379601	1.807622
7	2022-09-05	20	22.089480	2.089479	9.459161
8	2022-09-06	21	21.988176	0.988176	4.494126
9	2022-09-07	21	23.260126	2.260126	9.716741
10	2022-09-08	20	22.621630	2.621630	11.589040
11	2022-09-09	20	19.767303	0.232697	1.163483
12	2022-09-10	21	19.700119	1.299881	6.189909
13	2022-09-11	18	19.916002	1.916002	9.620416
14	2022-09-12	16	18.539553	2.539553	13.698026

```
Total Error percentage :  3.2013326875213104
Model Accuracy :  96.79866731247868
```

(a) PM2.5 using LSTM   (b) PM2.5 using Modified LSTM

Fig. 28.7 Actual and predicted values for PM2.5 using LSTM and Modified LSTM

Source: Made by Author

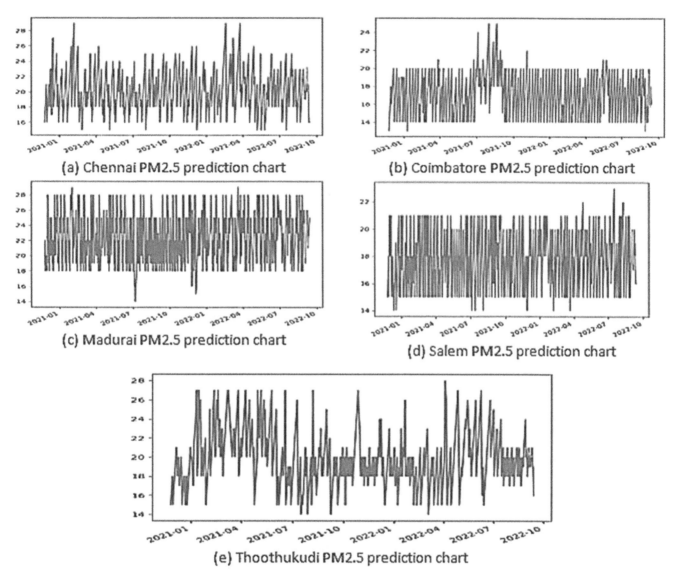

Fig. 28.8 Prediction chart of PM 2.5 for 5 cities

Source: Made by Author

Table 28.2 Comparison of Prediction accuracy of pollutants

Pollutants/ City	PM$_{2.5}$		SO$_2$		PM$_{10}$		NO$_2$	
	LSTM	MLSTM	LSTM	MLSTM	LSTM	MLSTM	LSTM	MLSTM
Chennai	92.18	96.7	85.83	93.8	95.40	96.7	92.18	93.87
Coimbatore	91.69	95.77	90.37	94.55	92.10	95.31	93.72	95.42
Salem	93.18	94.78	90.86	92.55	93.59	94.64	92.42	94.4
Madurai	94.50	96.01	93.3	94.62	94.98	96.54	95.90	96.80
Thoothukudi	94.97	95.49	92.96	93.93	94.86	95.34	89.31	93.77

Source: Made by Author

Table 28.2 exhibits the forecast accuracy of the pollutants PM2.5, SO2, PM10 and NO2 for 5 cities. For all the pollutants the Modified LSTM method has shown higher prediction accuracy compared to LSTM method.

Table 28.3 Execution time for normal and modified LSTM model for 5 cities

Model/City	Chennai	Madurai	Thoothukudi	Coimbatore	Salem
LSTM	100.28 s	99s	110s	112s	115s
MLSTM	44.6 s	43s	79.025s	84.62s	87.8s

Source: Made by Author

Table 28.3 Exhibits the execution time for LSTM and Modified LSTM for 5 cities. Execution has considerably reduced in Modified LSTM method compared to LSTM method which shows higher performance of Modified LSTM

6. Conclusion

In this study, the absorption of the air contaminants $PM_{2.5}$, PM_{10}, SO_2, and NO_2 were predicted for the five cities of Chennai, Salem, Thoothukudi, Madurai, and Coimbatore using the Modified LSTM (MLSTM) approach. The LSTM approach and the Modified LSTM method were compared in terms of performance. When compare to the LSTM method, the Modified LSTM method predicts outcomes more accurately. In addition, the Modifier LSTM technique executes faster than the LSTM method.

References

1. "Unique identification authority of India," May 2020, [Online;accessed 16-July-2020]. [Online]. Available: https://uidai.gov.in/images/state-wise-aadhaar-saturation.pdf.
2. "6 of the world's 10 most polluted cities are in India," August 2020, [Online; accessed 22- August-2020].[Online]. Available: https://www.weforum.org/agenda/2020/03/6-of-the-world-s- 10-most-polluted-cities-are-in-india/.
3. Y. Bai, B. Zeng, C. Li, and J. Zhang, "An ensemble long short term memory neural network for hourly PM2.5 concentration forecasting," Chemosphere, vol. 222, pp. 286–294, 2019.
4. J. Zhao, F. Deng, Y. Cai, and J. Chen, "Long short-termmemory—fully connected (LSTM- FC) neural network for PM2.5 concentration prediction," Chemosphere, vol. 220,pp. 486–492, 2019.
5. A. Masih, "Application of random forest algorithm to predict the atmospheric concentration of NO2," in Proceedings of the 2019 Ural Symposium on Biomedical Engineering, Radio electronics and Information Technology (USBEREIT), pp. 252–255, IEEE, Yekaterinburg, Russia, April 2019.
6. Q. Wu and H. Lin, "A novel optimal-hybrid model for daily air quality index prediction considering air pollutant factors," Science of the Total Environment, vol. 683, pp. 808–821, 2019.
7. K. Xiao, Y. Wang, G. Wu, B. Fu, and Y. Zhu, "Spatiotemporal characteristics of air pollutants (PM10, PM2.5, SO2, NO2, O3, and CO) in the inland basin city of Chengdu, southwest China," Atmosphere, vol. 9, no. 2, p. 74, 2018.
8. V. Reddy, P. Yedavalli, S. Mohanty, and U. Nakhat, "Deep air: forecasting air pollution in Beijing, China," 2018, http://arxiv.org/abs/1804/1804.07891.
9. U. Pak, C. Kim, U. Ryu, K. Sok, and S. Pak, "A hybrid model based on convolutional neural networks and long short-term memory for ozone concentration prediction," Air Quality, Atmosphere & Health, vol. 11, no. 8, pp. 883–895, 2018.
10. H. Liu, X.-w. Mi, and Y.-f. Li, "Wind speed forecasting method based on deep learning strategy using empirical wavelet transform, long short term memory neural network and Elman neural network," Energy Conversion and Management, vol. 156, pp. 498–514, 2018.
11. C.-J. Huang and P.-H. Kuo, "A deep CNN-LSTM model for particulate matter (PM2.5) forecasting in smart cities," Sensors, vol. 18, no. 7, p. 2220, 2018.
12. M. Rezamand, M. Kordestani, M. Orchard, R. Carriveau, D. S. K. Ting, and M. Saif, "Improved remaining useful life estimation of wind turbine drivetrain bearings under varying operating conditions (VOC)," IEEE Transactions on Industrail Informatic, vol. 17, no. 3, pp. 1742–1752, 2020.
13. Abdelhadi Azzouni and Guy Pujolle. "A Long Short-Term Memory Recurrent Neural Network Framework for Network Traffic Matrix Prediction". In: CoRR abs/1705.05690 (2017). arXiv: 1705.05690. URL: http://arxiv.org/ abs/1705.05690.
14. Xiang Li et al. "Deep learning architecture for air quality predictions". In: 23 (Oct. 2016).
15. Xiang Li et al. "Long short-term memory neural network for air pollutant concentration predictions: Method development and evaluation". In: Environmental Pollution 231.Part 1 (2017), pp. 997–1004. ISSN: 0269-7491. DOI: https://doi. org / 10 . 1016 / j .envpol .2017 .08 .
16. URL: http : / / www . sciencedirect .com / science / article/pii/S0269749117307534.
17. Bing-Chun Liu et al. "Urban air quality forecasting based on multi-dimensional collaborative Support Vector Regression (SVR): A case study of Beijing-Tianjin-Shijiazhuang". In: PLOS ONE 12.7 (July 2017), pp. 1–17. DOI: 10.1371/ journal.pone.0179763. URL: https://doi.org/10.1371/journal.pone.0179763.

18. Samira Shamsir et al. Applications of Sensing Technology for Smart Cities. Aug. 2017.
19. Adams, Matthew D., and Pavlos S. Kanaroglou. "Mapping real time air pollution health risk for environmental management: Combining mobile and stationary air pollution monitoring with neural network models." Journal of environmental management 168 (2016): 133–141.
20. Al-Rfou, Rami, et al. "Theano: A Python framework for fast computation of mathematical expressions." arXivpreprint arXiv:1605.02688 (2016).
21. Che, Zhengping, et al. "Recurrent neural networks for multivariate time series with missing values." arXiv preprint arXiv:1606.01865 (2016).
22. http://www.tnpcb.gov.in/air-quality.php.

Human Machine Interaction in the Digital Era – Prof. J. Dhilipan et al. (eds)
© 2024 Taylor & Francis Group, London, ISBN 978-1-032-54998-9

An Effective Analysis of Feature Selection Methods for High Dimensional Data—A Survey

29

M. Kalaivani[1]
Research Scholar, Department of Computer Science, Vels Institute of Science,
Technology & Advanced Studies (VISTAS) Chennai, Tamil Nadu

K. Abirami[2]
Assistant Professor, Vels Institute of Science,
Technology & Advanced Studies.

K. Dharmarajan[3]
Associate Professor, Vels Institute of Science,
Technology & Advanced Studies

Abstract In the field of genomics, High dimensional data is primarily utilized to detect the essential genes that play a vital role in determining the disease diagnosis using expression levels. The number of features in the High dimensional dataset is extremely very high when compared to the samples present in the dataset. The features in the dataset are usually given as input to a learning algorithm for classification of diseases. However, in the High dimensional data most features are redundant and irrelevant or noisy which will decrease the learning accuracy. To solve these problems, Feature selection technique is employed a significant role. Feature selection is one of the important preprocessing step for prediction and classification of disease. It aims to find informative features, selecting a small subset of relevant features from the original set of features by removing the redundant and irrelevant features from the dataset which can reduce the computational time and improving the classification accuracy. Due to increase in dimensionality of High dimensional data imposes a significant challenge to many existing feature selection methods in terms of prediction and accuracy of the model. This research work analyses about the use of various Feature selection methods that can select prominent attributes from the High dimensional dataset for classification of diseases.

Keywords Classification, Feature selection, High dimensional data, Prediction, Preprocessing

1. Introduction

High dimensional data occur when the feature count (P) higher than the number of samples (N), represented as P>N. and contains a large number of redundant and irrelevant attributes. To select the significant features from the data is a major challenging task. The main application of High dimensional data is in the field of Genomics such as Clinical decision support system for classification of the type of disease and facilitates the clinical analysis of those genes that are responsible for a particular disease especially Cancer Classification. In particular Deoxyribonucleic Acid (DNA) Microarray is an effective tool that supports to monitor the level of gene expression in an organism. The microarray dataset are represented as images that are converted into two dimensional matrices where the rows represent genes, and the columns represent samples (Brazma

---
[1]mkvanimca@gmail.com, [2]abiramidharmarajan@gmail.com, [3]dharmak07@gmail.com

DOI: 10.1201/9781003428466-29

and Vilo, 2001). The total number of samples in the dataset is significantly lesser when compared to the number of features (genes) commonly known as "curse of dimensionality" (Hamim et al., 2021). Genes are now taken into consideration with the development of biomedical research for the classification of a disease, particularly cancer, for prognosis or diagnosis of disease at an early stage. It is necessary to reduce the dimensions prior to classification because a high number of features may lead to increase in computational time and memory consumption. In an attempt to overcome these issues, Feature selection algorithms have been applied prior to classification. This research work analyses the different Feature Selection methods applied in High dimensional data used to remove redundant and irrelevant features for better classification of the model.

The rest of this paper is organized as follows. Section II discusses about the Role of Feature selection methods. Section III explores the different Feature Selection methods. Analysis of Feature Selection methods applied for High dimensional data by different researchers is discussed in section IV. Finally, Section V concludes the suitability of Feature Selection methods in High dimensional data.

2. Role of Feature Selection Methods

Dimensionality reduction is one of the primary pre-processing method in machine learning to determine the significant attributes in the data and can be broadly classified into two ways (Sahu et al., 2018). Feature selection and Feature extraction. Feature selection, selects the subset of features from the original set whereas Feature extraction is to generate new features from the existing set. Feature selection techniques can be broadly classified into three important methods. Supervised, Unsupervised and Semi Supervised method. A Framework for feature selection methods are shown in Fig. 29.1. In Supervised Feature Method the availability of class label information allows feature selection algorithms to efficiently select discriminative and relevant features to distinguish features from different classes.

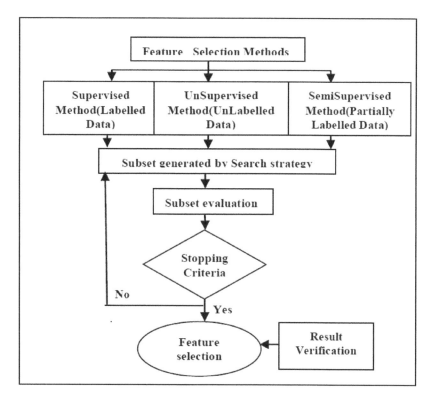

Fig. 29.1 Framework for feature selection methods

Source: Subset evaluation—Made by Author

In Unsupervised Feature Selection method, new features are created by combining the original features. The absence of class label information allows selecting discriminative features and produces high performance without prior knowledge is required and the limitation is to neglect the feasible correlation between features. Semi supervised method is the combination of Supervised and Unsupervised methods. It learns the data from the combination of both labeled and unlabeled data to evaluate the features (Sechidis & Brown 2018). It selects features by utilizing unlabeled data when there is limited number of labeled data. According to the subset generated by search strategy, the commonly used search methods are Sequential, Exponential and Random Search. In sequential search, feature selection methods can be used Sequential Backward Elimination, Sequential Forward Selection and Bidirectional search (Cai et al., 2018). Feature evaluation criteria is mainly categorized into distance, consistency, dependency, correlation and information. Stopping Criteria indicates the end of the process. Finally, evaluate the accuracy of the method.

3. Classification of Feature Selection Methods

In machine learning, Feature selection method plays an important role in the field of biomedical data analysis particularly cancer classification with an increase in the number of data dimensions. Feature selection methods are classified into the following categories: Filter ,Wrapper ,Embedded, Hybrid and Ensemble method. Filter method consists of selecting attributes based on inherent characteristics of the data without using classifier and generally involves non-iterative computation. Each attribute is analyzed individually by employing its basic statistical properties and can be classified as Univariate and Multivariate filters. Univariate filters evaluate the individual feature whereas multivariate filters estimate the whole feature subset. Wrapper approaches arc using a specific learning algorithm to select subsets of features (Panthong & Srivihok,2015) which is more accurate than filter methods. The number of operations required to obtain the feature subset and run each time a new learning method is applied. Computationally difficult method when compared to filter method. In Embedded methods, the feature selection algorithm is incorporated into the learning process (Jovic et al., 2015) and takes the advantages of its own attribute selection process and performs Feature selection and Classification at the same time. The hybrid method integrates the benefits of Filter, Wrapper and Embedded methods. Filter method is used as an initial step to reduce the attribute dimensions particularly when dimensions in the feature space are high and remove irrelevant and similar features with less computation cost. In the next step of process, Wrapper or Embedded method is applied to the selected features and evaluated to determine the best feature subset. An ensemble method is based on two steps. Initially, two or more component learners are trained either sequential method or parallel method and finally aggregating their predictions based on algorithms.

4. Analyzing Feature Selection Methods

High dimensional dataset normally consists of a huge number of features, but all the features are not contributing to the goodness of classification. Due to redundant and noisy features of High-Dimensional dataset, Feature selection is one of the essential pre-processing step which is used to identify the significant attributes present in the dataset and produce higher performance accuracy with less processing time. Various Feature Selection methods are used to select the significant attributes in the High dimensional data. This section discusses about different Feature selection methods are applied by many researchers in High dimensional dataset. D.M. Deepak Raj et al. (2020) proposed a new feature weighting algorithm called boundary margin relief (BMR) to predict the feature weights through the metric of local hyper plane to determine the set of the closest hits and misses and calculate the features weight by increasing the hyper plane margin. The proposed method identified non redundant features and produced higher accuracy value.

Dewi Pramudi Ismi et al. (2016) developed a model using k-means partitional cluster method. The model solves the curse of dimensionality by partitioning the data into clusters and removed similar and irrelevant features. Abdulrauf Garba Sharifai et al. (2020) have designed the multiple filter method with correlation based redundancy CBRMFA. In this method, Mutual Information, Symmetric Uncertainty and Euclidean distance was applied to select the relevant features from each filter method and aggregating the features using the union operator. The top N ranking features are selected to form a new set of features based on the threshold value. Correlation between features was computed and finally sequential forward search was applied to select the optimal set of features. Manikandan et al., (2017) proposed a wrapper based feature selection approach using symmetrical uncertainty method to calculate the feature weight based on Entropy and information gain values. The selected features are arranged in descending order and given as input to the classifier and the accuracy of each attribute is compared with the previous attribute and finally selected the higher accuracy features. Liuzhi Yin et al. (2013) discussed class imbalanced data using feature selection and developed two approaches. In the first method, large classes are partitioned into pseudo-subclasses with equal sizes and find the efficiency of features with the partitioned data to reduce the biased value of the imbalanced class. In the next step, Hellinger distance was applied to select the essential features. Compared the results with Pearson Correlation, Mutual information gain and Fisher Score methods. Yongbin Zhu et al. (2022) have developed a method referred as HFIA based on AI Optimization technique. Fisher Score and clonal section algorithm are applied to explore the feature subset. The result showed that the proposed method produced 91.67% of accuracy rate.

Aiguo Wang et al. (2017) have described a novel method to combine the MB method into WBFS. The developed method, eliminated the redundant and noisy attributes using a classifier which helps to speed up the process and select the significant features. Jamshid pirgazi et al. (2019) have designed a new metaheuristic attribute selection technique and implemented in two phases. In the first phase, reliefF was employed to select the attributes based on the rank and SFL,IW Subset Selection relevance algorithm was applied in the second stage and selected effective features in the dataset to increase the rate of accuracy. Amirreza Rouhi et al. (2017) have presented an algorithm using hybrid and ensemble method. FCBF filter method was applied to reduce the features space and different metaheuristic optimization algorithms are applied independently to the selected attributes and finally combined the important attributes. Chaonan Shen et al. (2021)proposed a model that combined MLP and IGWO. Network was trained with Lasso method to evaluate the hidden layer neurons using the weights. In the next step, IGWO was applied to reduce the size of the features space and also selected an optimal number of features. The results showed that 93.05% of accuracy for eleven tumor dataset. Rania Saidi et al. (2019) have proposed a hybrid selection technique using Pearson Correlation Coefficient and Genetic algorithm. In the initial stage, GA was applied for the Ionosphere dataset using Random Forest as evaluation function. Pearson Correlation Coefficient was applied in the next stage and combined the resulting attribute subsets to increase the classification accuracy. Mohammed Loey et al., (2020)have designed an algorithm consisting of three processing stages to reduce the feature space. At the initial stage, Information Gain (IG) was applied in the dataset and selected the most relevant features based on the weight value. To optimize the features using GWO ptimization and SVM classifier was employed to evaluate the features. Abdulrauf Garba Sharifai et al. (2021) have presented multiple filter approach with hybrid Grasshopper optimization and Simulated Annealing (HGOASA) technique. According to the threshold value, the top ranked attributes from each filter are selected. In the next stage, optimization approach was applied based on global search method to select the prominent set of features. Annavarapu & Dara ,2021 have described a cluster based feature selection method. Partitional kmeans cluster analysis with snr rank method was applied to reduce the feature dimension space of the dataset. CLA combined with ant colony optimization (CLACO) was applied on the selected attributes to obtain the significant feature subset and analysed using several classifiers.

5. Conclusion

Feature selection is an essential pre-processing step for the discriminate analysis of very High-Dimensional data because it contains a large number of attributes and limited sample sizes. Due to curse of dimensionality it is vital to implement Feature selection algorithm to reduce the high dimensional feature space to low dimensional feature space and select the most significant features. The primary goal of the method is to enhance the high accuracy of the model by the elimination of redundant, irrelevant and noisy features. In this process, Filter methods are very fast and simple, whereas each feature is measured separately and thus it does not take into account of the dependencies among the features. The Wrapper methods using exhaustive search to generate optimal solutions, whereas its limitation is that a risk of over fitting. Embedded methods are lower risk of over fitting but faster running time as compared to a wrapper method. Hybrid feature selection method can combine the benefits of Filter, Wrapper and Embedded methods which can significantly improve the performance by increasing the rate of accuracy and decreasing the execution time for classification of diseases. Ensemble methods are less prone to over fitting and certain regularization are more suitable to specific types of learners. The future work determines to develop an effective feature selection algorithm for identifying significant features as well as remove redundant and irrelevant attributes present in the High dimensional data, analysing the merits and demerits of the surveyed methodology.

References

1. Brazma, A., & Vilo, J. (2001). Gene expression data analysis. *Microbes and Infection, 3*(10), 823–829.
2. Hamim, M., El Mouden, I., Ouzir, M., Moutachaouik, H., & Hain, M. (2021). A novel dimensionality reduction approach to improve microarray data classification. *IIUM Engineering Journal, 22*(1), 1–22.
3. Sahu, B., Dehuri, S., & Jagadev, A. (2018). A study on the relevance of feature selection methods in microarray data. *The Open Bioinformatics Journal, 11*(1).
4. Sechidis, K., & Brown, G. (2018). Simple strategies for semi-supervised feature selection. *Machine Learning, 107*(2), 357–395.
5. Cai, J., Luo, J., Wang, S., & Yang, S. (2018). Feature selection in machine learning: A new perspective. *Neurocomputing, 300*, 70–79.
6. Panthong, R., & Srivihok, A. (2015). Wrapper feature subset selection for dimension reduction based on ensemble learning algorithm. *Procedia Computer Science, 72*, 162–169.
7. Jovic, A., Brkic, K., & Bogunovic, N. (2015). A review of feature selection methods with applications. In 2015 38th international convention on information and communication technology, electronics and microelectronics (pp. 1200–1205). IEEE.

8. Raj, D. D., & Mohanasundaram, R. (2020). An efficient filter-based feature selection model to identify significant features from high-dimensional microarray data. *Arabian Journal for Science and Engineering*, *45*, 2619–2630.

9. Ismi, D. P., Panchoo, S., & Murinto, M. (2016). K-means clustering based filter feature selection on high dimensional data. *International Journal of Advances in Intelligent Informatics*, *2*(1), 38–45.

10. Sharifai, A. G., & Zainol, Z. (2020). The correlation-based redundancy multiple-filter approach for gene selection. *International Journal of Data Mining and Bioinformatics*, *23*(1), 62–78.

11. Manikandan, G., Susi, E., & Abirami, S. (2017). Feature selection on high dimensional data using wrapper based subset selection. In *2017 Second International Conference on Recent Trends and Challenges in Computational Models* (pp. 320–325). IEEE.

12. Yin, L., Ge, Y., Xiao, K., Wang, X., & Quan, X. (2013). Feature selection for high-dimensional imbalanced data. *Neurocomputing*, *105*, 3–11.

13. Zhu, Y., Li, T., & Li, W. (2022). An Efficient Hybrid Feature Selection Method Using the Artificial Immune Algorithm for High-Dimensional Data. *Computational Intelligence and Neuroscience*, *2022*.

14. Wang, A., An, N., Yang, J., Chen, G., Li, L., & Alterovitz, G. (2017). Wrapper-based gene selection with Markov blanket. *Computers in biology and medicine*, *81*, 11–23.

15. Pirgazi, J., Alimoradi, M., Esmaeili Abharian, T., & Olyaee, M. H. (2019). An Efficient hybrid filter-wrapper metaheuristic-based gene selection method for high dimensional datasets. Scientific reports, 9(1), 18580.

16. Rouhi, A., & Nezamabadi-pour, H. (2017). A hybrid feature selection approach based on ensemble method for high-dimensional data. In *2017 2nd conference on swarm intelligence and evolutionary computation (CSIEC)* (pp. 16–20). IEEE.

17. Shen, C., & Zhang, K. (2021). Two-stage improved Grey Wolf optimization algorithm for feature selection on high-dimensional data. *Complex & Intelligent Systems*, 1–21.

18. Saidi, R., Bouaguel, W., & Essoussi, N. (2019). Hybrid feature selection method based on the genetic algorithm. *Machine learning paradigms: theory and application*, 3–24.

19. Loey, M., Wajeeh Jasim, M., El-Bakry, H. M., Hamed N. Taha, M., & Khalifa, N. E. M. (2020). Breast and colon cancer classification from gene expression profiles using data mining techniques. *Symmetry*, *12*(3), 408.

20. Sharifai, A. G., & Zainol, Z. B. (2021). Multiple filter-based rankers to guide hybrid grasshopper optimization algorithm and simulated annealing for feature selection with high dimensional multi-class imbalanced datasets. *IEEE Access*, *9*, 74127–74142.

21. Annavarapu, C. S. R., & Dara, S. (2021). Clustering-based hybrid feature selection approach for high dimensional microarray data. *Chemometrics and Intelligent Laboratory Systems*, *213*, 104305.

Human Machine Interaction in the Digital Era – Prof. J. Dhilipan et al. (eds)
© 2024 Taylor & Francis Group, London, ISBN 978-1-032-54998-9

Management Information System for Operation, Planning and Control of Wind Energy Farms

30

E. Kirubakaran[1]

Research Scholar, Department of Management Studies,
Karunya Institute of Technology and Science, Coimbatore, India

K. Karthikeyan

Assistant Professor, Department of Management Studies,
Karunya Institute of Technology and Sciences Coimbatore,India

Abstract Management information system (MIS) is basically a computerized information system for providing appropriate and timely information related to the various functional activities of the concerned organization. Internal data generated within the organization as well as related external data are processed and made available as relevant information to the management. This information helps the various levels of the management to take appropriate and timely action for effective operation and control of the various functional activities of the organization. In this paper, renewable wind energy generation is chosen as the domain organization and an appropriate MIS is developed for supporting the operation and maintenance activities of these wind turbine generators (WTG). The internal data generated from the supervisory control and data acquisition (SCADA) system and the external data from the environment in which these WTGs are operating are processed and stored in the database system. This information is made available to human machine interface (HMI) and also to the various functional agencies for managing the operation, planning and control of these WTGs. The main advantage of this MIS is knowing the conditions of the various components of the WTG and the overall performance, for taking timely action which eventually increases the operational efficiency. The scope of enhancing the existing MIS by incorporating the emerging trends in Information technology, namely the Industry 4.0 standards are also discussed.

Keywords Wind turbine generator, SCADA, MIS, Database, Networking, Information processing, Industry 4.0

1. Introduction

Wind energy is one of the fastest growing renewable energy technologies around the world. The wind energy is captured by the device called Wind Turbine Generators (WTG), which converts the kinetic energy of wind into clean, renewable electricity. In India Wind energy is growing very fast. The installed capacity has grown from 10,925 MW in 2009 to 40,067MW in 2021. In India, Tamil Nadu, Gujarat and Karnataka has a number of wind belt areas, facilitating installation of a number of WTGs. The capacity of these WTGs varies from 250 KW to 4000 KW, height varies from 50 to 150 meters and age from the recently installed ones to more than 30 years. The wind belt area of Mupandal pass located at Kanyakumari district of Tamil Nadu in India, houses a good number of Wind Turbine Generators. (Figure 30.1). Generally owners of the WTG outsource the operation and maintenance to companies specialized in the work of WTG maintenance. The working of these companies are complex

---
[1]ekirubakaran@gmail.com, [2]karthikeyan@karunya.edu

DOI: 10.1201/9781003428466-30

Fig. 30.1 Wind farm at Aralvaimohi, in TN, India

Source: https://www.kanyakumarians.com/aralvaimozhi-wind-farm-kanyakumari-district

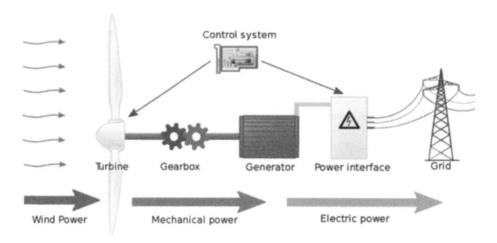

Process of converting wind energy into electric energy

Fig. 30.2 Basic functional components of WTG

Source: https://www.felsics.com/5-wind-turbine-drivetrain-components-and-functions-discussed/

in nature basically because of the difficulties in accessing the components of these WTG due to the height and the remote location of installation of these WTGs. A good Management Information System (MIS), showing the data and information arising out of the various components of the WTG is the basic requirement for effective management of the Wind turbine operations. This Paper deals about the salient features of the MIS and way it facilitates the operations of WTG. A company called M/S RS Windtech Engineers Pvt Ltd, located at Kanyakumari district is doing the operation and maintenance of about 1000 Wind Turbine Generators in India. In order to facilitate the operations, a good web based Management Information System is developed for this company. A study is made about the operation and MIS of this company. The data/ information from one of the client WTG namely ' 847Version' , is taken and illustrated in this paper.

2. Wind Turbine Generators (WTG)

The basic functional activities of WTG are conversion of wind power to mechanical power and then the mechanical power to electric power (Fig. 30.3). These functions are controlled and co-ordinated through the control system using various types of sensors and connectivity. All these components (except the blades) are enclosed inside the Nacelle of the WTG at the top of the wind turbine tower.

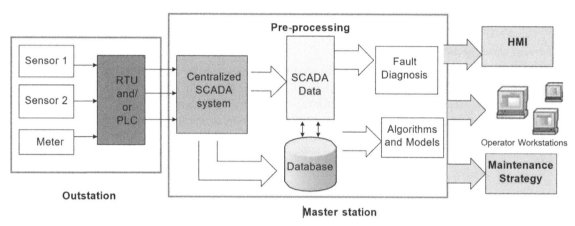

Sensors feed the signals to SCADA, processed and stored in database, made accessible to operators and management

Fig. 30.3 Basic building blocks of management Information system [1]

Blades (Wind energy): The blades of the turbine, are directly exposed to wind. On attaining the required wind speed, the blades starts rotating. Also beyond a particular speed of the wind, for safety reasons, the blades are automatically stopped by control system .

Shaft and Gear (Mechanical energy): The blades are attached to the main shaft. The rotational speed of the main shaft is very less (less than 20 rpm). In order to increase the speed to the desired value (around 1500 rpm) as required by the electric generator, suitable gear system is used.

Generator and Power Interface (Electric energy): The fast shaft running shaft is connected to the electric generator, where the mechanical energy is converted to electric energy. The power interface helps to connect the electricity thus generated to be fed to the electric grid.

Control System: This is basically the Supervisory Control and Data Acquisition (SCADA) system , which receives inputs from the sensors (like temperature sensors, pressure sensors, gas sensors etc.) from the various components of the WTG. Based on the signals received, the Controller controls the functional activities like starting and stopping the wind turbine, movements for the rotation of blade about its own axis, rotation of the nacelle etc ...

3. Management System for Wind Turbine

3.1 Infrastructure for Building Management Information System

The working of the MIS infrastructure in shown in Fig. 30.3. Reference (Khairy Sayed Et-al 2021). Various data like the ambient temperature, wind speed and turbulence is captured by the appropriate sensors and fed to SCADA through the Remote Terminal unit (RTU) or the Programmable Logic Controller (PLC). The sensors inside the nacelle directly send the operating data of WTG to SCADA. Pre-processing of the data is done and then fed to the Data Base. From the SCADA data, the fault that has occurred (like high temperature, over speed, high turbulence ..etc) are communicated using appropriate Human Machine Interface (HMI) and to the various devices. As per the requirement, data in the database are further processed using algorithms and models to get the desired information that are used to formulate appropriate maintenance strategy and operating plans. These strategy and plans help to avoid break down maintenance by using appropriate preventive maintenance schedules.

3.2 Information in Management Information System

The following are the salient management information provided in the Web based MIS System developed for the various WTGs for which the Operation and Maintenance is done by M/S RS Windtech Engineering, Pvt. Ltd.

- Over view of current status of all WTG maintained by the company
- Detailed performance parameters for a specific date for a WTG
- Month wise summary performance information for a WTG
- Comparison of design performance vs actual performance of a WTG at different seasons

Over view of current status of all WTG

Figure 30.4 is the screen shot showing the status of the various WTG maintained by M/S RS Wintech Engineers Pvt. Ltd. The various status like whether the WTG is presently doing Generation or Free Wheeling (blades are running but due to in adequate wind speed power is not generated) or Brake is applied (for reasons like high speed or specific input from sensors) or not under operation due to Grid Drop (because the electric grid is not in a position to accept the electricity generated) or under Service (where the wind turbine is under maintenance) or Manual Stop. The status are indicated by differ colours. Apart from this the present power generation, the present wind speed and the power generated for the day till the time of view is also shown.

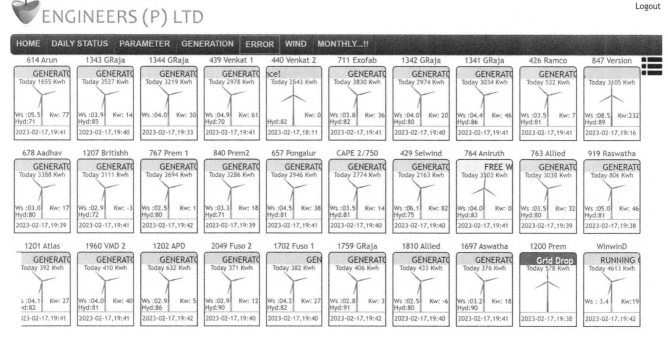

Fig. 30.4 Screen showing the status of the various WTG operated maintained by the company

Source: www.rooktec.in user : RS Windtech Engineers - opening page

Detailed performance parameters on a specific date for a specific WTG

The various parameter details for every 5 minutes on the specific date for the specific Wind Turbine is shown in Fig. 30.4. The various parameters like the wind speed, ambient temperature, temperature at various locations of the generator, current and voltage at various phases, output power and also the status of the WTG are shown.

Summary of Performance Information

Figure 30.6 shows the date wise summary of performance information for a particular WTG for a particular month. The various indictors are: the lull hours (where wind speed is not suitable), grid drop (grid is not prepared to receive the power), uptime (WTG is ready to work when the wind speed is sufficient), import (power drawn from the grid when the WTG is not working), stoppage(wind mill forcefully stopped).

TURBINE PARAMETER REPORT

Select Turbine 847 Version ⌄ Select Date 01-02-2023 ▦ Show   Export to excel

Turbine Parameter Report

For Date : 01-02-2023

Turbine : **847 Version**   HTSC No : **847**   Site : **Alaganeri**

| Temperature | | | | | | RPM | Freq | Hydraulic | Nacelle | Vol | | | Amp | | | Status | Wind | KW | KVr | Time |
Am	GG	Gg	GO	GB	Nc	Rot/Gen			Position	L1	L2	L3	L1	L2	L3		Speed			
030	035	035	044	047	034	00/0000	49.9	81	232	391	389	391	000	000	000	FREE WHEELING	01.1	0	00	23:56
030	035	035	044	047	034	00/0000	49.9	81	232	393	389	388	000	000	000	FREE WHEELING	01.5	0	00	23:52
030	035	035	044	048	034	00/0000	50.0	81	232	394	388	388	000	000	000	FREE WHEELING	01.8	0	00	23:50
030	036	036	044	047	034	00/0000	50.0	81	232	392	388	391	000	000	000	FREE WHEELING	02.0	0	00	23:48
030	036	036	045	047	034	00/0000	49.9	81	232	393	390	388	000	000	000	FREE WHEELING	02.0	0	00	23:44
030	036	036	045	048	034	00/0000	49.9	81	232	392	388	389	000	000	000	FREE WHEELING	02.2	0	00	23:40
030	036	036	045	049	034	00/0000	50.0	82	232	391	389	389	000	000	000	FREE WHEELING	02.0	0	00	23:36
030	036	036	045	049	034	00/0000	49.9	82	232	392	389	388	000	000	000	FREE WHEELING	02.5	0	00	23:32
031	036	036	045	049	034	00/0000	50.0	82	232	392	387	389	000	000	000	FREE WHEELING	02.0	0	00	23:28
031	037	037	045	049	034	00/0000	49.9	82	232	391	387	388	000	000	000	FREE WHEELING	02.4	0	00	23:24
031	037	037	045	049	034	00/0000	49.9	82	232	389	388	388	000	000	000	FREE WHEELING	02.2	0	00	23:20
031	037	037	045	050	034	00/0000	50.0	83	232	391	386	389	000	000	000	FREE WHEELING	01.3	0	00	23:16
031	037	037	046	050	034	00/0000	49.9	82	232	392	386	386	000	000	000	FREE WHEELING	01.6	0	00	23:12
031	037	037	046	050	034	00/0000	49.9	83	232	389	386	388	000	000	000	FREE WHEELING	02.0	0	00	23:08

Fig. 30.5 Showing the various performance parameters for a day for specific WTG (for every 4 minutes)

Source: User rswindtech Turbine selected '847 Version' date '01-02-2023'

INDIVIDUAL TURBINE MONTHLY REPORT

Select Turbine 847 Version ⌄ For the Month January 2023 Show to Excel

Turbine Monthly Report

Turbine : **847 Version**   HTSC No : **847**   Site : **Alaganeri**

Date	Generation	Lull Hrs	Grid Drop	UpTime	Import	Stoppage
01-01-2023	1787	03:10		100	-5	
02-01-2023	2772	00:04	02:48	99.86	-1	00:02
03-01-2023	3621	00:06	00:21	99.79	0	00:03
04-01-2023	5510	00:07	00:02	99.58	0	00:06
05-01-2023	5192			100	0	
06-01-2023	5008	00:03	04:13	99.17	-1	00:12
07-01-2023	5081	00:04	03:00	97.43	0	00:37
08-01-2023	6064			100	0	
09-01-2023	4009	00:11	00:28	99.38	-2	00:09
10-01-2023	3744	00:04	06:36	97.92	-4	00:30
11-01-2023	2942	00:41	01:00	99.79	-3	00:03
12-01-2023	2768	00:12		94.31	-3	01:22
13-01-2023	1779	04:31	00:23	97.43	-8	00:37
14-01-2023	2563	01:19		100	-2	
15-01-2023	1687	00:39		100	-9	
16-01-2023	2110	00:02		100	-1	
17-01-2023	3054	00:09		100	-4	
18-01-2023	5410			100	-1	
19-01-2023	5296	00:09	00:04	99.72	-1	00:04

Fig. 30.6   Is the screen shot showing the date wise summary for a specific month for specific WTG

Source: User rswindtech Turbine selected '847 Version', month 'January 2023'

Comparison of design performance and actual performance

Power graph is the graph showing the design performance (power generation vs wind speed) and this is provided by the original equipment manufacturer (OEM). On this theoretical power graph, the power that is actually generated for various samples of speed are plotted and thus actual power graph is arrived. The comparison of the design power graph and actual power graph for two selected months one for a high wind season and another for a low wind season (Fig. 30.7 and Fig. 30.8) are compared and overall performance inference could be arrived.

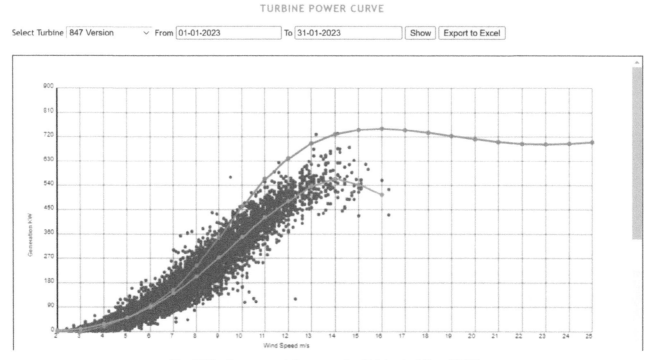

Fig. 30.7   Power graph for a month of high wind (Aug 2022)

Sourece: User rswindtech Turbine selected '847 Version', month 'aug 2022'

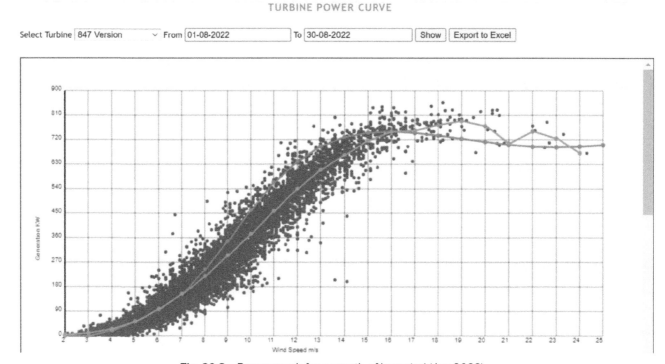

Fig. 30.8   Power graph for a month of low wind (Jan 2023)

Source: User rswindtech Turbine selected '847 Version', month 'jan 2023'

From Fig. 30.7 and Fig. 30.8, it could concluded that if the low wind season is of long duration, then the over all performance of this turbine ('847 Version') is unsatisfactory and it calls for a detailed diagnostic of the various components of the WTG.

4. Conclusion

It could be seen that the basic operating parameters (detailed data) is generally required for the operating level staff, summarized data for the higher level (summarized information) is required for the managerial staff and the top level summary and comparison information is required for the top level management and owners. This basic MIS is developed using the Internet based web system, The Internet has advanced as Industry 4.0 standards from the recent period from 2012 onwards. New technologies in sensing and networking (like Iot, IIot, Edge and Fog computing), virtualization techniques like (Virtual reality and Augmented reality), Artificial Intelligence techniques (like Machine Learning and Deep Learning), Big Data, Cloud and Virtualization has emerged. Incorporating these technologies, new paradigm can be evolved in this area. One such thing is the Condition based maintenance, where the conditions of the various components are constantly monitored and the appropriate Operation and maintenance could be done. Thus the Operation of the Wind Turbine Generators could be made as a 'Smart Factory'.

5. Acknowledgment

The authors thank M/S RS Wind Tech Engineers Pvt., Ltd., Aralvaimozhi for the permission and support to make a detailed study on Operation, Maintenance, Planning of the Wind Turbine and the impact of MIS on these activities.

Reference

1. Khairy Sayed, Ahmed G. Abo-Khalil & Ali M. Eltamaly 6 Wind Power Plants Control Systems Based on SCADA System (2021)., https://link.springer.com/chapter/10.1007/978-3-030-64336-2
2. Huan Long, Long Wang, Zijun Zhang, Zhe Song, and Jia Xu "Data-Driven Wind Turbine Power Generation Performance Monitoring" IEEE TRANSACTIONS ON INDUSTRIAL ELECTRONICS (2015)
3. Srbinovski, B., Conte, G., Morrison, A. P., Leahy, P., & Popovici, E. (2017, March). ECO: An IoT platform for wireless data collection, energy control and optimization of a miniaturized wind turbine cluster: Power analysis and battery life estimation of IoT platform. IEEE international conference on industrial technology (2017)
4. An, J., Zou, Z., Chen, G., Sun, Y., Liu, R., & Zheng, L. An IoT-Based Life Cycle Assessment Platform of Wind Turbines. Sensors, 21(4), 1233(2021).
5. IRENA : Innovation Landscape Brief: Artificial Intelligrnce and Big Data, International Renewable Energy Agency, Abu Dhabi (2019)
6. Chenggen Wang, Qian Zhou, Mingzhe Han, Zhan'ao, Xiao Hou, Haoran Zhao, Jing Bu, Review of Recent Development of Dynamic Wind Farm Equivalent Models Based on Big Data Mining, 2018 Asia Conference on Energy and Environment Engineering (ACEEE 18), 2018
7. Mikel Cañizo, Enrique Onieva, Angel Conde, Real-time predictive maintenance for wind turbines using Big Data frameworks, 2017 IEEE International Conference on Prognostics and Health Management (ICPHM), June 2017
8. Xu Andy Sun, Nagi Gebraeel, Murat Yildirim, Integrated predictive analytics and optimization for wind farm maintenance and operations, IEEE Transactions on Power Systems PP(99), February 2017
9. Yingying Zhao, Dongsheng Li, Ao Dong, Dahai Kang, Qin Lv, Li Shang, Fault Prediction and Diagnosis of Wind Turbine Generators Using SCADA Data, Energies 2017
10. evin LeahyKevin Leahy, R. Lily HuIoannis, KonstantakopoulosIoannis, Konstantakopoulos, Diagnosing wind turbine faults using machine learning techniques applied to operational data, 2016 IEEE International Conference on Prognostics and Health Management (ICPHM), June 2016
11. Anindita Roy, Amita Rathod, G.N. Kulkarni, "Challenges to diffusion of small wind turbines in India", 2nd IET Renewable Power Generation Conference, 2013, DOI: 10.1049/cp.2013.1840
12. Chenggen Wang, Qian Zhou, Mingzhe Han, Zhan'ao, Xiao Hou, Haoran Zhao, Jing Bu, "Chenggen Wang, Qian Zhou, Mingzhe Han, Zhan'ao, Xiao Hou, Haoran Zhao, Jing Bu", ACEEE- Asia Conference on Energy and Environment Engineering (2018)

Human Machine Interaction in the Digital Era – Prof. J. Dhilipan et al. (eds)
© *2024 Taylor & Francis Group, London, ISBN 978-1-032-54998-9*

Deep Learning Procedures to Detect Breast Cancer in Mammogram and MR Images and Relative Exploration of Images

31

P. Kanimozhi[1]

PhD Scholar in CIS, Annamalai University,
Tamilnadu, India

S. Sathiya[2]

Assistant Professor in CSE, Annamalai University,
Tamilnadu, India

Abstract Breast cancer is one of the deadly disorders found among female population widely world around. Cancer attacks human cells without any symptoms until it reaches advanced stage. So proper screening and diagnosing methods are needed to identify this disease in earlier stage. Several image processing techniques are in use but they have both advantages and limitations. This paper focuses two diagnosing methods currently in practice, namely Magnetic Resonance Imaging and Digital Mammogram method. For both the images, classification is done by deep learning methods like CNN model, Inception V3 model and Residual Network model and accuracy is shown highest in Mammogram images with 86.34% for Resnet50 model.

Keywords Magnetic resonance imaging, Mammogram, CNN model, Inception V3 model, Residual Network model

1. Introduction

In human body, breast cancer originates in the ducts or lobules of the breast tissues and slowly spreads to surrounding areas and adjacent lymph nodes. Throughout the world women are facing this problem for past one century. Regarding a recent survey, over 2.3 million women have the symptoms of breast cancer and more than six lakhs death occur in the world. [1]. The popular procedure in medical field used recently is the Magnetic Resonance Imaging method to produce the images of the tissues and structures inside the body [2]. They are more advantageous when compared with other imaging modalities. Mammogram is considered as the best test to diagnosis breast cancer even without symptoms. Though mammogram shows more sensitivity it is unsuitable for dense breast tissues and radiologists find hard for visual perception.

2. Related Work

According to recent analysis [3] using multi image modalities breast cancer is diagnosed using deep learning methods. Using well known data bases segmentation, feature extraction and classification steps are carried out. In another research paper Ahmet Hasim Yurttakal et al. used new method to detect cancer in MRI images with only pixel information a multi-layer CNN architecture was designed.[4]. Mahmuda Rahman et al. proposed a novel technique of automated identification of breast tumors

---

[1]khani.pappi@gmail.com, [2]sathiya.sep05@gmail.com

DOI: 10.1201/9781003428466-31

from MRI images using otsu's thresholding segmentation method and the results were compared with other existing methods [5].

3. Proposed Methodology

The main aim of this work is to distinguish lesions from tumor using both MRI and Mammogram images. The steps involved in MRI images are segmentation using Gaussian Mixture Model and the segmented images are classified as benign lesions and malignant tumor using deep learning techniques like Inception V3 model and Residual Network model. In Mammogram method the images are segmented using K-Means algorithm as normal and abnormal images. The abnormal images are then classified as benign, malignant and micro calcification using CNN model and Res net50 model. Accuracy is compared for both the images using classification models.

4. Preprocessing of Images

Both MR and Mammogram images are converted from gray scale into RGB color model for segmentation and divided into three corresponding channels. After segmentation the images are again converted into gray scale model. Mammogram images are converted from 3D to 2 dimensional for accurate segmentation and all the images are rescaled to standard size.

5. Segmentation of Images

5.1 MRI Segmentation

Segmentation of Magnetic Resonance images is done by first converting the gray scale image into RGB color model simplifying for differentiation of brighter region from the dark back ground area. Segmentation is done by machine learning unsupervised technique called Gaussian Mixture Model which is similar to K-means except that it shows mixed representation of various probability distribution. GMM segments the images into definite number of classes. The results are used for classification purpose.

5.2 Gaussian Mixture Model

The MR images are segmented using GMM method with Expectation-Maximization algorithm which is an iterative procedure for finding the maximum likelihood. GMM belongs to soft clustering method. GMM is defined as a probabilistic model with combination of individual gaussians of unknown parameters. EM algorithm automatically discover all these parameters. This method is used to cluster the data into several parts and each cluster forms a Gaussian distribution leading to analysis based on this model. Gaussian is a bell shaped curve with center as mean and width of the bell as standard deviation. The probability density function of a Gaussian distribution is defined as

$$G\ (X|\mu,\ \sigma) = 1/(\sigma\sqrt{2\Pi})\ exp - (x-\mu)2/2\sigma^2$$

Where μ and σ^2 are mean and variance of the distribution.

The unknown parameters are guessed using EM algorithm which works best for missing data problems.

5.3 Expectation-Maximization Method

EM is an algorithm for Gaussian Mixture Model which is iterative that starts from some initial estimate value and then updates iteratively until convergence is derived. Each iteration consists of E- step and M-step

E-step: Denote current parameter values and based on these values cluster the data points.

M-step: Using the data calculate new parameter values to generate best clusters.

Thus EM method is used for guaranteed convergence to local minima values where the maximum likelihood values are calculated. [6]. In the proposed segmentation method the number of clusters k is assigned as value 3 and the parameters are calculated.

5.4 Mammogram Images Segmentation

Mammogram images are segmented to acquire cancerous region using K- Means clustering method. It belongs to hard clustering type where clusters do not overlap and elements either belong to a cluster or not. K means clustering find similarity between data points to form clusters. Here K represents the number of clusters. The given image is converted for easy clustering. Digital Mammogram which is in 3D representation is converted into 2 dimensional image for clustering.

5.5 K-Means Clustering Method

The method involves three steps. They are the algorithm used, the metrics for calculating the distance and the method to find the number of clusters. Euclidean distance method and Elbow Method are used for last two steps.

Algorithm

1. Preprocess the images for segmentation
2. Find the number of clusters using Elbow method. Calculate Sum of Squared Error for the data points. Here the number of clusters is taken as 3.
3. Initialize the centroids randomly in the data points.
4. Compute the distance of every point from centroid using Euclidean distance method and partition the data into 3 groups.
5. Calculate the mean of the clusters and averaging each cluster, move the center point to the calculated position.
6. Find any points near to the new center position from the other cluster. If so move the point to the near cluster.
7. Repeat these steps again and again until all the data points are in correct cluster. That is no points change their cluster anymore and stop the process .

6. Experimental Results

The segmented MR images and Mammogram images are shown in Figs 31.1 and 31.2.

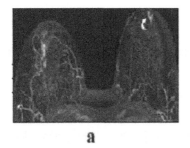 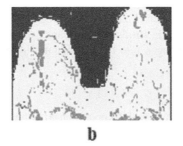 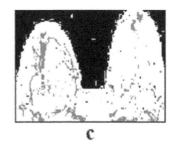

a   b   c

Fig. 31.1   MRI segmentation of cancer affected areas

Source: Made by the Authors (Primary)

Fig. 31.2   Mammogram segmentation-Malignant, Micro calcification, Benign segmentation

Source: Made by the Authors (Primary)

7. Classification

Several machine learning and deep learning methods are proposed for classification of images into numerous classes but the recent study proves deep learning approaches to be more suitable for breast cancer images. For classification of MR images and mammogram images deep learning models like CNN architecture, Inception V3 and Res net50 models are used. Res net model is used to classify both the images and accuracy is compared.

7.1 Inception V3 Model

Classification of Magnetic Resonance Images is done using two models with transfer learning method as the base. One of the model is the Inception model which is called Image recognition model with 48 layer architecture for image identification and classification. The basic architecture is based on the fact that wider network is better than deeper network and two modules are defined.

I module: Parallel filters with different sizes are used for different scaling purpose in same layer to recover all the information.

II module: Reduce the parameter number thus dimension is also condensed to avoid problem of overfitting.

7.2 Architecture of the Model

Inception model must be trained first and then used for classification purpose. For training ImageNet data set is used which can be used for classifying 1000 images. Some modifications should be used for the basic model like changing the output layer and change the size of the input image used. The basic architecture contains building blocks like

BLOCK A- Factorization into smaller convolutions.(i.e.) 5×5 convolution layer is replaced to two 3×3 convolutions.

BLOCK C- Factorization into asymmetric convolutions.(i.e.) 3×3 is replaced by 1×3 and 3×1 layers.

BLOCK B- Factorization of 7×7 convolutions. (i.e.) $7\times7 = 1\times7$ and 7×1 convolution After factorization put it in asymmetric convolution.(i.e.) replace $7x7$ to a series of $3x3$ convolution.

Classification is done as follows. A combination of convolution layers and max pooling layer is performed followed by Block A and C. Next, either Block c with Global average pooling method performed, given to Relu and Sigmoid classifer. Or Average pooling and convolution layer filter size 1×1 done then Batch normalization and sigmoid classifier.

In the proposed method the modification is done in the last layer. New layer is added to the basic architecture suitable for the data set provided. Then data set is divided into 2 classes namely benign lesion and malignant tumor.

7.3 Residual Network Model

Another CNN model using transfer learning method to classify MR images is the Residual Network model used for solving vanishing gradient problem in deeper networks. In this model, skip connections allow the gradient to be directly back propagated to earlier layers.

7.4 Basic Architecture

Resnet 50 model contains 50 layers with two important blocks as identity block and convolution block. The predefined layer 7×7 is used, in Identity block convolution function is performed with 1×1 and 3×3 filters and final values are added. In convolution block when input is not equal to output size, additional convolutional layer in shortcut path is added to the output to equal the input size. The basic architecture contains the following steps

- Input image is given to the layer followed by batch normalization and max pooling layer function is performed
- Identity block is implemented followed by convolution block. Four such operations are performed
- Finally global average pooling function is performed and dense layers are added and the final value is given to sigmoid classifier for classification of two classes
- The modification made in the proposed architecture is appending the bottom layer to match the problem of image classification for the data set. Using the basic architecture with new last layer, MRI images are classified.

8. Experimental Results

Performance analysis is done for both the methods using MR images and the accuracy is derived. Inception model shows accuracy up to 76.01% while ResNet model shows accuracy for 83.53%.

8.1 CNN Model

Mammogram images are classified using deep learning techniques for accurate result into three classes namely benign, malignant and micro calcification. The two models used for classification are CNN architecture and Res Net50 model. Both feature extraction and classification are done using CNN model

8.2 Basic Architecture

The main architecture consists of two measures namely Convolution tool that extracts the features from the image for analysis. The features are edge, blob, texture and parts. A fully connected layer predicts the class the image belongs using the features extracted. The basic steps are as follows

- CNN architecture contains 3 convolution and 3 max pooling layers. No of filter is 64 with size 4x4 used to give various features for class prediction.
- Convolution layer extract   the features from the input image and the activation function Relu adds non linearity with the dropout layer avoiding over fitting problem.
- Global Max pooling function is performed to generate onefeature map for each category and the resulting vector is fed to the dense layer (i.e.) fully connected layer1.
- In this layer ANN is added and dense layer receive whole feature vector and combine to form attributes of the image to predict the class.
- Finally in fully connected layer2 three neurons to predict 3 classes are produced.

8.3 ResNet 50 Model

Using Residual network model mammogram images are classified using the basic architecture. The images are classified into three classes as benign, malignant and micro calcification. The proposed architecture uses basic 7x7 convolution layer with identity and convolution block for classification. Batch normalization and activation functions are used. Finally dense layers and classifier are used to classify the images into three desirable classes.

8.4 Experimental Analysis

CNN model shows accuracy up to 78.32% while Resnet50 model shows accuracy up to 86.34%.with loss 0.49 and 0.40.

9. Conclusion

From the analysis done for detection of breast cancer using deep learning methods, both MRI images and mammogram images are showing better accuracy. Comparing the performance of both the methods Mammogram shows more accuracy than MRI images in classifying both the images as benign and malignant. Further enhancement can be done to improve the quality and resolution of these images to get more accurate results. This work can be extended to get optimum results in future.

References

1. Breast cancer-https://www.who.int/news-room/fact-sheets/detail/breast-cancer.
2. Medical News Today- https://www.medicalnewstoday.com/articles/146309.
3. Tariq Mahmood, Jianqiang Li, (Senior Member, Ieee), Yan Pei (Senior Member, Ieee), Faheem Akht, (2020), A Brief Survey on Breast Cancer Diagnostic With Deep Learning Schemes Using Multi-Image Modalities, *IEEE*, Volume 8.
4. Ahmet Hasim Yurttakal,Hasan Erbay, Turkan Ikizceli, Seyhan Karacavus, (2021), Detection of breast cancer via deep convolution neural networks using MRI images, *Multimedia tools and Applications* .
5. Mahmuda Rahman, Md Gulzar Hussain, Md Rashidul Hasan , Babe Sultana , Shamima Akte,(2020),Detection and Segmentation of Breast Tumor from MRI Images Using Image Processing Techniques, *researchgate.net*.
6. Malve Shravya & Nagaraja Rao P PM ,Breast Cancer Detection from Mammography Images Using GMM Method. Sri Venkatesa Perumal College of Engineering and Technology, RVS Nagar, K. N Putt.
7. Adnan M. Khan , Hesham El-Daly, Nasir M. Rajpoot, (2013), A Gamma-Gaussian Mixture Model for Detection of Mitotic Cells in Breast Cancer Histopathology, *pubmed.ncbi.nlm.nih.gov*.
8. Luana Conte, Benedetta Tafuri, Maurizio Portaluri , Alessandro Galiano , Eleonora Maggiulli and Giorgio De Nunzio, (2020), Breast Cancer Mass Detection in DCE–MRI Using Deep-Learning Features Followed by Discrimination of Infiltrative vs. In Situ Carcinoma through a Machine-Learning Approach. *Appl.Sci*, 10,6109.
9. Nesma Elsokkary, A.A. Arafa, A.H. Asadand H.A. Hefny, (2019), Computer-Aided Detection System for Breast Cancer Based on GMM and SVM, *Arab journal of Nuclear Sciences and Application,* 10.21608.
10. Abdel hafiz, D., Yang, C., Ammar, R. and Nabavi, S., (2019)**,** Deep convolutional neural networks for mammography: advances, challenges and applications., *BMC Bioinformatics*. 2019 Jun 6; 20 (Suppl 11): 281. doi: 10.1186/s12859-019-2823-4.

Human Machine Interaction in the Digital Era – Prof. J. Dhilipan et al. (eds)
© *2024 Taylor & Francis Group, London, ISBN 978-1-032-54998-9*

Usage and Adoption of Technology Among Christians During Covid-19: Based on UTAUT2 Model

32

Fr. Marvin Paul Frank[1]

Research Scholar, Department of Commerce,
CHRIST (Deemed to be University), Bangalore, India

Ginu George[2]

Assistant Professor, Department of Commerce,
CHRIST (Deemed to be University), Bangalore, India

Abstract Modern technology and devices such as smartphones, tablets, and gadgets have become integral to everyday life. These technologies are also applied in different types of businesses, education institutions, and all other service-related activities. This survey-based study used UTAUT 2 model to perceive the use and adoption of technology for a spiritual purpose among Christians. The adapted questionnaire was used to collect data and received 200 responses from Christians living in India through an online questionnaire. The study revealed that Effort Expectancy, Social Influence, and Habit Behaviour significantly impact technology use. Using technology for spiritual purposes has created a habit among Christians, leading them to use it post-COVID-19. The study found that the highest user of technology for spiritual purposes is in the age group of 20 to 39, around 66.30%. The study recommends that Church authorities highlight the importance of spiritual activities by effectively using technology to develop one's faith, helping the believers to remain connected to God through extensive use of technology by providing various spiritual activities even post-COVID-19.

Keywords Adoption, COVID-19, Christians, UTAUT2, Usage

1. Introduction

The outbreak of COVID-19 that began in China in 2019 spread worldwide within no time. The authorities of Asian countries, such as South Korea, China, Japan, and Singapore, successfully brought down the gravity of the spread. Still, European countries like the USA, France, and Germany failed to control the pandemic. COVID-19 had an extreme impact on the functioning of every aspect of society. On 25 March 2020, the nation closed down all institutions, organizations, and corporates, including Mosques, Churches, and Temples. Religious communities suffered in terms of spiritual and religious activities. Spirituality is the whole essence of every religion. One cannot speak about religion without spirituality, and spirituality cannot be expressed without the idea of religion. The spirituality of every religion plays several vital roles in people's everyday lives. The long-stretched pandemic hindered many people's faith, particularly among Christians who regularly attend the Church's obligatory spiritual exercises. This led many religious communities to move from physical practices to virtual means and modes. Some accepted the shift from a physical method of worship to a virtual manner. Still, some found it difficult, and a few believers'

[1]marvin.frank@res.christuniversity.in, [2]ginu.george@christuniversity.in

DOI: 10.1201/9781003428466-32

opinions were that they lost connection with God eventually (Sułkowski, 2020). People shifted from the physical mode of worship to the virtual way to bring continuity in their spiritual life, particularly Christians in India.

The believers and worshippers adopted modern technology to fulfill their routine spiritual activities. Campbell (2016) highlighted that linking religion with new media by emphasizing that technology in the religious setup and spiritual exercises is a constraint to expressing one's belief system. There was no other option than to shift to online mode. Further in his study, it was assumed that digitizing rituals might lead to conflict within religious communities. According to Campbell and Lovheim (2011), it was stated that the internet changes spiritual practices, communities, and ideologies and has become part and parcel of everyday life. COVID-19 has affected, and many have experienced depression, anxiety, and panic. To tackle the impact of the pandemic through spirituality, many sought their way of participating in spiritual activities. Many believers started adopting online modes like YouTube, Google Meet, Zoom, Webex, and Microsoft Teams for spiritual activities during the pandemic. Even the authorities of spiritual services, like clergy members and pastors, kept the faithful occupied with many online spiritual activities. Clergy members and pastors started recording videos and telecasting them to their flocks (Fullana et al., 2020).

1.1 Research Problem

The COVID-19 pandemic has shaken the entire world and has affected every aspect of human life. It has affected individuals physically, emotionally, and spiritually. The physical activities were shifted to virtual mode. Religious activity is one of the activities which moved to virtual mode. This shift affected every religious community and, in particular, every individual. The most affected individuals were those who were regular with the religious activities of their respective religious institutions. Temples, Mosques, Churches, and other worshipping centers were shut for a lengthy period. Few individuals went into downheartedness, and some managed to cope with themselves spiritually. Some sought spiritual benefits virtually, and others took time to accept the virtual mode of spirituality. There was internal spiritual despair among a few and external chaos due to the pandemic. Every religion underwent a kind of spiritual disorderliness. Christian religious communities gathered virtually rather than physically for obligatory worship and religious activities. Most believers and worshippers accepted the change, but few were reluctant. Therefore, this research studies technology usage and adoption among Christians during COVID-19.

2. Literature Review

2.1 Evolution of the UTAUT2 Model

The UTAUT2 Model has undergone several years to evolve fully. The Unified Theory of Acceptance and Use of Technology (UTAUT) combines or integrates eight existing theories from past studies (Venkatesh et al., 2003). The below table explains the evolution of the UTAUT2 Model.

Table 32.1 Evolution of the UTAUT2 Model

1	Theory of Reasoned Action (TRA)	This theory was developed in the social psychology field in 1980. The approach predicted individual behaviour based on behavioural intention and pre-existing attitudes.
2	Technology Acceptance Model (TAM)	TAM was developed in 1989 by Fred Davis to understand the users' acceptance and use of technology and the factors affecting them.
3	Motivational Model (MM)	Davis developed MM to study ICT adoption and use in 1992. The motivational Model concludes that one's behaviour is based on extrinsic and intrinsic motivation.
4	Theory of Planned Behaviour (TPB)	TBP is an extension of TRA by the same author in 1985. TPB was designed to explain and predict one's behaviour by considering the effects of the social system and roles of self and institutional level (Ajzen, 1991).
5	Planned Behaviour/Technology Acceptance Model (C-TPB-TAM)	C-TPB-TAM is the Technology Acceptance Model (TAM) and Theory of Planned Behavior (TPB) developed by Tylor and Todd in 1995 to explain and predict user behavioral intentions toward using technology.
6	Model of PC Utilization (MPCU)	Thomson developed the MPCU model in 1991 to study the PC Utilization behaviour of the user and measure the attitudes, social norms, habits, and expected consequences of their behaviour.
7	Innovation Diffusion Theory (IDT)	Rogers designed IDT in 1995. It highlights the conception of new ideas, processes, or practices infused into society through diffusion.
8	Social Cognitive Theory (SCT)	Albert Bandura designed SCT in 1986. According to this theory, learning occurs in a social environment through personal, behavioural, and environmental interactions.

The UTAUT model overcomes all the challenges, difficulties, and limitations of previous theories and models. The UTAUT Model of Venkatesh (2003) has four primary constructs influencing user behaviour and behavioural intention while using technologies. These primary constructs are again moderated by four key factors age, gender, experience, and voluntariness of use. Later three more constructs were added to the original UTAUT Model and named UTAUT2 in 2012. The added constructs to the original UTAUT Model were Hedonic motivation, Price Value, and Habit. (Venkatesh et al., 2012).

Table 32.2 Constructs of the UTAUT2 model (Venkatesh, 2012)

Constructs of UTAUT2	Definition
Performance Expectancy (PE)	"The degree to which using a technology will benefit consumers."
Effort Expectancy (EE)	"Measures the relationship between the hard work and the effort in position, the performance achieved from the action, and the benefit or reward received from the hard work and effort."
Social Influence (SI)	"Degree to which using the technology is appreciated in the social network important to the individual."
Facilitating Condition (FC)	"The consumers' perception of available support when using the consumer system."
Hedonic Motivation (HM)	"The fun or pleasure or entertainment derived from using a technology."
Price (PV)	"Way in which buyers react to price changes."
Habit (HBT)	"The extent to which people exhibit behaviour automatically because of learning accumulated from their experience using certain technology."
Behavioral Intention (BI)	"A measure of the strength of an individual's intention to perform a specified behaviour."
User Behaviour (UB)	"The Use Behavior measures from the actual frequency of a particular technology use"

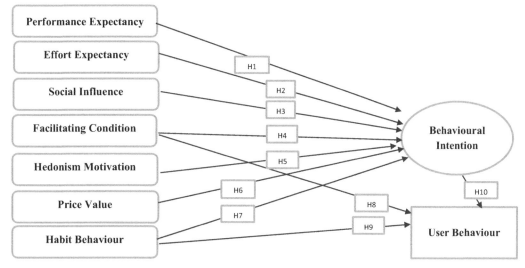

Fig. 32.1 UTAUT 2 model (Venkatesh, 2012)

2.2 Research Hypothesis

Based on the UTAUT2 Model, the research hypothesis is tested in the current study.

Table 32.3 Research hypothesis

	Research Hypothesis
H1	Performance Expectancy has a significant positive effect on the Behavioural Intention of Christians.
H2	Effort Expectancy has a significant positive effect on the Behavioural Intention of Christians.
H3	Social Influence has a significant positive effect on the Behavioural Intention of Christians.

	Research Hypothesis
H4	Facilitating Conditions have a significant positive effect on the Behavioural Intention of Christians.
H5	Hedonism Motivation has a significant positive effect on the Behavioural Intention of Christians.
H6	Price Value has a significant positive effect on the Behavioural Intention of Christians.
H7	Habit has a significant positive effect on the Behavioural Intention of Christians.
H8	Facilitating Conditions have a significant positive effect on the User Behaviour of Christians.
H9	Habit has a significant positive effect on the User Behaviour of Christians.
H10	Behaviour Intention has a significant positive impact on the Use Behaviour of Christians.

3. Research Methodology

3.1 Survey Instrument

The UTAUT2 questionnaire (Venkatesh, 2012) was adapted for the current study. The convenience method was used to collect data. The data was collected through an online questionnaire mode. The first part of the questionnaire consisted of demographic factors like age, gender, and religion. The factor religion again sub-divided into priests, religious brothers, religious sister, believers, worshippers, and other religions. The latter part of the questionnaire had queries relating to model constructs measured on a five-point Likert scale from strongly disagree to agree strongly. The study focused on the intention to use technology for spiritual activities. The sample for the quantitative survey had a total of 224 respondents. This study eliminated 24 of the responses due to non-Christian respondents.

3.2 Discussion and Implication

Demographic Description

Table 32.4 Demographic description

Factors	Percentage
1. Gender	
Males	44.20%
Females	55.8%
2. Age	
20-29	56.70%
30-39	19.60%
40-49	13.40%
50-59	07.60%
60-69	02.70%
3. Religion	
Priest	26.30%
Religious brother	7.10%
Religious sister	8.00%
Believers and worshippers	48.70%
Other religion	09.80%

The respondent's profile of our results suggested that most of the respondents were males (55.8%). The ages 20 to 29 were the highest (56.70%) of the respondents. Believers and worshippers were highest (48.70%), and Catholic priests formed 26.30% of the responses.

3.3 The Testing of Hypothesis

Table 32.5 The path analysis

			Unstandardized (B)	Standardized (β)	SE	CR.	P	Results
PE	--- >	BI	-.083	-.080	.051	-1.630	.103	Reject
EE	--- >	BI	.472	.287	.108	4.349	***	Accept
SI	--- >	BI	.213	.200	.058	3.649	***	Accept
FC	--- >	BI	-.118	-.084	.078	-1.519	.129	Reject
HM	--- >	BI	.017	.023	.037	.467	.641	Reject
PV	--- >	BI	.098	.112	.044	2.215	.027	Reject
HB	--- >	BI	.662	.804	.072	9.210	***	Accept
FC	--- >	UB	.027	.019	.064	.422	.673	Reject
HB	--- >	UB	.279	.327	.070	3.976	***	Accept
BI	--- >	UB	.687	.664	.106	6.479	***	Accept

The relationships between different variables are measured through standardized regression. In table 5, the regression coefficient shows that when independent variables go up by 1 unit, the dependent variable goes up by the unit of the respective estimate and vice versa. The above table evaluates the hypothesis for the perceived use of technology among Christians in India during COVID-19. The data results indicated that half of the hypotheses were supported among the variables evaluated. The path coefficient between EE and BI for H2 (β = 0.472, p < 0.05, C.R = 4.349); path coefficient between SE and BI for H3 (β = 0.213, p < 0.05, C.R = 3.649); path coefficient between HB and BI for H7 (β = 0.662, p < 0.05, C.R = 9.210); path coefficient between HB and UB for H9 (β = 0.279, p < 0.05, C.R = 3.976); path coefficient between BI and UB for H10 (β = 0.687, p < 0.05, C.R = 6.479). However, the insignificant path coefficients that did not support the hypothesis include PE and BI (H1); FC and BI (H4); HM and BI (H5); PV and BI (H6); FC and BI (H8). The study's findings conclude that there is a positive impact on the behavioural intention of Christians when the effort is put into learning, interacting, and using technology for a spiritual purpose. The study's findings also highlight that even the Influence of society positively impacts the behavioural intention of Christians to use technology for spiritual purposes. As the COVID-19 continued to persist for many months, the habit of using technology for spiritual purposes among Christians accelerated. The findings determine that using technology positively affects Christians' behavioural intention and user behaviour. Thus, behavioural intention has a significant positive impact on the user behaviour of Christians.

Table 32.6 Model fit with standardized estimates

Fit statistics values	Standard Values	Actual Values	Fitness
CMIN/DF	5	4.895	Acceptably fit
GFI	> 0.90	.983	Good fit
AGFA	0 to 1	.913	Accepted
IFI	0 to 1	.898	Accepted
TLI	0 to 1	.666	Good Model fit
CFI	> 0.90	.900	Good fit
RMSEA	< 0.05	.020	Accepted

Model fitness indices are typically used to evaluate the appropriateness of SEM. The Model exhibits a reasonably fit with the value of 4.895 when the Minimum discrepancy per Degree of Freedom (CMIN/DF) lesser than 5 indicates a reasonably fit.

4. Limitations and Further Research

This study is done in India, where the Christian population is around 2.4%. The respondents of the survey study were just 200, and we cannot generalize that the use of technology in spiritual activities has created a habit of using technology even

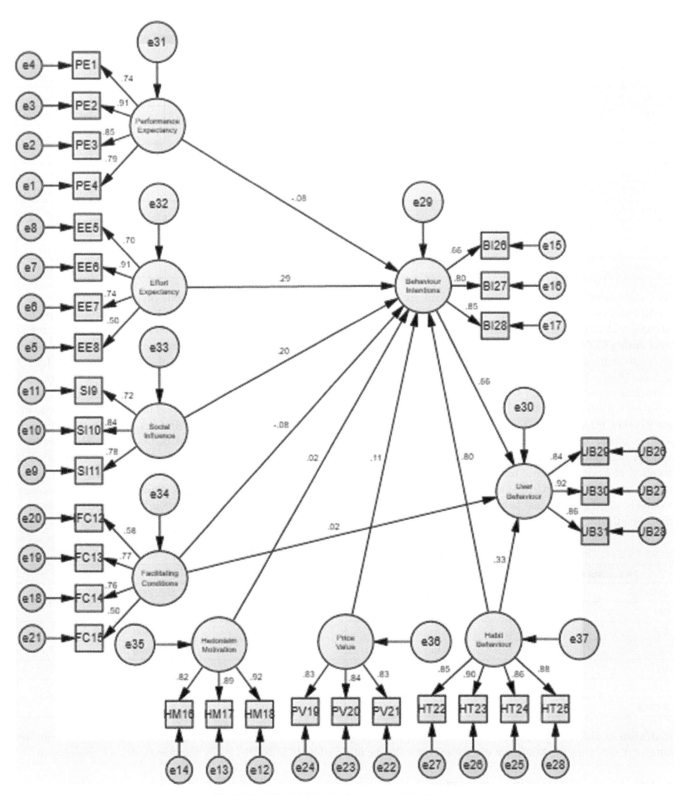

Fig. 32.2 Model fit with the standardized estimate

after COVID-19. Age, experience, and gender are the moderating variables of the UTAUT2 Model, but this study ignored the moderating variables, which can be further researched. This study keeps open for further research with a large sample size covering most of India's states. This study is limited only the Christianity. Similar analyses can be done in other religions to determine technology's impact.

5. Conclusion and Recommendations

Though COVID-19 affected human routineness, technology came as a substitute the human routineness. Even though COVID-19 pandemic disturbed routine physical and spiritual activities, humanity found technology to keep spiritual activities alive. The study's analysis and findings can conclude that the created habit of using technology for spiritual purposes can be made more effective. People who could not physically attend spiritual or religious activities can now participate virtually. The usage and adoption of technology have become an eye-opener to the officials of the Church to extend their services in physical and virtual modes. Therefore, the study recommends that Church leaders use the technology extensively. The study suggests that the Church authorities highlight the importance of spiritual activities. This study urges the Church authority to use technology effectively to develop one's faith and remain connected to God. This study also suggests conducting many more spiritual activities, community-building activities, competitions, seminars, spiritual sessions, retreats, and motivational talks from international personalities through modern technology, even post-COVID-19.

6. Acknowledgment

The author gratefully acknowledges the valuable time and advice of the research supervisor Dr. Ginu George, for her valuable time and timely suggestions.

References

1. Ajzen, I. (1991). The Theory of Planned Behavior. *Organizational Behavior and Human Decision Processes*, 50(2), 179–211. https://doi.org/10.1016/0749-5978(91)90020-t
2. Campbell, H. A. (2016). Surveying theoretical approaches within digital religion studies. *New Media & Society*, 19(1), 15–24. https://doi.org/10.1177/1461444816649912
3. Campbell, H. A., & Lövheim, M. (2011). Introduction. *Information, Communication & Society*, 14(8), 1083–1096. https://doi.org/10.1080/1369118x.2011.597416
4. Fullana, M. A., Hidalgo-Mazzei, D., Vieta, E., & Radua, J. (2020). Coping behaviors were associated with decreased anxiety and depressive symptoms during the COVID-19 pandemic and lockdown. *Journal of Affective Disorders*, 275. https://doi.org/10.1016/j.jad.2020.06.027
5. Sulkowski, L. (2020). Covid-19 Pandemic; Recession, Virtual Revolution Leading to De-globalization? *Journal of Intercultural Management*, 12(1), 1–11. https://doi.org/10.2478/joim-2020-0029
6. Venkatesh, V., Morris, M. G., Davis, G. B., & Davis, F. D. (2003). User Acceptance of Information Technology: Toward a Unified View. *MIS Quarterly*, 27(3), 425–478. https://doi.org/10.2307/30036540
7. Venkatesh, V., Thong, J. Y. L., & Xu, X. (2012). Consumer Acceptance and Use of Information Technology: Extending the Unified Theory of Acceptance and Use of Technology. *MIS Quarterly*, 36(1), 157. https://doi.org/10.2307/41410412

Note: All the figures and tables in this chapter were made by the Author.

Human Machine Interaction in the Digital Era – Prof. J. Dhilipan et al. (eds)
© 2024 Taylor & Francis Group, London, ISBN 978-1-032-54998-9

Artificial Intelligence (AI) Enabled Decision Framework for Supply Chain Resilience (SCRs)

33

Manikandan Rajagopal[1] and Sreerengan V. R. Nair[2]
Lean Operations and Systems, School of Business and Management,
CHRIST (Deemed to be University), Bengaluru, Karnataka

Ramkumar Sivasakthivel[3]
Department of Computer Science, School of Sciences,
CHRIST (Deemed to be University), Bengaluru, Karnataka

Abstract Artificial Intelligence (AI) based systems are normally data driven applications, where the model is trained to think on its own based on the external circumstances. Systems and literature of the past shows that AI based technologies are promising in intelligent supply chain management (SCM) and building resilient SCMs. There is a gap in literature which addresses on the framework for decision support systems in SCM and application of AI methods for building a robust Supply Chain Resilience (SCRs) leading to more exploration on the topic. In this paper, a decision framework is proposed by incorporating fuzzy logic and Recurrent Neural Networks (RNN) for disclosing the patterns of various AI enabled techniques for SCRs. The proposed analysis involved data from leading literatures to determine the most adoptable and significant applications of AI in SCRs. The analysis shows that techniques such as Fuzzy programming, network based algorithms and Genetic Algorithms has large impact on building SCRs. The results help in decision making by exhibiting an integrated framework which can help the AI practitioners for developing SCRs.

Keywords Artificial intelligence (AI), Supply chain management (SCM), Supply chain resilience (SCR), Fuzzy programming and multi-level decision support (MLDS) framework

1. Introduction

The world has seen a testing period on the effectiveness of research carried out in the Supply Chain management. The plans which were found effective and robust for business continuity were proven wrong.[1-3]. For a robust supply chain resilience, the companies have to adopt an effective mechanism for risk identification, risk assessment and mitigation for all possible unlikely events [4-5]. Multi-sourcing was also adopted by certain companies instead of having one source as supply. It is evident that standardization of elements for a range of products, specifically which are not vital and visible for a customer can minimize the overheads in sourcing which naturally increases the degree of SCRs. The SCM ecosystem is finding partnerships with small manufacturers and outsourced logistics which again play a vital role in resilience [6]. The potentials of Artificial Intelligence in SCM and SCRs have grabbed attention of many companies and researchers. Traditional decision making systems could not effectively address these problems which make it a mandate for coming up with potential enhancement of those particularly to

---
[1]manikandan.rajagopal@christuniversity.in, [2]sreerengan.vr@christuniversity.in, [3]ramkumar.s@christuniversity.in

DOI: 10.1201/9781003428466-33

address the complex relationships. There are only very few hybrid approaches that considers both the ability of AI and strategies of SCRs and to build a decision making framework [7-8]. This work presents a two-folded method where in first, the potential AI methods and SCRs strategies are identified from intense literature study and second to build an integrated framework for determining the application patterns of AI in SCRs. The data for the study is obtained from [9].

2. Literature Review

2.1 SCR Strategies

SCRs refer to the ability of a SCM system to adapt and efficiently handle at the time of disruptions. Some researchers have also conducted in depth study and has come out with broader strategies [10-11]. Resource balancing, Robustness and Re-inventing supply chain network are the major strategies found to be followed in Multi-Sourcing principle. It had become the choice of many of the organizations based on the location decisions. The prominent factors that affect this are the size of the firm and industry [12-15] It is evident that this is followed and taken as an important strategy for building SCRs irrespective of the demography and business domain. Data sharing, revenue sharing and agile contracts have been observed as key strategies in near-shoring strategies. Operational flexibility is the key strategy for inventory buffer which are again supported by other strategies such as IT intervention and flexibility in operations. Platform Harmonization is the mechanism of regionalizing the supply network for seamless supply within a network or a region [16-17] and achieved through. Table 33.1: lists out the top SCRs found in literature.

Table 33.1 SCRs principles and strategies

SCRs principles	SCR Indicator	SCRs strategies
Multi-Sourcing	SCR1	Resource Balancing
	SCR2	Robustness
	SCR3	Re-inventing supply chain network
Near Shoring	SCR4	Data Sharing
	SCR5	Revenue sharing
	SCR6	Agile contracts
Platform Harmonization	SCR7	Business continuity
	SCR8	Social concern
	SCR9	Dynamic revenue management
Ecosystem Partnership	SCR10	Risk Management
	SCR11	Risk mitigation
	SCR12	Knowledge transfer
Inventory Buffers	SCR13	Visibility
	SCR14	IT interventions
	SCR15	Operations flexibility

2.2 Potential AI Techniques for SCRs

Artificial Intelligence has been largely adopted in SCM and building SCRs. AI techniques are broadly classified into reactive theory, limited memory, and mind theory and self-learning. Owing to data uncertainty and ambiguous data, it becomes more difficult for the SCM systems to achieve the required degree of resilience and companies were making short focused decisions which increased their risk. From the literature, different techniques used in AI are identified and grouped in a categorical manner and presented in Table 33.2:

Table 33.2 AI categories and Techniques for SCRs

AI Category	AI Techniques
Reactive Machines	Linear Regression
	Logistic regression
	Decision Tree
Limited Memory	SVM
	NB classifier
	KNN
Mind Theory	K Means
	Random forest
	Dimension reduction
Self-Learning	Fuzzy
	Genetic Algorithm
	Neural Networks

3. AI Based Decision Framework

AI based Decision framework is proposed for evaluating various AI methods in context of the ability to be used in SCRs strategies. Figure 33.1 shows the proposed decision framework. The proposed framework has two phases (i) to evaluate the SCRs strategies and (ii) for evaluating the different AI methods adopted in respect to the capacity for supporting SCRs.

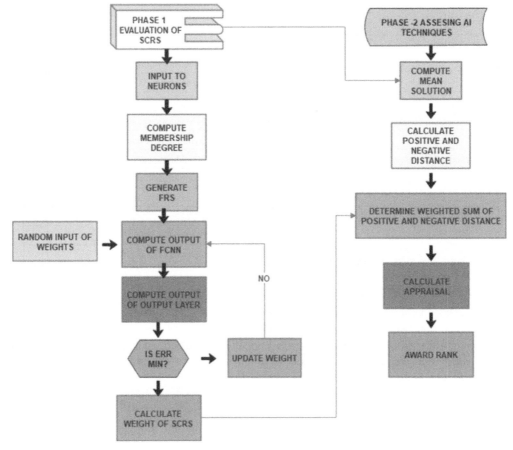

Fig. 33.1 Proposed architecture

3.1 SCRs Strategy Evaluations

The proposed phase uses Fuzzy logic driven wavelet neural network (FWNN) for calculating the weights for a given SCRs strategy. The proposed FWNN follows unsupervised learning which merges the objective function of the wavelet, the neural network and the fuzzy inference system. Fuzzy systems are largely used when there is an uncertainty in data and situation, the exact environment where SCRs are required. In order to achieve more generalization, the wavelets and neural networks are combined. This also achieves the model to learn fast and has minimal errors during approximation than the regular Artificial Neural Networks.

3.2 FWNN Implementation

The proposed FWNN has five layers; the foremost is input. The variables taken as input are put under the process of fuzzifiction in 2nd layer. Here, each of the neuron comprises of a fuzzy set. The output hence directly reflects on the value of membership to the appropriate input, which is mapped to a certain Euclidean function $\varphi j(yi)$, which can be represented as

$$\varphi j(yi) = e^{-\left(x_i - d_{ij}/e_{ij}\right)^2}, i = 1,2,\ldots\ldots, m, j = 1,2,\ldots,n \tag{1}$$

Here, *Dij* and *Eij* denote the appropriate centers and corresponding breadth of memberships function which gets updated in each step. The next layer receives the degree of membership of the corresponding input for the layers above for generating the front part of the fuzzy interface. The output of this layer is computed as

$$\varphi_j(y) = \prod_{i=1}^{n} n_i(y_i), \quad i = 1,2,\ldots\ldots,n \tag{2}$$

Evaluation of AI Aimed for determining the rank of a AI technique, the phase is based on weights arrived for SCRs strategy in the previous phase. EDAS method is followed for evaluation. This makes use of vectors coming out from FWNN and then assigns a final rank to the AI technique.

Evaluation based on EDAS

The EDAS undergo a five step process

Step 1: Computer the mean of all solution components for each criteria using the below equation

Step2: Computer the Positive and Negative distances and for all the elements present the matrix using the equation,

Step 3: compute the weighted sum of positive and negative distance from their corresponding averages

Step 4: Compute the appraisal score for each of the alternative

Step 5: Assign ranks according to the appraisal scores.

4. Result Analysis

4.1 SCRs Evaluation

The initial step to evaluate the strategies of SCRs is to introduce a fuzzy inference engine for normalizing the decision matrix and to arrive at normalized weights and then to assess the AI techniques. The weighted summation was then computed and normalized from their averages. At last the final appraisal score is calculated and the same is presented in Table 3. Overall, it is identified that Fuzzy, Genetic and Neural Network are the most predominant in SCRs and Linear and Multiple regression has the lowest rank with minimal impact on SCRs.

4.2 Sensitivity

A sensitivity study was done to analyze the robustness of the proposed model by testing for its variations. Three models (RNN, CNN and SVM) were taken for consideration and were compared with the original method for showing the proposed model is robust. Spearman's method for correlation is used for identifying the statistical difference among the ranks that are obtained from the proposed and other model taken for comparison. Figure 33.2 shows the sensitivity analysis and performances. The numbers 1, 2... 1 denote the corresponding supply chain resilience strategies discussed in section 2 and AI1, AI2, AI12 denote the corresponding AI techniques. It is seen obvious that Self Learning Algorithms have greater and likely impact in almost all of the SCR and the final impact matrix is shown in Table 33.3.

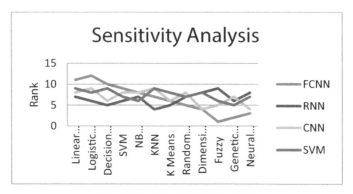

Fig. 33.2 Sensitivity Analysis

Table 33.3 Final Impact

AI category	AI Technique	Multi-Sourcing			Near Shoring			Platform Harmonization			Eco System partnership			Inventory Buffer		
		1	2	3	4	5	6	7	8	9	10	11	12	13	14	15
Reactive	AI1	0.62	0.71	0.82	0.01	0.01	0.01	0.69	0.58	0.67	0.02	0.01	0.01	0,01	0.02	0.03
	AI2	0.54	0.59	0.63	0.01	0.01	0,01	0.02	0.03	0.04	0.03	0.03	0.01	0.01	0,01	0.02
	AI3	0.62	0.71	0.82	0.01	0.01	0.01	0.69	0.58	0.67	0.02	0.01	0.01	0,01	0.02	0.03
Limited Memory	AI4	0.01	0.01	0.01	0.69	0.58	0.67	0.02	0.01	0.01	0,01	0.02	0.03	0.04	0.03	0.03
	AI5	0.03	0.02	0.03	0.05	0.64	0.47	0.01	0.01	0.01	0.02	0.03	0.04	0.05	0.06	0.07
	AI6	0.03	0.05	0.06	0.47	0.58	0.69	0.01	0.01	0,01	0.02	0.03	0.04	0.03	0.03	0.01
Mind Theory	AI7	0.01	0.03	0.04	0.05	0.05	0.06	0.47	0.45	0.41	0.01	0.03	0.04	0.05	0.05	0.06
	AI8	0.01	0.01	0,01	0.02	0.03	0.04	0.41	0.56	0.49	0.06	0.07	0.08	0.01	0.02	0.03
	AI9	0.01	0.01	0,01	0.02	0.03	0.04	0.61	0.72	0.61	0.07	0.08	0.01	0.02	0.03	0.07
Self-Learning	AI10	0.62	0.71	0.82	0.69	0.58	0.67	0.47	0.45	0.41	0.1	0.2	0.12	0.14	0.02	0.01
	AI11	0.54	0.59	0.63	0.54	0.64	0.47	0.41	0.56	0.49	0.23	0.24	0.25	0.14	0.31	0.05
	AI12	0.62	0.71	0.82	0.47	0.58	0.69	0.61	0.72	0.61	0.73	0.45	0.65	0.71	0.79	0.81

5. Conclusion

The research was aimed to find out the various patterns of AI techniques for building SCRs by providing a holistic view of the impact and promising avenues of AI in SCRs. AI techniques such as genetic algorithm and network based algorithms also have a predominant impact in building SCRs. The sensitivity analysis proves that the proposed decision making framework has the highest correlation with R reaching the value of 1. The generalization should be done more cautiously as different business domain has different values and virtues. The future scope of the research shall include the interrelationships among the AI algorithms and SCRs principle through modern techniques such as Analytics Network processing (ANP). It is concluded that AI promises to fill the gap of industry which has not got adequate SCRs implemented.

References

1. Brockman, P., French, D. and Tamm, C. (2014).REIT organizational structure, institutional ownership, and stock performance.J. Real Estate Portf. Manag. 20(1): 2136.
2. Cella, C. (2009).Institutional investors and corporate investment. United Sates: Indiana University, Kelley School of Business.
3. Chuang, H. (2020). The impacts of institutional ownership on stock returns.Empir. Econ. 58(2): 507533.
4. Clark, G. L. and Wójcik, D. (2005). Financial valuation of the German model: the negative relationship between ownership concentration and stock market returns, 1997–2001. Econ. Geogr. 81(1): 1129.
5. Dasgupta, A., Prat, A. and Verardo, M. (2011). Institutional trade persistence and long-term equity returns. J. Finance.66(2): 635653.

6. Demsetz, H. and Lehn, K. (1985). The structure of corporate ownership: causes and consequences. J. Polit. Econ. 93(6): 11551177.
7. Dyakov, T. and Wipplinger, E. (2020). Institutional ownership and future stock returns: an international perspective.Int. Rev. Finance. 20(1): 235245.
8. Gompers, P.A. and Metrick, A. (2001).Institutional investors and equity prices. Q. J. Econ. 116(1):229259.
9. Han, K.C. and Suk, D.Y. (1998). The effect of ownership structure on firm performance: Additional evidence.Rev. Financ. Econ.7(2):143155.
10. Kennedy, P. (1985). *A Guide to Econometrics*, MIT Press, Cambridge.
11. La Porta, R., Lopez-de-Silanes, F., and Shleifer, A. (1999). Corporate ownership around the world. J. Finance. 54(2): 471517.
12. Manawaduge, A. S., Zoysa, A., and Rudkin, K. M. (2009).Performance implication of ownership structure and ownership concentration: Evidence from Sri Lankan firms. Paper presented at the Performance Management Association Conference. Dunedin, New Zealand.
13. McNulty, T. and Nordberg, D. (2016). Ownership, activism and engagement: institutional investors as active owners. Corp. Gov.: Int. Rev.24(3): 346358.
14. Othman, R., Arshad, R., Ahmad, C.S. and Hamzah, N.A.A. (2010). The impact of ownership structure on stock returns, In 2010 International Conference on Science and Social Research (CSSR 2010) (pp. 217–221). IEEE.
15. Ovtcharova, G. (2003). Institutional Ownership and Long-Term Stock Returns. Working Papers Series. Available at SSRN: https://ssrn.com/abstract=410560.
16. Shleifer, A. and Vishny, R.W. (1986).Large shareholders and corporate control. J. Polit. Econ. 94(3): 461488.
17. Sikorski, D. (2011).The global financial crisis.*The Impact of the Global Financial Crisis on Emerging Financial Markets. Contemporary Studies in Economic and Financial Analysis*,ed. A. Jonathan Batten, and G PeterSzilagyi, 93:1790. United Kingdome:Bingley:Emerald Group Publishing.
18. Singh, A. and Singh, M. (2016). Cross country co-movement in equity markets after the US financial crisis: India and major economic giants. J. Indian Bus. Res. 8(2): 98121.

Note: All the figures and tables in this chapter were made by the Author.

Human Machine Interaction in the Digital Era – Prof. J. Dhilipan et al. (eds)
© 2024 Taylor & Francis Group, London, ISBN 978-1-032-54998-9

Prognosticating Mobile Application Usage Pattern with Modified Genetic Algorithm and Dynamic Density Updated Clustering Approach

34

P. Priyanga[1]

Research Scholar, Department of Computer Science,
Alagappa University, Karaikudi, Tamilnadu, India

Padmapriya Arumugam[2]

Professor, Department of Computer Science,
Alagappa University, Karaikudi. Tamilnadu, India

Abstracts In recent years, mobile applications have made crucial contributions to day-to-day lives. At present, these applications are being utilized by almost all individuals. Including social media applications, messaging, browsers, etc., on Google playstore, users can select millions of mobile applications that correspond to their devices. Therefore, detecting the individual's application usage pattern is vital for the current data marketers to comprehend the demand and needs of the users. Concurrently, ML (Machine Learning) based algorithms have gained significance in this area due to their innate capability to identify patterns easily. Existing research has endeavored to use different ML-based algorithms to determine application usage patterns. However, such studies have been ineffective in overfitting and low prediction rate. This study proposes MGA (Modified Genetic Algorithm) for selecting relevant features to attain high pattern prediction. As the conventional GA can get trapped into a local optimum with an objective function and similarity in parameter inversion, an elitist and multi-objective optimization approach is introduced to resolve the inverse issue.

Further, the research proposes MK-MA (Modified K-Means Algorithm) based on the Dynamic Density Updated Isolated Forest. In this process, the density updation is considered to solve the challenges of unstable prediction outcomes caused by the random partition of dataset features in Isolation Forest. The overall performance is comparatively evaluated by performance metrics that expose the efficacy of the proposed system in pattern recognition.

Keywords Mobile applications, Pattern usage recognition, Machine learning, Genetic algorithm, Isolated forest, Density updation

1. Introduction

Mobile phone technology has penetrated every life aspect of people. Smartphones have become essential for rapid participation and communication in online activities [1]. The count of mobile-application has progressed exponentially and is utilized daily, covering broader areas such as entertainment, finance, health care, games, e-commerce, etc. [2]. These smartphones possess various embedded sensors, including an accelerometer microphone, Bluetooth, camera, gyroscope, and GPS. The mobile phone's ubiquity and its maximizing usage have generated a larger volume of behavior-associated data gained from

[1]priyangaresearch@gmail.com, [2]padmapriya@alagappauniversity.ac.in

DOI: 10.1201/9781003428466-34

users [3]. These smartphones, both non-structured and structured information, consist of richer information and can offer insights beneficial for various categories of businesses like application developers, online retailers, and network providers [4].

The explosive progress of cellular networks and smart-devices users have witnessed in the recent decade. Cisco's white paper argued that mobile devices and global connections attained a range of eight billion in 2016, and it is forecasted to reach upto 11.60 billion by the end of 2022 [1]. Meanwhile, more traffic seems to be generated when the users interact with different mobile applications. The users should traverse the core network or backhaul, and information is fetched out and stored in the remote application servers. This, not alone, aggravated the core network framework. However, it incurs a long delay in accessing exhibited application content. It exposes that mobile application traffic loads distribute non-uniform data and differ across the time across the (BS) base stations [5]. The prevalence of Mobile-applications (Apps) introduced a prominent evolutionary difference in people's lives. In 2014, Google Play and Apple Application store hosted more than two million applications, having been downloaded by users more than a hundred billion times [6]. Different applications were also hosted through marketplaces like the Samsung app store, Amazon app store, Fetch, and F-Droid. In certain regions where these native application marketplaces were not accessible, the third-party (marketplaces) contributed to the users in determining, downloading, and managing the mobile applications. In order to aid the user in determining and exploring the high-qualified application, many application market-placed collects the application's user-input ratings in the form of numerical ratings, free-text comments, and like votes or dislike votes. More research was performed to assess those ratings as well [7].

Smart device proliferation prompts explosive mobile app usage that increases network traffic loadwork. Characterizing applications' traffic and usage patterns from individual perceptions seems valuable for content providers and operators to bring out business and technical strategies. The patterns discovery regarding how, where, and when the users interact with those mobile apps deliberated significant insights for providers of mobile services. As a consequence of this, various mobile apps were downloaded and installed on the smartphone. It seems more general that users of smartphones run and simultaneously open multiple applications.

Nevertheless, this limited memory and battery capacity also turn out to be a bottleneck in the smartphone's performance since various applications running simultaneously would occupy more resources and memory consumption unnecessarily [8, 9]. Further, the application usage and execution of more applications would maximize the reaction time and impact the mobile user experience. Hence it seems crucial to manage the usage of mobile applications efficiently, to enhance the performance of smartphones and the corresponding app investors. Meanwhile, letting the user perform this job themselves seems impossible as it would necessitate more user interventions. In this scenario, profiling the user based on app usage patterns could leverage the profiles to develop an automatic-app management approach. The appropriate application marketplaces could take out action immediately. When those user ratings were uncommon, the app-management activities usage patterns were the most robust and effective user preference indicators. Even though using single patterns, the apps quality could be accurately ranked, which is specifically valuable for high-quality and new applications.

Multi-time-aware sequential patterns [10]~,.G could be integrated with different approaches like (ML)Machine-learning algorithms to enhance the accuracy of ranking the attributes to predict the usage patterns [11]. More significantly, to save data flow and protect user privacy, it seems desirable to mine usage patterns of apps on user's smartphones rather than to transit raw context data to the back-end server and performs mining later. In this approach, the smartphone would effectively be aware of which app the user is currently using and in responding prior to user-request. Through logs of mobile-application usage, usage time series could be expediently generated. Forecasting time series of app usage is another suitable approach in judging whether an application could be used or not [12]. However, anyway, the issue of predicting the usage patterns of mobile application were not addressed by the conventional time-series methods [13], as the models were generally more time-consuming and complex to get adopted to make out real-time prediction. Hence, the need to bring out a new effective and the simple prediction model is desperately needed. However, limited smartphone resources demand companies and researchers to pay great attention to the unique approach to managing and estimating app usage patterns. The primary challenge relies on assessing how mobile-user application usage patterns can be modeled to improve performance [12]. In order to overcome the above-discussed issues in the model overfitting while in the training phase, instability in selection and prediction outcomes, lacking accurate prediction of usage patterns, the ML algorithms are implemented in the present study for designing efficient mobile app-Usage patterns prediction model proposed Modified K-Means algorithm in accordance with Dynamic Density Updated Isolation Forest. In coping with this requirement,

The major contribution of the study is elucidated below:

- Select relevant features using the proposed MGA (Modified Genetic Algorithm) by introducing elitist and multi-objective optimization approaches to avoid getting trapped into local optima.

- To perform clustering using the proposed MK-MA (Modified K-Means Algorithm) in accordance with Dynamic Density Updated Isolation Forest for solving the unstable prediction outcomes caused by the random partition of dataset features in Isolation Forest.
- To validate the proposed system by comparison with existing studies in accordance with accuracy, calinski score, and silhouette score.

1.1 Problem Identification

In the smartphone era, mobile applications were ubiquitous, with these mobile applications, communication with one another in multiple pathways like voice mails, videos, images, and text entertain people with the learning of new things, shopping, and navigating the network traffic. Since the number of smartphone applications increases, the prediction phase of the popular or more application usage by different users worldwide is becoming increasingly complex [14]. The day-to-day usage analysis of the applications, for instance, the user's behavior, can assist the application investors and makers in making process simplifications and enhancement of existing app priorities. However, it is more tedious for mobile service providers to estimate future application usage by relying on large-implicit app-usage records. Hence the investors or the app makers can invest in applications having potential, bringing out targeted advertising and offering the consumers personal app-based services. In line with the necessity for app usage prediction, three primary parameters, app-specific characteristics, user-specific priorities, and aggregate behavior depending on associated users, were considered. In fact, the mobile-relate data in handling them may have challenges characterized by their variety, velocity in the network, and larger volume (3Vs). This is prone to the complexity of the model to select the attributes of user app data and predict user data and application behavioral patterns.

In the recent period, various models were developed in modeling exogenous variance having multi-explicit features and social relationships with attributes of one another. However, in different scenarios, abundant features and explicit information, in its selection and classifying of the selected features efficiently with good accuracy, is tedious to obtain in the anonymous and large-scale dataset [15]. For instance, according to the usage records data of smartphones obtained from local operators, it became difficult to get these usage patterns of different networks. However, users' call logs [16] are retrieved as features. Another researcher Yan, applied contextual information, including temporal access patterns and location aids in the prediction of application usage-of users. However, the precise outcomes are not yielded. Besides this, Rebwar Bakhtyar Ibrahim then used trace-based assessment in predicting the future application usage data of different users in the count of many user participants [17].

1.2 Paper Organization

Section I states the introductory concepts of research and the problem statement of the research. Section II enumerates the review of literature studies and their gaps. Section III deliberated on the research method. Finally, section IV illustrates the results of the research implementation, and section V explicates the conclusive statement of the study.

2. Review of Literature

The below section enumerates the review analysis of existing approaches and methods to predict the application usage patterns. In addition, the challenges that adhered to the respective prediction model were also discussed in the section.

The usage information patterns discovered regarding when, how, and where the users interact with its mobile applications explicates significant insights for mobile service providers. One such study by Silva, A. C. Domingues exploited the large-scale and real-time dataset, for the first time, that consists of mobile application usage records, with 5343 user data. This data was gathered through a software agent installed on users' smartphones. This software agent monitors the application usage. In the first phase, the software looks out for the patterns of how the users have access to certain mobile applications to data traffic, duration, diversity, and frequency [18]. After this phase, the dataset is mined to derive the temporal patterns, like how this accessibility occurs. In the final phase, the access location to determine users' points of interest and the location-based communities occurs. Based on outcomes, the model is designed, that generate synthetic data sets with mobile apps usages. The solutions are also evaluated in the next application prediction. The group of implications of mobile advertisement, smart cities, and telecommunication services is discussed.

The marketplaces of application host million counts of mobile applications that are downloaded billions of times. To investigate the approach to managing the mobile application in day-to-day life paves a peculiar opportunity to understand mobile device-user preferences and behavior [19]. This infers the app's quality and enhances the user experience. Unfortunately, the past literature offers constrained knowledge related to activities of app management because of lacking application usage information [20]. An article by Tang, G. Huang, and F. Feng took the initiative to assess the management logs of a large application gathered using the Android application marketplace.

Likewise, the volume and variety of applications accessed by younger children towards an understanding of difficulties in using mobile apps. Given the significance of high-quality learning experiences, it is essential that application usage in schools is understood better. In line with this, the research explores applications used by real children using larger aggregated Australian datasets in primary schools [21]. The data was automatically collected from around 15 thousand android devices over the past three years. For this app usage pattern prediction, data mining techniques of association rules and clustering analysis were utilized to pick out the app usage patterns. The outcomes of the research deliberated the prominent five distinct app usage patterns. The inferences yield significant insights into multiple application-usage complexities within the classroom. The different app-usage pattern's implications associated with teaching and learning were discussed.

Health departments like the (NHS) National health-service were looking forward to utilizing digital well-being technologies like self-management health applications. Hence logging into the user interaction could permit great insights into the user necessities and might offer ideas to enhance the digital intervention, for instance, by using enhanced personalization. If this NHS promotes the log-analysis [22] and health application seems insightful, then there prevails a necessity for standardization to increase the recorded event-logs utilities for healthcare data analysis. When parent downloads applications, they possess access to different easy-to-use tools like mood tracker, location tool, symptoms checker, and keys to find the required resources closer to them [23].

In the recent period, mobile applications have been utilized widely by Smartphone users. In this context, understanding the usage of mobile applications assists in drawing out predictions upon their tendency in development. Meanwhile to this, it also enhances the experiences of users. For the prediction of usage of mobile applications, a novel technique considering two categories of a user-related network app-usage similarity-network and call-log based-network of users is developed. This model also explores the correlation between the app functions and user characteristics. The model was tested along with dataset information consisting of 25376 data users obtained from large connected components of a local-operator user in regions of chine. The outcomes revealed that this model integrated with combined-network attains the best outcomes with 60 percent of recall and precision metric). Furthermore, the model is compared to conventional methods without network data or on a single network, such as application-usage similarity and call-logs-based network [14].

Similarly, the researcher who worked on a small-scale dataset had dependency typically on the user characteristics and logs of the application usage to estimate the future application usage patterns. For instance, Secci, L. Tabourier, and B. Tebbani assessed the Markov model in predicting app-usage patterns. Specifically, the relation between smartphones and apps (referred to as explicit feature) is modeled through the usage logs gained from selected app users. The researchers constructed the delivery network comprising application temporal ordering (called implicit features). Those explicit and implicit features are used in modeling the prediction method. In this research, Liao gathered time-based information obtained from sensors installed on fifty different smartphones to predict applications' usage. However, it might be tedious to construct an explicit design, impacting the factors of time-varying usage behavior of the application[24]. To overcome this issue, a DeepAPP (Deep reinforcement-learning model) learns predictive neural networks by utilizing the application usage data. Nevertheless, the model should also overcome the data sparsity and minimize the time taken for feature prediction by a lightweight personalized agent [24].

Different studies have posed certain assessments related to the application usage patterns prediction and its data through smartphones. The researchers were distinct, from present existing researchers, in an approach of data collection from users' smartphones and assessing the individual's usage behavior on the application [25]. Another side, smartphone app usage is analyzed through anonymized-dataset from the different cellular networks and internet service providers that are more abundant and diversified. The correlation among the parameters like subscribers, access time, applications, and data volume are evaluated. This sort of assessment aids in increasing the provider's QoS (Quality-of-Service). However, the dataset utilized by the researchers consists of certain similarities. The research distinguishes other existing literature from other methods like to have focus as profiling, prediction of individual application usage, and clustering method of various mobile apps through temporal characteristics at a high fine-grained range. In the current research, the clustering data-consumption behavior method, according to what category of application users at a particular period.

The application prediction system gets the benefits of reducing its launching time and search time as the launched application could preload in memory before its usage in the market. Even though few researchers proposed an app usage analysis issue, the prediction model recommends the applications for the users based on app usage frequencies. In this research, an association between temporal and spatial user behaviors and application utilization demands might be stronger. In this research, the (STAR) Spatial and Temporal-app Recommender is proposed, and a new framework is employed for the prediction and recommendation of mobile users' applications underneath a smartphone ecosystem's environment. This STAR model comprises four primary modules. At first, the semantic and meaningful location movements from GPS geographical data are determined through the Spatial Relation-mining module. The next phase generated the appropriate temporal feature segments through the Temporal Relation-Mining module. After this segment, the (STAUP-Mine) Spatial and-Temporal Application usage-patterns mine algorithm is designed efficiently to discover the (STAUP) Spatial and Temporal-application usage-patterns of mobile users. Further, the Application usage-demand prediction phase propounded to estimate the demands of app usage by the temporal/spatial and discovered STAUP [26].

2.1 Research Gaps

One of the major limitations in the prediction phase is the utilization of the association rule method to assess the clusters of data. The association rules determine only the frequently occurring data patterns, but the small feature patterns are not visible. Further to this, association rules depend on the categories of information. As a result, the outcomes bring out the most frequently occurring data categories for every cluster. However, the model does not point out the most population application data in every cluster. Further assessment is required for this variation, and data collection phenomena ought to be improved [21].

More research must examine and enhance the usage patterns of an application using a similarity network. For instance, studies should concentrate on different feature characteristics to gain out implicit networks, like the node correlation and distribution degree, etc.. Additionally, more networks need to be learned in utilities in the characterization of aggregate impacts on users' behaviors. Since the user may differ in frequency of application usage, that is applied to construct a frequency-similarity network. The users could also be grouped through the specific utilization of gaming and shopping apps to construct a unique similarity network. However, all those networks might have sparsity and overfitting issues in their training and attribute selection phase. However, Users can also be grouped. Further, research ought to be employed with large-scale users dataset, and models need to be designed to rectify the challenges associated with time-intensive similarity score calculation in large-scale network data, dealing with clustering methods through the application-usage patterns [27].

The traffic pattern of the app usage could vary, for example, due to upgrades of version or sudden changes in popularity, trends may be prone to the complexities in determining applications usage patterns in terms of downloads, installation, user profiles, and rating attributes. In this scenario, a method must permit varying classes to match the new user traffic patterns detected. Although significant research was developed to exploit the Spatio-temporal mobile user's behaviors, the research on data usage patterns, application usage, and web access seems to be limited. Information in profiling and clustering techniques guarantees a simplified method for Wi-Fi usage. However, the efficient feature classification for ease of prediction is still lacking, but the heterogeneous clustering method yields out highly efficient prediction accuracy level [25].

3. Proposed Methodology

The research endeavors to determine the application usage patterns in Google playstore through suitable ML (Machine Learning) algorithms. Though existing research intended to accomplish this, they need to improve the prediction rate due to ineffective feature selection. In addition, conventional models also resulted in overfitting. To eliminate such negative effects, this study proposes an appropriate ML-based model that follows certain sequential processes, as depicted in Fig. 34.1.

As exposed in Fig. 34.1, the dataset is initially loaded. Following this, the pre-processing phase is performed, wherein the missing values are checked. This phase assists in minimizing ineffective prediction. Then, selecting relevant features is performed using the proposed MGA (Modified Genetic Algorithm). This is later fed into the train and test split with 80% training and 20% testing. Subsequently, clustering is performed by the proposed MK-MA (Modified K-Means Algorithm). In this process, Isolated Forest with dynamic density updation is fed into MK-MA. Finally, the pattern prediction is performed. Lastly, performance metrics are regarded for evaluating the effectiveness of the proposed system. Additionally, the overall process involved in clustering is exposed in Fig. 34.2.

As exposed in Fig. 34.2, the data points and clusters are initialized. Then, the K-Centroids are initialized. Following this, the distant object to individual centroids with the Isolated Forest approach is performed. Subsequently, the dynamic density is updated. Lastly, clustering is performed under a minimum distance to accomplish better results. The algorithms considered in this study are discussed below.

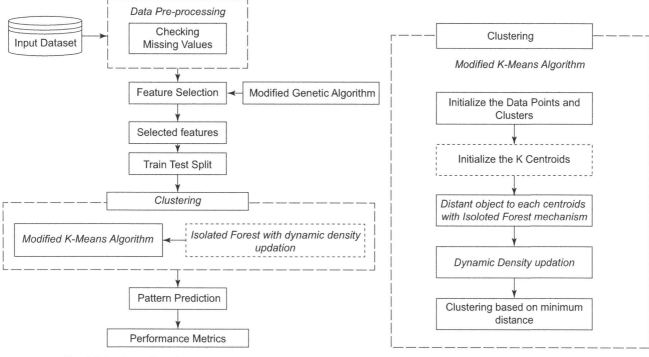

Fig. 34.1 Overall outline of the proposed system

Source: Made by Author

Fig. 34.2 Clustering process using MK-MA

Source: Made by Author

3.1 MGA (Modified Genetic Algorithm)

GA (Genetic Algorithm) is typically a methodology to resolve unconstrained and constrained optimization issues which rely on the natural-selection process which initiates biological evolution. This algorithm repetitively alters the population of distinct solutions. This algorithm possesses certain innate merits like robustness and optimization upon a huge space state. Unlike conventional AI, GA does not halt slight alterations in input. Fundamental components of GA include the representation of chromosomes, selecting fitness values, and operators with biological inspiration. Generally, chromosomes consider strings in binary form. In chromosomes, the individual locus (particular chromosome position) possesses two probable alleles (various gene forms) as 1 and 0. The chromosomes are regarded as points within the solution space. Such a process is performed through genetic operators, which is undertaken by iterative replacement of its population.

Moreover, FF (Fitness Function) is employed for assigning a value to all the chromosomes in a population. Biologically inspired operators include S (Selection), M (Mutation), and C (Crossover). In the selection phase, chromosomes are chosen based on fitness value to perform further processing. Following this, a random locus is selected in crossover operation, which alters the subsequence amongst the chromosomes for offspring creation. Finally, a few chromosome bits are flipped randomly based on the probability of the mutation operation. The corresponding process is described in Algorithm-I.

Algorithm-I: Genetic Algorithm
$\mathbf{Max_{generation}}\text{-}\to 100{,}000$
Initialize **population**;
res-$\to$ min(**population**)
for population generation-$\to 1$ to $\text{Max}_{generation}$
randomly select two parents;
offspring-$\to$ BLX_a crossover (p_1, p_2)
offspring into population
p-$\to$a random real number between 0.0 and 1.0
if $p \leq 0.2$

```
mutate (offspring);
remove the worst vector from the population;
res-→min(res,offspring);
return res
```

Despite various merits of GA, the typical GA possesses the possibility of getting trapped into local optima with an objective function and similarity in parameter inversion. This makes it complex to create appropriate composite operation as the decision taken could not circulate suitable weights of an individual target. On the other hand, existing GA possesses the ability for enhancement and employment for which an elitist and multi-objective optimization approach is introduced to resolve the inverse issue compared to fundamental single objective GA.

The objective function (f) in the equation, one is stated as the proportion of RMSE (Root Mean Square Error) and SD (Standard Deviation).

$$\text{Objective function (f)} = \frac{\text{RMSE}}{\text{SD}} = \frac{\sqrt{\frac{1}{n-1}\sum_{i=1}^{n}(P_i - R_i)^2}}{\sqrt{\frac{1}{n-1}\sum_{i=1}^{n}(R_i - R')^2}} \tag{1}$$

In equation 1, R_i indicates specific computations at various times, P_i indicates the respective model identification data, and n denotes the overall computation sets.

In this case, the overall population is partitioned into various subcategories with similar sizes, and corresponding values sorted by the individual group are computed from the respective objective functions. Few optimal individuals might be chosen as elitist individuals. These can be stored in the subsequent generation, but few must be discarded. Besides, other individuals are mixed, implemented crossover and variational operations. Random re-distribution is performed for supplementing individual sub-group in the subsequent generation. During iteration, crossover and variational operations are not undertaken on elitist individuals to avoid destruction and loss. All of the elitist individuals encompassed optimal genetic data. These are copied to individual sub-population of subsequent generations and assessed by the varied objective function. Such a process could expose the merit of multiple objective optimizations, which attains the ideal local solution approaching the ideal global solution. Comprehensive stages of the population cycle updation stage through the enhanced GA approach in such inverse problem are described below:

(i)   Partition the overall population into sub-groups ($k(sub_g)$) uniformly. Respective varied objective functions assess the individuals in each of the sub-groups.

(ii) Sort the individuals in each of the sub-group. Then, each of the sub-group is partitioned as an Elitist sub-group (E_k), Middle sub-group (M_k), and Discarded sub-group (D_k) per the respective FF.

(iii) Integrate all Elitist and middle individuals in distinct sub-groups into an Elitist group (E) and Middle group (M) and reject all the unwanted individuals.

(iv) Undertake crossover operation and variation operation upon individuals in M. Create copies (k_{copies}) of E, Redistribute them to attain a sub-group in the subsequent generation that encompasses individuals in E and certain individuals in M.

The main fact is that total individual consistency must be managed between two generations. Precisely, overall individuals have to be complemented through random individuals during the reduction in the overall amount. Additionally, certain redundant individuals within M have to be discarded when there is an increase in the overall amount. The overall process of MGA is discussed in Algorithm II,

Algorithm-II: Modified Genetic Algorithm
$Max_{generation}$-→ 100,000
Initialize population;
res-→ min(**population**)
for population generation-→ 1 to $Max_{generation}$
$Max_{generation} = k(sub_{g1}, sub_{g2}, \ldots\ldots, sub_{gk})$

sorted$_{\text{indivduals 1}}$ = S(E$_k$) → **Elitist subgroup**

sorted$_{\text{indivduals 2}}$ = S(M$_k$) → **Middle subgroup**

sorted$_{\text{indivduals 3}}$ = S(D$_k$) → **Discarded subgroup**

randomly select two parents from sorted$_{\text{indivduals}}$;

merge → S(E$_k$) – m(E$_k$)

merge → S(M$_k$) – m(M$_k$)

discard → S(D$_k$)

perform crossover and variation operations in k

k$_{\text{copies}}$(E)

k$_{\text{copies}}$(m(E$_k$ + M$_k$)

offspring -→ BLX$_a$ crossover **m**(p$_1$, p$_2$)

offspring into population

m(p)-→ a random real number between 0.0 and 1.0

if m(p) ≤ 0.2

mutate (offspring);

remove the **Discard individuals** from the population;

res-→ min(res, offspring);

return res

3.2 MK-MA (Modified K-Means Algorithm)

Generally, K-Means is an algorithm that is categorized under unsupervised learning. It is used for solving clustering issues in data science or machine learning. In this case, K claims the overall pre-defined clusters that must be developed in this process. For instance, when K = 2, it represents that there are 2 clusters, while when K = 3, it represents that there are 3 clusters. The main intention of the approach is to alleviate the overall distances between data-point and their clusters. Traditional K-means needs only certain stages. The initial stage involves the random selection of K centroids wherein K is the overall selected cluster. In this case, centroids are the data points indicating the cluster's center. The main component of this algorithm functions on two major stages termed expectation-maximization. In the expectation stage, data points are individually assigned to their neighboring centroid. Subsequently, the maximization stage calculates the mean of each point for a distinct cluster and then sets a new centroid. This study regards the K-means algorithm as it encompasses certain innate advantages like simple execution, scaling to huge datasets, guaranteeing convergence, acclimatizing to new instances, and generalizing to the clusters of varied sizes and shapes like elliptical clusters. This clustering approach aims to divide n objects into k clusters where the individual object pertains to clusters with neighbouring mean. Such methodology affords k clusters of maximum probable distinction. Optimal clusters (k) resulting in the greatest separation seem unknown and have to be calculated from corresponding data. The overall process of the K-means approach is exposed in Algorithm III.

Algorithm-III: K-Means Clustering
Begin
assign the specific number of k cluster
initialize the k centroids in a random format
repeat
expectation: assign an individual point to its nearest centroid
maximization: compute the mean (new centroid) of each other the main intention of this clustering seems to alleviate the overall intra-cluster variance or squared-error functions that are given by the objective function, $$\text{objective function}: f = \sum_{j=1}^{k}\sum_{i=1}^{n} \parallel x_i^j - c_j \parallel^2$$
f – objective function, k – number of clusters, n-number-of-cases, x-case of i, c-centroid for j cluster
Until the positions of the centroid do not alter

Initially, a specific number of clusters are assigned. Then, the centroids are initialized in a random format. Following this, expectation and maximization are performed. Subsequently, the intention of the clustering approach seems to alleviate the complete intra-cluster variance/squared error operations. The operation is performed until all the centroid operations seem to get unaltered. Moreover, this study focuses on solving the issues of unstable prediction outcomes and low efficacy caused by the random partition of dataset features in an Isolation Forest. To accomplish this, the study proposes a Modified K-Means algorithm in accordance with Dynamic Density Updated Isolation Forest.

3.3 Isolation Forest with Dynamic Density Updation

Isolation Forest typically is an unsupervised learning approach that averts the requirement for the label. This algorithm constructs a binary tree ensemble for a specific dataset. This algorithm also converges faster with minimum trees, and subsampling permits one to accomplish effective outcomes while possessing computational efficacy. The overall process of Isolation Forest and Density Updation is depicted in Algorithm IV.

Algorithm-IV: Isolation Forest
Algorithm: isolation forest (X, t, δ)
input: X– $\rightarrow$ **input** data, t– $\rightarrow$ number of trees, δ – $\rightarrow$ **sub – sampling size**
output: a set of t isolated trees
initialize the forest
set the limit l=attributes in data frame ($\log_2 δ$)
for i = 1 to t **do**
A^1– $\rightarrow$ samples (A, δ)
Forest – $\rightarrow$ F ∪ i T (A^1, 0, l)
end for
return Forest
//Density Updation
Main (A, MinPts, ϑ) ε– $\rightarrow$ ϑ
for j = 1, …… t **do**
while [cp, |CP|] = ClusterPoint[A, ϑ, It$_0$]
//corepoint-clusterpoint
|CP| ≥ MinPts, **do**
From the cluster (π) consisting of all densities – reacheable points from the point C$_p$
Add the cluster (π)to partition 2, and A = A\(π)
end while
end for
output: [2]

In the present research, the classical Isolation Forest approach is comprehensively assessed and augmented by introducing a K-Means-based Density Updated Isolation Forest approach. This method is capable of permitting the construction of a search tree with several branches that contradict with two regarded in the actual method. Accordingly, K-Means clustering involves the prediction of various divisions on individual DT (Decision Tree) nodes. The proposed system is also capable of working efficiently and permits a user to intuitively find anomaly scores corresponding to individual records for the evaluated dataset. This system's main merit is that it can fit data at the phase of DT construction. Besides, it returns intuitively appealing anomaly values. The overall process is depicted in Algorithm-V.

Algorthm-V: Isolation Forest with Density Updation
begin
algorithm: isolation forest (X, t, δ) φ – – $\rightarrow$ δ
input: A – $\rightarrow$ **input** data, n$_t$ – $\rightarrow$ number of trees, δ – $\rightarrow$ **sub-sampling size** X $\rightarrow$ A, t $\rightarrow$ n$_t$

output: a set of n_t isolated trees(ist)

initialize the forest

set the limit l = attributes in dataframe ($\log_2 \boldsymbol{\delta}$)

for i = 1 to t **do**

$X^1 - \rightarrow$ samples $(X, \boldsymbol{\delta})$

Forest $- \rightarrow F \cup i\ T (A^1, 0, l)$

density updation in an isolated forest of tree formation

while [cp, |CP|] = CorePoint[A, ϑ, It$_0$]

|CP| $\geq$ MinPts, do

$F \cup i\ T (A^1, 0, l) + A(\pi)$

end for

return Forest with MinPts

assign the specific number of k of the cluster to return the forest

k=5

randomly initialize k centroids.

repeat

expectation: assign each to its closet centroid

maximization: compute the new centroid (mean)of each other

objective function: $\mathbf{f} = F \cup i\ TA\left(A^1, 0, 1\right) + \sum_{j=1}^{k}\sum_{i=1}^{n} \| \mathbf{x}_i^j - \mathbf{c}_j \|^2$

end while

end for

end

4. Results and Discussion

The outcomes procured through the execution of the proposed system are discussed in this section with a description of the considered dataset, performance metrics, exploratory data analysis, empirical outcomes, and performance analysis results.

4.1 Dataset Description

The study is assessed by considering the applications of google playstore that are available on kaggle [28]. The considered dataset comprised nearly 10,840 applications with reviews and rankings. It possesses several variables like app name, id, category, rating, reviews, number of installations, application size, kind of application, rating of content, genres, updated version, current version, and final updated version.

4.2 Performance Metrics

The performance of the study is evaluated through significant metrics, which are discussed in this section.

Accuracy

It is claimed as the calculation of overall correct classification performance and is given by equation 2.

$$\text{Accuracy}\left(A\right) = \frac{\text{True}_{\text{Negative}} + \text{True}_{\text{Positive}}}{\text{True}_{\text{Positive}} + \text{True}_{\text{Negative}} + \text{False}_{\text{Negative}} + \text{False}_{\text{Positive}}} \tag{2}$$

Silhouette Score

Silhouette score or silhouette coefficient is defined as a metric for computing the efficacy of the clustering approach. The value ranges between -1 and 1.

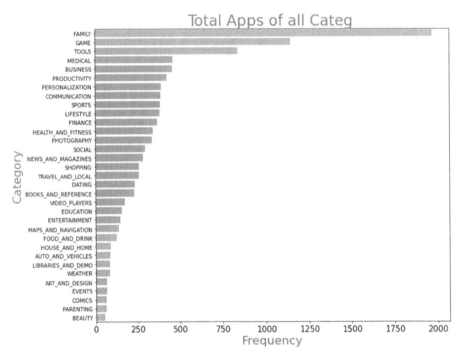

Fig. 34.3 Total applications of all categories

Source: Made by Author

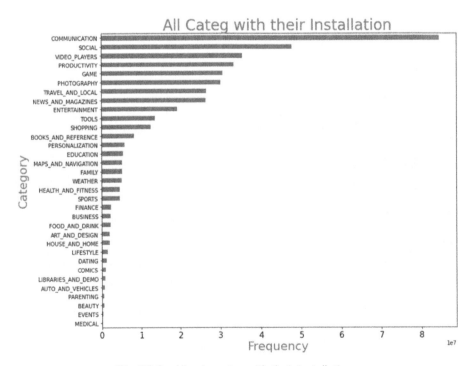

Fig. 34.4 All categories with their installations

Source: Made by Author

Calinski Score

Calinski score, also termed variance ratio criteria, is computed as the proportion of the totality of intra-cluster dispersion and the totality of inter-cluster dispersion for all the clusters. In this case, dispersion involves the totality of the squared distances.

4.3 Exploratory Data Analysis

Exploratory data analysis is a critical process involved in initial evaluations of the data for discovering patterns, spotting anomalies, and validating the assumptions with the assistance of graphical representations and summary statistics. Initially, the total applications corresponding to all category is exposed in Fig. 34.3, while all the application categories with their corresponding installations are exposed in Fig. 34.4.

Following this, the paid applications are visualized in Fig. 34.5. In contrast, all the application categories with their corresponding ratings are exposed in Fig. 34.6. In contrast, the installations and ratings of the applications in Google playstore are explored in Fig. 34.7.

Moreover, a correlation matrix is attained, typically a table that exposes the correlation coefficient for diverse variables. In this process, the matrix denotes the correlation amongst all the probable value pairs. It is generally a robust tool for summarizing large datasets and determining and visualizing the patterns for specific data. These metrics measure if there exists any relationship between the two variables. The correlation graph is explored in Fig. 34.8.

As exposed in Fig. 34.8, the correlation between variables is perceived. In this case, rating, reviews, price, size, and installations are variables. A better correlation is found for the considered variables, with one as correlating value.

4.4 Experimental Results: Pattern Prediction Analysis

The results attained through the execution of the proposed system are exposed in Fig. 34.9. In this case, Fig. 34.9a explores the clustering results of Modified K-Means, which shows 2 clusters, similarly 3 clusters are found in Fig.34.9b, while 4 clusters are explored in Fig. 34.9c and Fig. 34.9d, whereas, 5 clusters are exposed in Fig. 34.9e, finally, a single cluster is explored in Fig. 34.9f.

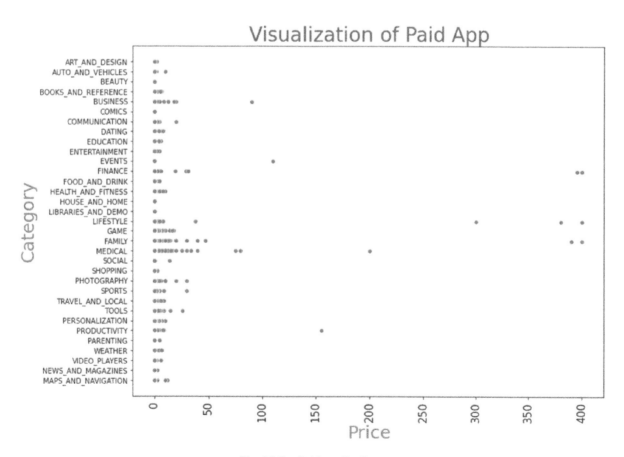

Fig. 34.5   Paid applications

Source: Made by Author

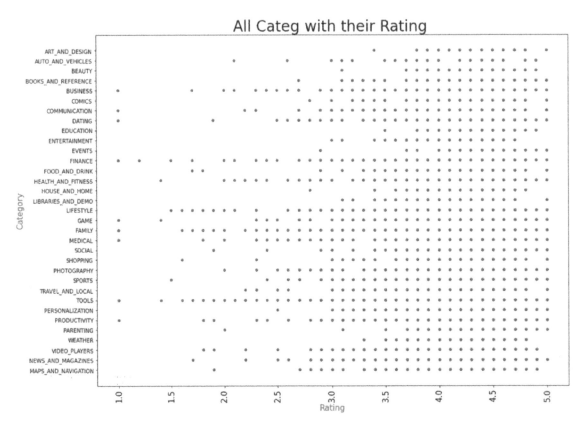

Fig. 34.6 All application categories and their ratings

Source: Made by Author

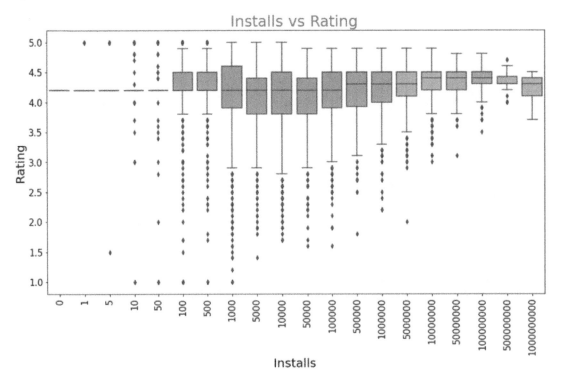

Fig. 34.7 Installations and ratings of the applications in Google playstore

Source: Made by Author

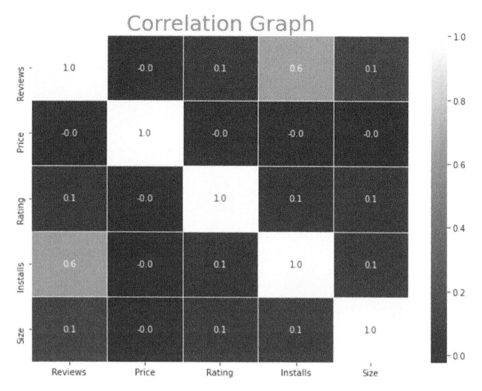

Fig. 34.8 Correlation matrix

Source: Made by Author

4.5 Performance Analysis

The Calinksi and Silhouette scores have assessed the performance of the proposed system. The corresponding outcomes are exposed in Fig. 34.10 and Fig. 34.11.

Figure 34.10 shows that the overall Calinksi score of the proposed system is 0.94%. On the other hand, from Fig. 34.11, it is found that the Silhouette score of this system is 0.94, and this effective rate is attained when the overall clusters are 3. Hence, it is clear that, among different clustering results, better outcomes are attained when 3 clusters are formed.

4.6 Comparative Analysis

The existing system has been comparatively assessed for accuracy, and the corresponding outcomes are discussed in this section. At first, the proposed system was comparatively evaluated with conventional algorithms, namely RF (Random Forest), SVR (Support Vector Regression), LR (Linear Regression), K-Means clustering, and K-NN (K-Nearest Neighbour). The outcomes are exposed in Table 34.1, with its equivalent graphical representation in Fig. 34.12.

From Table 34.1 and Fig. 34.12 show that existing approaches like RF have exposed 73.55% accuracy, SVR has explored 76.49%, LR has exposed 72.45%, K-Means Clustering has exposed 69.56%, and KNN has revealed 92.22%. In contrast, the proposed system has exposed 94% accuracy. Hence, it is found that the proposed system has shown better performance than conventional algorithms with 94% accuracy. In addition, the comparison has been performed with existing methods [30] and the proposed system. The corresponding outcomes are exposed in Table 34.2, with its graphical representation in Fig. 34.13.

From Table 34.2 and Fig. 34.13 show that the existing method has explored 0.9 as accuracy, while the proposed system has revealed 0.94 as accuracy. Hence, it is clear that the proposed system has exposed better and outstanding performance than conventional works. The proposed MGA possesses innate advantages like optimization and robustness. At the same time, MK-MA resolves the problems of unstable prediction outcomes and low effectiveness caused by the random partition of dataset features in Isolation Forest with Dynamic Density Updated Isolation Forest. These advantages have made the proposed system expose optimal outcomes confirmed through the results.

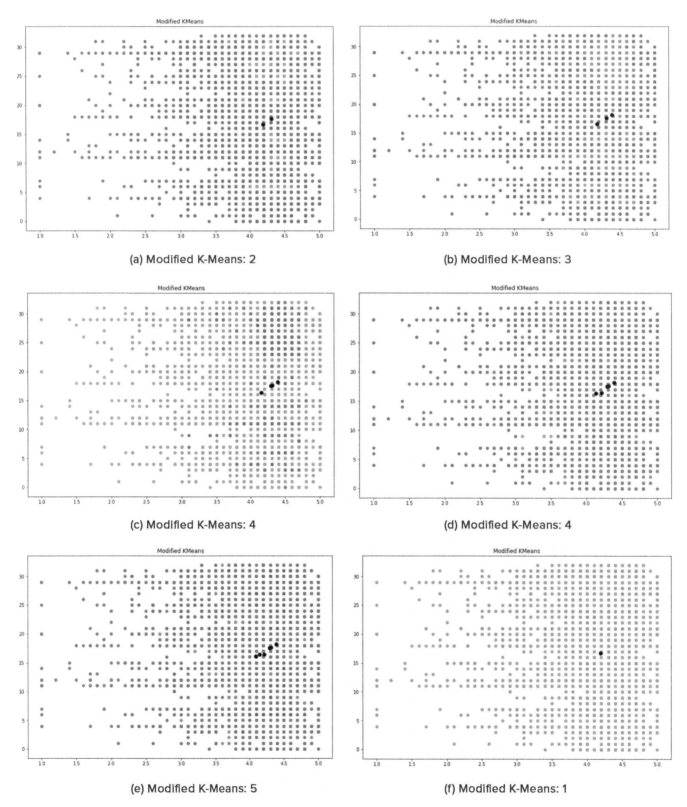

Fig. 34.9 Experimental clustering results

Source: Made by Author

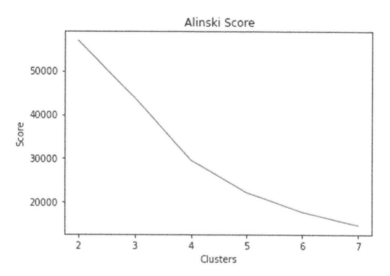

Fig. 34.10   Calinksi score (Accuracy = 0.94%)

Source: Made by Author

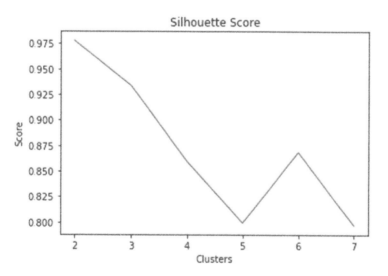

Fig. 34.11   Silhouette Score (Total clusters:3, optimal Silhouette Score during clustering=0.94)

Source: Made by Author (Values Got from - https://www.kaggle.com/code/chiragsaini97/google-playstore-apps-dataset.)

Table 34.1   Evaluation of outcomes about accuracy [29]

Algorithm	Accuracy
Random Forest	73.55%
SVR	76.49%
Linear Regression	72.45%
K- Nearest Neighbor	92.22%
K-Means Clustering	69.56%
Proposed algorithm	94.00%

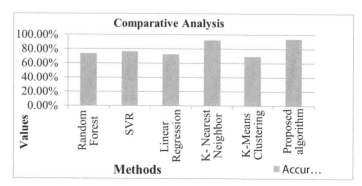

Fig. 34.12   Comparative analysis of the proposed and conventional study [29] about the accuracy

Table 34.2 Evaluation of accuracy [30]

Methods	Accuracy
Existing methods	0.9
Proposed Method	0.94

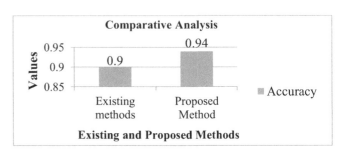

Fig. 34.13 Comparative analysis of the proposed and conventional study [30] about the accuracy

5. Conclusion

The present study intended to determine the Google playstore application usage patterns by clustering using suitable ML-based algorithms. MGA was proposed for feature selection, while MK-MA was employed for clustering. An exploratory data analysis was undertaken to expose the patterns and to assess the assumptions through graphical depiction and summary. In addition, a correlation matrix was attained for the proposed system to explore the correlations amongst the considered variables (price, reviews, rating, size, and installations). Finally, the system's overall performance was assessed in accuracy, silhouette score and calinski score. The results showed that the calinksi score was exposed to be 0.94, while the silhouette score was revealed to be 0.94 when 3 clusters were formed.

Further, to confirm the effectiveness of the proposed system, a comparison was undertaken with two existing researches. Comparative evaluations revealed the outstanding performance of the proposed system with 94% accuracy. In the future, other hybrid ML algorithms can be regarded for enhancing the accuracy rate.

References

1. I. H. Sarker, M. M. Hoque, M. Uddin, and T. Alsanoosy, "Mobile data science and intelligent apps: concepts, AI-based modeling, and research directions," *Mobile Networks and Applications,* vol. 26, pp. 285–303, 2021.
2. L. Cruz and R. Abreu, "Catalog of energy patterns for mobile applications," *Empirical Software Engineering,* vol. 24, pp. 2209–2235, 2019.
3. R. Xu, R. M. Frey, E. Fleisch, and A. Ilic, "Understanding the impact of personality traits on mobile app adoption–Insights from a large-scale field study," *Computers in Human Behavior,* vol. 62, pp. 244–256, 2016.
4. X. Lu, B. Rai, Y. Zhong, and Y. Li, "Cluster-based smartphone predictive analytics for application usage and next location prediction," *International Journal of Business Intelligence Research (IJBIR),* vol. 9, pp. 64–80, 2018.
5. M. Zeng, T.-H. Lin, M. Chen, H. Yan, J. Huang, J. Wu, *et al.,* "Temporal-spatial mobile application usage understanding and popularity prediction for edge caching," *IEEE Wireless Communications,* vol. 25, pp. 36–42, 2018.
6. H. Cho, D. Ippolito, and Y. W. Yu, "Contact tracing mobile apps for COVID-19: Privacy considerations and related trade-offs," *arXiv preprint arXiv:2003.11511,* 2020.
7. J. Feng, X. Chen, R. Gao, M. Zeng, and Y. Li, "Deeptp: An end-to-end neural network for mobile cellular traffic prediction," *IEEE Network,* vol. 32, pp. 108–115, 2018.
8. W.-T. Wang, W.-M. Ou, and W.-Y. Chen, "The impact of inertia and user satisfaction on the continuance intentions to use mobile communication applications: A mobile service quality perspective," *International Journal of Information Management,* vol. 44, pp. 178–193, 2019.
9. J. Cui, Y. Zhang, Z. Cai, A. Liu, and Y. Li, "Securing display path for security-sensitive applications on mobile devices," *Computers, Materials and Continua,* vol. 55, p. 17, 2018.
10. D. Nguyen, W. Luo, T. D. Nguyen, S. Venkatesh, and D. Phung, "Sqn2vec: Learning sequence representation via sequential patterns with a gap constraint," in *Joint European Conference on Machine Learning and Knowledge Discovery in Databases,* 2018, pp. 569–584.
11. M. Gan and C. Tan, "Mining multiple sequential patterns through multi-graph representation for next point-of-interest recommendation," *World Wide Web,* pp. 1–26, 2022.
12. M. Cai, M. Pipattanasomporn, and S. Rahman, "Day-ahead building-level load forecasts using deep learning vs. traditional time-series techniques," *Applied Energy,* vol. 236, pp. 1078–1088, 2019.

13. C.-L. Liu, W.-H. Hsaio, and Y.-C. Tu, "Time series classification with the multivariate convolutional neural network," *IEEE Transactions on Industrial Electronics,* vol. 66, pp. 4788–4797, 2018.

14. Y. Jiang, X. Du, and T. Jin, "Using combined network information to predict mobile application usage," *Physica A: Statistical Mechanics and its Applications,* vol. 515, pp. 430–439, 2019.

15. D. Yu, Y. Li, F. Xu, P. Zhang, and V. Kostakos, "Smartphone app usage prediction using points of interest," *Proceedings of the ACM on Interactive, Mobile, Wearable and Ubiquitous Technologies,* vol. 1, pp. 1–21, 2018.

16. A. A. Mubarak, H. Cao, and W. Zhang, "Prediction of students' early dropout based on their interaction logs in an online learning environment," *Interactive Learning Environments,* vol. 30, pp. 1414–1433, 2022.

17. P. Tak and S. Panwar, "Using UTAUT 2 model to predict mobile app based shopping: evidences from India," *Journal of Indian Business Research,* 2017.

18. F. A. Silva, A. C. Domingues, and T. R. B. Silva, "Discovering mobile application usage patterns from a large-scale dataset," *ACM Transactions on Knowledge Discovery from Data (TKDD),* vol. 12, pp. 1–36, 2018.

19. X. Liu, W. Ai, H. Li, J. Tang, G. Huang, F. Feng, *et al.*, "Deriving user preferences of mobile apps from their management activities," *ACM Transactions on Information Systems (TOIS),* vol. 35, pp. 1–32, 2017.

20. A. Dubov, L. Fraenkel, Z. Goldstein, H. Arroyo, D. McKellar, and S. Shoptaw, "Development of a smartphone app to predict and improve the rates of suicidal ideation among transgender persons (TransLife): Qualitative study," *Journal of medical internet research,* vol. 23, p. e24023, 2021.

21. S. K. Howard, J. Yang, J. Ma, K. Maton, and E. Rennie, "App clusters: Exploring patterns of multiple app use in primary learning contexts," *Computers & Education,* vol. 127, pp. 154–164, 2018.

22. D. Zhang, C. Yuntian, and M. Jin, "Synthetic well logs generation via Recurrent Neural Networks," *Petroleum Exploration and Development,* vol. 45, pp. 629–639, 2018.

23. R. Bond, A. Moorhead, M. Mulvenna, S. O'Neill, C. Potts, and N. Murphy, "Exploring temporal behavior of app users completing ecological momentary assessments using mental health scales and mood logs," *Behaviour & Information Technology,* vol. 38, pp. 1016–1027, 2019.

24. J. Wang, L. Zhao, J. Liu, and N. Kato, "Smart resource allocation for mobile edge computing: A deep reinforcement learning approach," *IEEE Transactions on emerging topics in computing,* vol. 9, pp. 1529–1541, 2019.

25. K.-W. Lim, S. Secci, L. Tabourier, and B. Tebbani, "Characterizing and predicting mobile application usage," *Computer Communications,* vol. 95, pp. 82–94, 2016.

26. E. H.-C. Lu and Y.-W. Yang, "Mining mobile application usage pattern for demand prediction by considering spatial and temporal relations," *GeoInformatica,* vol. 22, pp. 693–721, 2018.

27. H. Cao and M. Lin, "Mining smartphone data for app usage prediction and recommendations: A survey," *Pervasive and Mobile Computing,* vol. 37, pp. 1–22, 2017.

28. Dataset, "Google playstore apps dataset," 2019.

29. B. N. S Shashank1, "Google Play Store Apps- Data Analysis and Ratings Prediction," *International Research Journal of Engineering and Technology (IRJET),* vol. 7, pp. pp. 265–274, 2020.

30. A. Karim, A. Azhari, M. Alruily, H. Aldabbas, S. B. Belhaouri, and A. A. Qureshi, "Classification of Google Play Store Application Reviews Using Machine Learning," 2020.

Human Machine Interaction in the Digital Era – Prof. J. Dhilipan et al. (eds)
© 2024 Taylor & Francis Group, London, ISBN 978-1-032-54998-9

Differential Privacy for Set Valued Data Publishing to Avoid Illegal Attack

35

T. Puhazhendhi[1]

Research Scholar,Department of MCA,
Bharath Institute of Higher Education and Research (BIHER), Chennai

K. Rajakumari[2]

Associate Professor,Department of CS,
Bharath Institute of Higher Education and Research (BIHER), Chennai

Abstract Cloud computing is the most significant technology in the current dynamic area in information technology. It is known to be flexible, extensible and cost effectual in information technology operations which have made the media and the analysts to look around towards it. As well as it affords various approaches that manipulates the extensive range of organizational requirements by offering convincingly scalable servers and various applications. Cloud computing also establishes a new paradigm, and facilitates use of information storage for its users to with ability to create dynamic applications which can be accessed from anywhere by the internet. Cloud utilizes different storage services where data is distantly maintained, overseen and backed up. This facility can concede the end users to stores their data online. So that users can access the data from anywhere and any location using the internet. The benefits of using cloud storage are usability, accessibility, bandwidth disaster recovery, cost saving.

Utilizing a cloud environment raises several problems, chief among them being security and privacy. A third party provides the cloud's infrastructure control and data storage. Risk of sensitive data being given to cloud service provider can occur occasionally. Because of this, users always need useful, secure data storage. Here, the privacy and security of cloud data storage are primarily the focus. To protect the data, many algorithms and techniques are used. One of the most exciting aspects of data security technology is encryption. When if any algorithm is used to protecting data during transfer from one user to another, the data content is translated into a seemingly meaningless ciphertext, it is unreadable form.

Keywords Cloud computing, Privacy, Illegal attack, Usability, Accessibility

1. Introduction to Cloud Computing Model

Modern cloud computing with its emerging methodology provides benefits to both public and private organizations with cost saving computations and flexible services through Cloud Service Provider (CSP). CSP has to ensure confidentiality, security and integrity of outsourced data storage using a suitable following access policy (Pasupuleti et al. [113]). Cloud computing technology is a leading improved technology which provides paid access to software resources and hardware resources. Cloud computing handles enormous flexible information technology facilitates ability to deliver the service using internet resources to extrinsic users. Figure 1.1 displays a common cloud computing model that performs sharing services through the internet.

[1]drpugalbds@gmail.com, [2]rajakumari.mca@bharathuniv.ac.in

DOI: 10.1201/9781003428466-35

Proliferate growth of cloud benefits, through availability of the resources, provides automated tools, the configuration of data on demand, thereby overpowering the existing traditional methods. Enterprises can quickly deploy and use cloud services on a rental basis. It also provides fast network access on at cost with more scalability. Protection against network attacks, security controls and recovery from the disaster are some additional benefits of cloud services (Singh and Shrivastava [8]). CLOUD COMPUTING DEPLOYMENT MODELS

2. Application of Cryptography in Cloud Computing

Cryptography is a technique used for the transfer of data securely from a sender to legal recipients (Jaber and Zolkipli [10]). Cloud computing model allows accessing its resource from the remote location. It is essential to the business organization to secure its data when moved to cloud cost effectively. The three critical security goals are confidentiality, integrity and availability. Cryptography is the most used area to achieve confidentiality and integrity in cloud model (Singh and Gilhotra [9]).

The three cryptographic techniques to protect the data are

- Symmetric key algorithm.
- Asymmetric key algorithm.
- Hashing.

2.1 Cryptography Techniques

Some of the symmetric algorithms with shared encryption and decryption keys are DES, 3DES, RC5, RC6, Blowfish, and AES. An asymmetric algorithm uses a public key that is made public but a secret private key. Examples of algorithms are RSA and ECC. The private key is only communicated to authorized users, which is the primary benefit of public key cryptography. An older technique called Public Key Infrastructure (PKI) was employed in cloud computing and other grid environments. It is based on short-term public keys and proxy certificates with strong credentials that were later shown to be unsuitable for a dynamic environment (Yin et al. [127]).

Key creation is one of the cryptography algorithm's primary three phases.

- Encryption.
- Decryption.

3. Related Work

RSA is one of the favourite cryptographic techniques used for data security (Eberle et al. [35]). Elliptical Curve Cryptography (ECC) and Proxy Re- Encryption (PRE) are the other supporting techniques used for providing data security. Some of the works related to data security in the cloud are discussed in this section.

Hongbing et al. [13] have attempted to meet the challenges such as non- interactivity, collusion attack, non-transitivity, uni-direction access and multi- tenancy. The authors have improved the work of Canetti and Rosenberger [96] in which PRE based on the bilinear group is used. Boneh and Franklin [19] used the identity-based encryption scheme (IBE) which is a fully functional based scheme. It is based on Diffie-Hellman on elliptic curves on a natural analog.

4. Methodology

4.1 Rivest-Shamir-Adleman (RSA) Crypto Algorithm

Rivest et al. [101], Milanov [27] proposed public key cryptosystem for securing data over the internet. This asymmetric algorithm has been proposed by Ronald Rivest, Adi Shamir and Leonard Adleman in 1977. It uses public and private keys for encrypting and decrypting the data. This integer factorization one-way function is used in the distribution of key and digital signature process. It is trouble-free to compute but tough to do the inverse of the process. These terms decide the computation complexity. In the key generation step, each user generates public and the private key pairs by choosing two large prime numbers in random. Euler's theorem and square and multiply algorithm of exponentiation are used. The result is obtained on repeated squaring on the base number, and exponents are multiplied. The main steps involved in RSA are:

- Random prime number selection.

- Calculate the modulus of keys.
- Euler's function computation.
- Compute private key.
- Encrypt message.
- Broadcast the public key and transfer the message.
- Send private key to authorized client very much safely.

In the RSA algorithm in attacks such as brute force key attack which is done by testing all possible chance of private keys, mathematical attack of factoring out all equivalent product of two prime and timing offense during the decryption time are resolved using regular exponentiation and by adding delays in random. The RSA algorithm has the following advantage and disadvantage.

4.2 ECC Key Generation

Key generation plays a vital role in the generation of public and private keys. For encrypting the data content need receiver's public key and for decrypting need private key.

Here, let us choose a number **'a'** in the range of **'z'**.

Generation of the public key needs the use of the following equation

$$P = a * C$$

a = The random number that we have selected within the range of (**1 to z-1**). **C** is the point on the curve.

'P' is the public key and **'a' is the private key.**

Encryption

For representing data information on the curve, let 't' be the text content for sending. Consider *'m'* has the point *'M'* on the curve *'E'*. Randomly select 'r' from [1 – (n-1)]. Based on the process 2 cipher texts as generated let it be **CT1** and **CT2**.

$$CT1 = r*C \quad CT2 = M + r*C$$

CT1 and CT2 will be send.

Decryption

The message 't' that was sent has to be got back

$$M = CT2 - a * CT1$$

M is the original message to be sent.

5. Results and Discussion

The experiment has been implemented using Net beans IDE and cloud environment through use of the Clouds tool which offers modelling, simulation, cloud infrastructures and cloud services. The performance metrics used are throughput and execution time of encryption and decryption action. The proposed algorithm is compared with Identity Based Encryption (IBE).

5.1 Encryption and Decryption Process Time Evaluation

Table 35.1 shows the comparative analysis with the proposed and IBE algorithm. Figure 35.2 and Fig. 35.3 shows Encryption and Decryption time of IBE method respectively whereas the hybrid encryption and decryption shows better effective and efficient results.

Table 35.1 Comparative analysis with the proposed and IBE algorithm

S. No	Algorithm	Encryption time	Encryption time	Encryption time
1	Hybrid Encryption Method (RSA with ECC)	1.5	1.0	100
2	IBE algorithm	2.0	1.7	1050

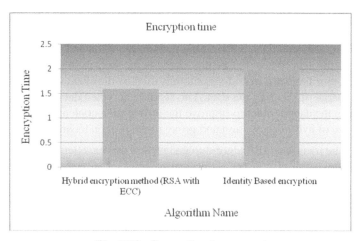

Fig. 35.1 Encryption time report

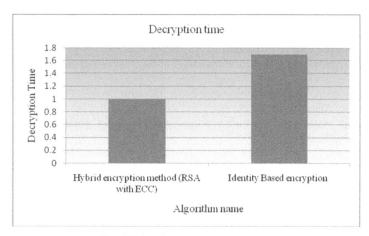

Fig. 35.2 Decryption time report

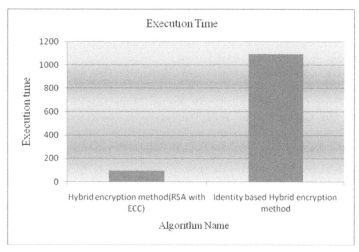

Fig. 35.3 Execution time report

5.2 Execution Time Evaluation

Figure 35.3 shows the execution time of proposed work and IBE. According to the evaluation, the execution time of hybrid algorithm is very good and decidedly less than the IBE system.

6. Conclusion

To achieve security for the cloud-outsourced data, a hybrid combination of RSA and ECC using the PRE algorithm has been proposed. The technique encrypts the data content before it is stored in the storage so that the data owner has complete control over how the data is protected. In this method, a hybrid algorithm is used by the client sender to encrypt the data payload. Both the sender and the CSP are in charge of securing the data. The sender encrypted data is combined with proxy re-encryption technique and used for encrypting the data in order to improve the information content of data security. Strongly transmission of the data is achieved through use of hybrid encryption and proxy re-encryption techniques.

The results are discussed among the traditional IBE algorithm and proved that excellent efficiency concerning encryption time, decryption time and execution time.

References

1. A. A. Falasi and M. A. Serhani, A Framework for SLA-based Cloud Services Verification and Composition, IEEE International Conference on Innovations in Information Technology, (2011), 287–292.
2. A. Gholami and E. Laure, Big Data Security and Privacy Issues in the Cloud, International Journal of Network Security and Its Applications, 8(1) (2016), 59–79.
3. A. Gholami and E. Laure, Security and Privacy of Sensitive Data in Cloud Computing: A Survey of Recent Developments, Computer Science and Information Technology, (2015), 131–150.
4. A. Ivan and Y. Dodis, Proxy Cryptography Revisited, Proceedings of the Symposium on Network and Distributed System Security, 2003.
5. A. Seccombe, A. Hutton, A. Meisel, A. Windel, A. Mohammed and A. Licciardi, Security Guidance for Critical Areas of Focus in Cloud Computing, v2.1. Cloud Security Alliance, (2009), 25.
6. A. Shamir, Identity-Based Cryptosystems and Signature Schemes, in Advances in Cryptology - Proceedings of CRYPTO'84, Lecture Notes of Computer Science (LNCS), 196 (2013), 47–53.
7. A. Sharif, S. Cooney, S. Gong and D. Vitek, Current Security Threats and Prevention Measures Relating to Cloud Services, Hadoop Concurrent Processing, and Big Data, IEEE International Conferenceon Big Data, (2015), 1865–1870.
8. A. Singh and M. Shrivastava, Overview of Attacks on Cloud Computing, International Journal of Engineering and Innovative Technololgy, 1(4) (2012), 321–323.
9. A. Singh and R. Gilhotra, Data Security using Private Key Encryption System based in Arithmetic Coding, International Journal of Network Security and its Applications, 3(3)(2011), 58–67
10. A. N. Jaber and M.F.B. Zolkipli, Use of Cryptography in Cloud Computing, IEEE International Conference on Control System, Computing and Engineering, (2013), 179–184.
11. B. Lin, W. Guo, N. Xiong, G. Chen, A.V. Vasilakos and H. Zhang, A Pretreatment Workflow Scheduling Approach for Big Data Applications in Multicloud Environments, IEEE Transactions on Network and Service Management, 13(3) (2016), 581–594.
12. C. Deyan and H. Zhao, Data Security and Privacy Protection Issues in Cloud Computing, IEEE International Conference on Computer Science and Electronics Engineering, 1 (2012), 647–651.
13. C. Hongbing, R. Chunming, H. Kai, W. Weihong and L. Yanyan, Secure Big Data Storage and Sharing Scheme for Cloud Tenants, China Communications, 12 (6) (2015), 106–115.
14. C. Liu, J. Chen, L.T. Yang, X. Zhang, C. Yang, R. Ranjan and R. Kotagiri, Authorized Public Auditing of Dynamic Big Data Storage on Cloud with Efficient Verifiable Fine-Grained Updates, IEEE Transactions on Parallel and Distributed Systems, 25 (2014), 2234–2244.
15. C. Liu, L. Zhu, L. Li and Y. Tan, Fuzzy Keyword Search on Encrypted Cloud Storage Data with Small Index, IEEE International Conference on Cloud Computing and Intelligence Systems (CCIS), (2011), 269–273.

Note: All the figures and tables in this chapter were made by the author.

Human Machine Interaction in the Digital Era – Prof. J. Dhilipan et al. (eds)
© 2024 Taylor & Francis Group, London, ISBN 978-1-032-54998-9

Recognition and Classification of Thyroid Nodule with Deep Recurrent Neural Network Using Ultrasound Images

36

V. Sandhiya[1]

Student, Department of Electronics and Communication Engineering,
Easwari Engineering College, Ramapuram, Chennai, India

R. Sri Malini[2]

Student, Department of Electronics and Communication Engineering,
Easwari Engineering College, Ramapuram, Chennai, India

Guided by,

A.T.Madhavi

Assistant professor, Department of Electronics and Communication Engineering,
Easwari Engineering college, Ramapuram, Chennai, India

Abstract Nodules on the thyroid gland are common and increases with age, it can either be classified as benign or malignant. To determine whether a nodule is benign or malignant, several techniques are used, including percutaneous biopsy and ultrasound imaging. However, existing methodologies can lead to medical errors. The proposed DRNN and ultrasound images work together to enhance nodule characterization and decrease biopsies. This ensemble DRNN implementation correctly classified thyroid nodules based on TI-RADS(Thyroid Imaging Reporting & Data System classifications. The experimental system was built utilising the Thyroid Digital Image Database (TDID) to validate its functionality. The pre-trained ImageNet dataset was used to evaluate the model setup in addition to the same dataset being used for training and testing. Recall, and F1-score were used to determine the ensemble network model's diagnostic performance.

Keywords TI-RADS, Ultrasound imaging, DRNN, Benign and malignant nodules, ImageNet database

1. Introduction

In the human body, the thyroid gland is a significant organ that is located towards the base of the neck. A connecting isthmus separates its two lobes, one on each side. Its major function is to secrete hormones, which regulate the temperature and heart rate. When these cells grow or spread beyond their normal borders, thyroid nodules are created. The severity of thyroid nodules varies from medullary (low to moderate) to anaplastic (severe to aggressive). Traditional methodologies such as fine needle aspirations, biopsies, and repeated CT scans, tend to miss segmented nodule areas for identification. While many strategies were employed to improve diagnostic performance, the system's overall performance remained unchanged. A further setback was the inability of the present Neural Network to offer the optimum diagnostic performance using images from a multicenter. The approach for reliably classifying nodule regions into low, medium, and high severity categories using deep learning architectures is put forward in this research.

[1]sandhiyavs02@gmail.com, [2]malini03r@gmail.com, [3]madhavi.t@eec.srmrmp.edu.in

DOI: 10.1201/9781003428466-36

2. Literature Review

(P. Yin, et al., 2022) automatically detect nodules in thyroid ultrasound images, researchers have introduced a CAD system named TND (Thyroid Nodule Detector). (J. Zielke et al., 2022) Region-based active contour has the advantage of improving thyroid image quality and segmenting the thyroid into its individual lobes. (S. Pavithra et al., 2022) The variable background active contour (VBAC) was developed for thyroid nodule classification.

(Ge-Ge Wu, et al., 2021) Deep learning is used to differentiate malignant and benign thyroid nodules based on TIRADS between TR4 and TR5 from B-mode ultrasound images. (Jingzhe Ma, et al., 2020) Texture features and edges were a major area of focus, and deep neural networks were designed to distinguish these features from nodules. (Tianjiao Liu, et al., 2019) A CAD-based deep-learning system was used for automated nodule detection and classification. Multiscale based detection network was designed to distinguish between benign and malignant nodules.

Data and variables

The dataset was collected from an open-access database for thyroid nodules called TDID (Thyroid Digital Image Database), which contains a total of 480 valid cases and the images in grayscale. Among the 480 cases with a TI-RADS score, 280 cases were diagnosed as malignant (TI-RADS scores 4 and 5) and 200 cases as benign (TI-RADS scores 2 and 3). The image augmentation process was used to produce 2000 datasets of images for training the convolutional neural network model. Among them, 1400 images were used for training, 400 images for validation, and the remaining 200 images for test sets.

3. Proposed Methodology

We proposed a Deep Recurrent Neural Network for categorising nodules into different TI-RADS levels ranging from TR1 to TR5. The proposed ResNet-50 model is used in two distinct modes: training and testing. In pre-processing, thyroid ultrasound images are resized to a standard 256 pixels on either side. Kirsch's edge detector is used to improve the pixels at the edges of a thyroid image, and noise removal is carried out by using the median filter as it is effective at removing outliers without compromising image clarity. For image enhancement, the adaptive equalisation technique is suitable for strengthening local contrast and enhancing edge definitions in an image. A top-hat model is used for performing morphological opening processes. Thyroid tumour identification and segmentation performance is improved by using fuzzy-K means (FKM) and MSROI (region of interest) to optimise features derived from the thyroid image.

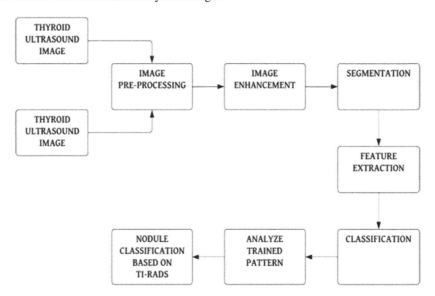

Fig. 36.1 A potential method for detecting and diagnosing thyroid nodules using ultrasound imaging

Source: Created by the authors using lucid chart tool

For detecting suspicious nodules, horizontal and vertical projections are used. Anatomical information is obtained, and the image is divided into three parts: the skin, thyroid area, and dark region. Otsu's thresholding separates the intensity information for both skin and nodule parts. Kaze employs nonlinear diffusion filters to detect and describe image features in a nonlinear

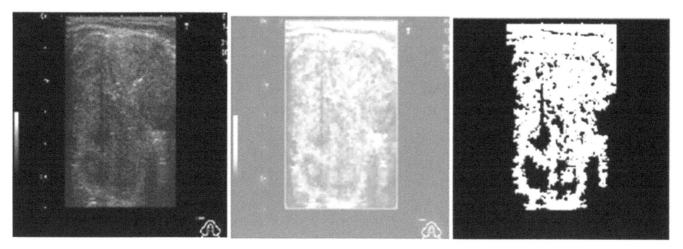

Fig. 36.2 The resulting region of interest using proposed methodology

Source: Created by the authors using MATLAB

scale space in order to extract features from the foreground. By lowering noise while maintaining object boundaries, we may locally adapt blurring to the picture data and achieve better localization accuracy and distinctiveness.

4. Results and Discussion

4.1 Parameter tuning (Hyperparameter for CNN)

Learning rate

The network model of AlexNet, GoogleNet, and CNN was trained through the ultrasound image of the training dataset by varying the learning rate between 0.1, 0.01 and 0.001. Training with 0.1, 0.01, and 0.001 learning rates resulted in low training accuracy and high validation losses, while training with 0.001 resulted in high accuracy and low validation losses.

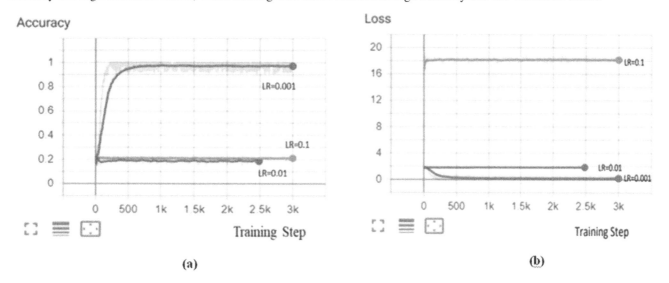

Fig. 36.3 Comparison of accuracy (a) and loss (b) graph of different learning rates

Source: Created by the Authors using MATLAB

Batch size

The ensemble network model of AlexNet, GoogleNet, and DRNN was trained through the ultrasound image of the training dataset, varying in batch size from 32, 64 to 128. Training with 32 and 64 batch sizes converged and resulted in high training and validation accuracy, whereas with 128 as batch size results high training and validation loss did not converge.

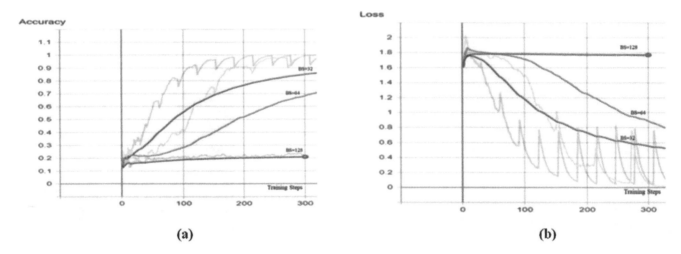

(a)   (b)

Fig. 36.4 Comparison of accuracy (a) and loss (b) graph of different batch sizes
Source: Created by the Authors using MATLAB

4.2 Test Results

The internal test datasets comprised 200 ultrasound nodule images, of which 64 were benign and 136 were malignant. Of the 64 benign nodules, 33 nodules were TR2, 31 nodules were TR3, and of the 136 malignant nodules, 104 were TR4 and 32 were TR5. The proposed nodule segmentation approach using deep learning technology acquired 97.21% precision, 97.65% specificity, 97.45% accuracy, 97.3% F1- score.

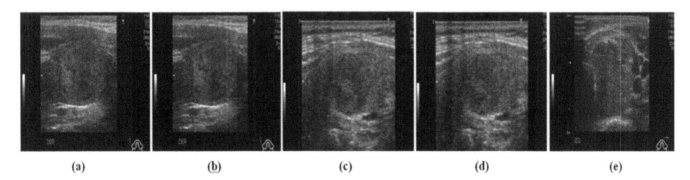

(a)   (b)   (c)   (d)   (e)

Fig. 36.5 Nodule is identified as benign (TR1) in (a), not suspicious (TR2) in (b), mildly suspicious (TR3) in (c), moderately suspicious (TR4) in (d) and highly suspicious (TR5) in (e) [11, 12, 13]
Source: Created by the Authors using MATLAB

The results of the diagnosis system's examination of ultrasound thyroid pictures are displayed in Fig. 36.5. Based on the assessments of a panel of radiologists, the nodule segmentation case photos are categorised as either "mild," "moderate," or "severe." As a result, the diagnostic rate for a mild case is around 97.15% (Fig. 36.5a, 36.5b), the diagnosis rate for a moderate case is approximately 97.2% (Fig. 36.5c, 36.5d), and the diagnosis rate for a severe case is approximately 94.1% (Fig. 36.5e). A proposed approach for detecting and diagnosing thyroid nodules has an approximate 97% overall diagnosis rate.

5. Conclusion

The proposed deep learning and machine learning-assisted method for identifying and segmenting nodule locations in ultrasound thyroid images achieved 97.21% precision, 97.65% specificity, 97.45% accuracy, 97.3% F1-Score, 96.2% recall. It diagnosed 130 pictures as abnormal and 70 as normal, with an overall diagnostic rate of approximately 97%. In future research, it is planned to apply this method of image classification to multicenter images for in-depth analysis of ultrasound images to gain a better comprehension result.

References

1. S. -X. Zhao, Y. Chen, K. -F. Yang, Y. Luo, B. -Y. Ma and Y. -J. Li, "A Local and Global Feature Disentangled Network: Toward Classification of Benign-Malignant Thyroid Nodules from Ultrasound Image," in IEEE Transactions on Medical Imaging, vol. 41, no. 6, pp. 1497–1509, June 2022, doi: 10.1109/TMI.2022.3140797.

2. X. Zhao et al., "Automatic Thyroid Ultrasound Image Classification Using Feature Fusion Network," in IEEE Access, vol. 10, pp. 27917-27924, 2022, doi: 10.1109/ACCESS.2022.3156096.

3. J. Lu et al., "GAN-Guided Deformable Attention Network for Identifying Thyroid Nodules in Ultrasound Images," in IEEE Journal of Biomedical and Health Informatics, vol. 26, no. 4, pp. 1582–1590, April 2022, doi: 10.1109/JBHI.2022.3153559.

4. M. Khairalseed et al., "H-scan ultrasound imaging for the classification of thyroid tumors," 2022 IEEE International Ultrasonics Symposium (IUS), Venice, Italy, 2022, pp. 1-3, doi: 10.1109/IUS54386.2022.9957380.

5. V. T. Manh et al., "Multi-Attribute Attention Network for Interpretable Diagnosis of Thyroid Nodules in Ultrasound Images," in IEEE Transactions on Ultrasonics, Ferroelectrics, and Frequency Control, vol. 69, no. 9, pp. 2611–2620, Sept. 2022, doi:10.48550/arXiv.2207.04219.

6. B. Shankarlal and P. D. Sathya, "Thyroid Tumor Diagnosis System using Spatial Fuzzy C-Means (SFCM) Classification Approach," 2022 International Conference on Edge Computing and Applications (ICECAA), Tamilnadu, India, 2022, pp. 1600–1604, doi: 10.1109/ICECAA55415.2022.9936189.

7. G. Zhang et al., "Ultrasound Super-Resolution Imaging for the Differentiation of Thyroid Nodules: A Feasibility Study," 2022 IEEE International Ultrasonics Symposium (IUS), Venice, Italy, 2022, pp. 1–4, doi: 10.1109/IUS54386.2022.9958286.

8. S. Pavithra, G. Yamuna and R. Arunkumar, "Deep Learning Method for Classifying Thyroid Nodules Using Ultrasound Images," 2022 International Conference on Smart Technologies and Systems for Next Generation Computing (ICSTSN), Villupuram, India, 2022, pp. 1–6, doi: 10.1109/ICSTSN53084.2022.9761364.

9. L. Ma, G. Tan, H. Luo, et al., "A Novel Deep Learning Framework for Automatic Recognition of Thyroid Gland and Tissues of Neck in Ultrasound Image," in IEEE Transactions on Circuits and Systems for Video Technology, vol. 32, no. 9, pp. 6113–6124, Sept. 2022, doi: 10.1109/TCSVT.2022.3157828.

10. J. Zielke, C. Eilers, B. Busam et al., "RSV: Robotic Sonography for Thyroid Volumetry," in IEEE Robotics and Automation Letters, vol. 7, no. 2, pp. 3342–3348, April 2022, doi: 10.1109/LRA.2022.3146542.

11. https://www.kaggle.com/datasets/azouzmaroua/algeria-ultrasound-images-thyroid-dataset-auitd

12. http://cimalab.unal.edu.co/applications/thyroid/thyroid.zip

13. http://cimalab.unal.edu.co/?lang=en&mod=program&id=5

Human Machine Interaction in the Digital Era – Prof. J. Dhilipan et al. (eds)
© 2024 Taylor & Francis Group, London, ISBN 978-1-032-54998-9

Recognizing Spectral Patterns of Phytophthora Infestans on Tomatoes Using Image Processing Techniques

37

Suresh D[1]

Research Scholar, Department of Computer Science, School of Computing Sciences, Vels Institute of Science, Technology and Advanced Studies (VISTAS), Chennai, Tamilnadu, India

Sree Kala T.[2]

Associate Professor, Department of Computer Science, School of Computing Sciences, Vels Institute of Science, Technology and Advanced Studies (VISTAS), Chennai, Tamilnadu, India

Abstract Tomatoes are one of the important crop cultivated every parts of the world. The tomatoes belong to the family solanaceae, which has more effective growth on summer season. The crops have deep history and originated from South and Central America. This crop is most important because of its usefulness in cooking and in many agro based industries. India is the second largest producer of tomatoes following China and majority of farmers of India are relaying on agriculture.

The tomato farmers all-round the world faces many kinds of economic barriers as well as they face many problems in growth of the crop. The growth of the crop is affected basically by many infections caused due to virus and bacteria. Most common tomato diseases are Early Blight, Late Blight, Septoria leaf spot, Leaf Mold, Bacterial Spot, Tomato Pith Necrosis, Buckeye Rot, Fusarium wilt, southern Blight, seedling Disease, Tomato spotted wilt and Tomato Yellow leaf curl disease.

Farmers face many problems in identifying the different level of infections and to manage with pesticides. Mostly they rely on agricultural officers for understanding the severity level of the infection, which causes time consuming process.

This research work might be useful for farmers and agricultural officers to understand severity level of the affected crop with in time duration. Digital Image Processing plays a major role in evacuating many problems in disease identification process. This research article explains various pre-processing techniques followed during the research work. The final part of the research work is useful in analyzing the affected part from unaffected part of the tomato with the usage of deep learning technique.

Keywords Bacterial spot, Tomato pith necrosis early blight, Late blight, Septoria leaf spot, Leaf mold

1. Introduction

Tomatoes are used for cooking and for many agro based industries all over the world. Tomatoes are mostly considered as functional food with more nutritious values and helps in preventing from many diseases. It is very useful in improving the health with proper supplely of phytochemicals substances known as Lycopene. Many agro industries are using the tomato crop as basic ingredients for many agro products. The usage of tomatoes is increasing day by day and farmers are in need in yielding the best tomato crop. The farming of tomato crop has to undergo many economic barriers created due to the various disease

[1]dsureshsuse@gmail.com, [2]sreekalatm@gmail.com

DOI: 10.1201/9781003428466-37

infections found on the tomato crop. The effects caused due to bugs and pests are also creating problem in proper growth of the crop.

Various researchers are working hard to solve many problems occurs in tomato crop. They use many latest technologies for solving many problems occurs in agricultural fields.

2. Tomato Crop Diseases

The infections found due to the effect of bacteria basically differs from stage to stage, which creates major problem in proper growth of the tomato crop. The infections on the tomatoes are beginning from roots and spread through fruit, suckers, stem, leaves and flowers of the crop. Most of the major tomato yielding problems is caused due to various virus problems. The following section explains the different type of virus affected in tomato crop.

2.1 Bacterial Spot

The bacterial spot are most common infections found on the tomato crop usually causes due to Xanthomonas perforans, which is one of the several bacterial species Xanthomonas. The bacterial spots are most common in affecting the green tomatoes more than red tomatoes. Few other crops like peppers are also infected with the bacterial spot diseases can causes severe damage in production. Predominantly the bacterial spots are most common on the wet seasons rather summer seasons. The conditions and circumstances for the growth of the bacterial spot remain good in wet seasons. Sunscalded fruit and defoliation are most common outcome of the disease, which gradually decrease the yielding and causes Sevier damage to farmer's economy. The bacterial spot has the symptoms of water spots on the leaves, angular to irregular, slightly raised scab spots on the fruits and numerous small fruits. The effected leaves might have yellow halo spot on the leaf surface. Finally, the center surface of the leaf got teared and dried out.

2.2 Tomato Pith Necrosis

The species of the soil born disease pseudomonas bacteria has subdivision of bacterial infections such as pseudomonas corrugate causes the severe damage on tomatoes and Pectobacterium carotovorum. Some of the tomato crops are affected with disease in the earlier stage of the season because of Tomato pith necrosis, which occurs in greenhouse high channel tomato product. In few conditions the infections spread even in rainy spring weather and cool climate. The infections are seen very common on the vegetable gardens and can have an impact the proper growth of the pepper plant. These types of bacterial effects are common on cool, cloudy and moister environment, which tormented effects the proper growth of the tomato crop.

3. Image Processing Techniques in Disease Identification

Image processing is most important technique followed for examining the disease affected parts of the tomatoes in various stages. The processing is more useful in accurate classification of effected and unaffected parts of the crop separately. The healthy parts of the tomato crop can be found with greenish color and fruit can be found with healthy red in color. Many discussed bacterial infections and viral infections are found on the stump, leaves and the fruits of the tomato crop. Even though there are many agro industries and researcher act to solve many issues in tomato crop, there is lot of gap in finding the result with accurate manner. The basic surface of fruit and leaves of the tomato crop are affected with infections are starts to change its color pigment into different colors. The color variations found on the surface of the tomato leaves or fruits are captured with digital image processing technology and used for identifying the type of the disease and severity level.

3.1 Formatting Leaf Images

The first section of the image processing starts with collecting the infected tomato crop from various from land. This research work uses plant village dataset for testing the tomato disease severity levels. The taken images may have full infected tomato crop images or only affected parts of the tomato crop. The collected tomato crop images are collected in the RGB color format initially and converted into gray scale if it is necessary. Many of the image processing research work are carried out gray scale images for clear examination and cross validations. This research work uses other models such as HSV, CMYK, RGB and LAB for analyzing process.

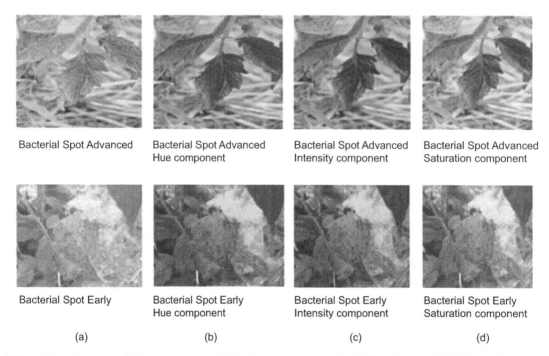

Fig. 37.1 (a) Input image, (b) Hue component (c) Intensity component of HSI color model (d) Saturation component

3.2 Pre-Processing of Tomato Crop Parts

The classification and enhancement technique followed in the image processing is very crustal and much needed one. Removing the irrelevancies in the collected parts of the crop images have to carried out for fixing the perfect framing. The noise removal technique is involved for removing the unwanted objects and irrelevant objects present in the collected images. The intensity of the collected images is increased for better clarity and enhancement of the collected images are also improved with various image processing techniques. Some of the image preprocessing technique followed for many image processing techniques is followed without losing the originality. Some of the basic image processing technique, such as resizing, applying filtering techniques and enhancing the quality of the banana leaves are commonly carried out in pre-processing stage. The basic image resizing of 512 X 512 size, is fixed for every images after converting it into gray scale images.

3.3 Thresholding in Tomato Spotted Wilt

Classifying the healthy part or unaffected parts from effected parts of tomato crop is the toughest part of the image processing. Image segmentation can be useful in segmenting the healthy and unhealthy parts of the tomato crop. Thresholding is one of the image processing techniques very useful in segmenting the healthy and unhealthy parts of the tomato crop. The segmentations are carried out with color difference found on the pixels and boundaries. The labeling is done with 0s and 1s for each and every segmented objects. The background color is given a value of 0 and foreground information are given as 1, which is basically known as effected parts. The procedure is followed not only for affected parts of the leaves, it also carried out for the infected tomato fruits and stumps.

3.4 Histogram for Leaf Mold

The intensity of pixels distributions can be analyzed with the use of Histogram equations. The graphs shown in the Fig. 37.3(a to d) demonstrate the histogram gathered for Leaf Mold Starting infection on tomato leaf and Leaf Mold advanced stage. The Histogram is tested for RGB, HSV, LCH and LAB, which clearly shows the distributions of color throughout the leaf parts.

The high peak in the graph denotes the high intensity range of the particular color. The lower peak found in the graph demonstrate the lightness of the color in leaf parts. The higher degree of acquisition can be identified with best accuracy rate with the usage of Histogram and segmentation.

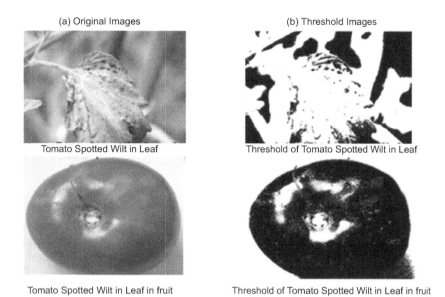

(a) Original Images

Tomato Spotted Wilt in Leaf

Tomato Spotted Wilt in Leaf in fruit

(b) Threshold Images

Threshold of Tomato Spotted Wilt in Leaf

Threshold of Tomato Spotted Wilt in Leaf in fruit

Fig. 37.2 Yellow leaf curl (a) Input yellow leaf curlimage (b) Output of threshold yellow leaf curl

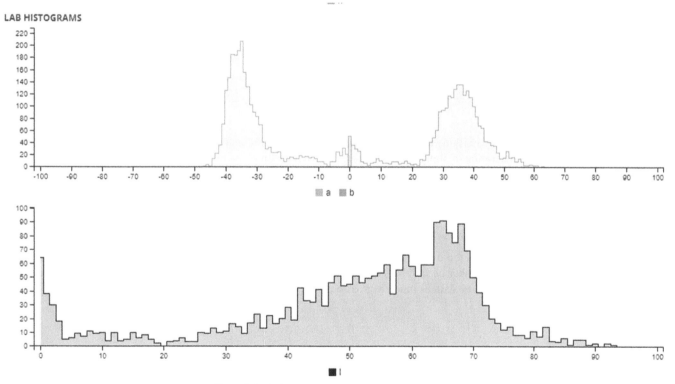

LAB HISTOGRAMS

Fig. 37.3 LAB histogram for leaf mold starting

Varity of leaf components can also be identified with the usage of Histogram equation. The variation is useful in extracting the affected parts from healthy parts very accurately. The procedure is carried out for rest of the infections also and the results are obtained. The LCH and LAB for all the collected infections are noted and taken for feature extraction process.

The Table 37.1 shows the color space and channel statistics of the leaf mold advanced stage. The color variations can be visible with positive numbers and negative numbers in the table. The Table 37.1 also shows the RGB, LCH, HSV and LAB color distribution of leaf mold advanced stage.

Table 37.1 RGB, HSV and LAB color distribution for leaf mold advanced stage

	avg	med	min	max
HSV:H	88 1.00	89	0	360
HSV:S	58	60	0	92
HSV:V	61	62	18	99
LAB:L	59	60	16	98
LAB:A	-27	-29	-41	5
LAB:B	39	41	-20	63

4. Pattern Classification

The classification procedure followed in the image processing technique is very essential part for segregating the healthy and unhealthy parts of the fruits, leaves and stumps of the tomato crop. The classification algorithm used for the segregating the affected and unaffected part of the tomato crop is done with the help of support vector machine, Neural Network and Principal Component Analysis, which are well known for the best classification techniques.

4.1 Neural Network

A neural network is a type of computer system made up of neurons, which are small processing units that are often arranged into layers and connected fully or partially. Receiving activation levels from its neighbours (the output of other neurons), computing an output based on its weighted input parameters, and sending that output to its neighbours are the three fundamental tasks performed by a neuron.

The Hopfield, BPNN, FFNN, MLFF, MLP, SOM, and PCNN neural networks are some of the most popularly used neural networks for picture segmentation. The two stages of neural network-based picture segmentation are pixel categorization and edge detection.

4.2 Support Vector Machine

SVM Supervised algorithm gives classifying data into different classes. To distinguish various classes, SVM creates a hyperplane in multidimensional space. Generation of applying best hyperplane is decision plane between various classes to reduce an error, and to find maximum marginal hyperplane. Data points are closest to hyperplane are the support vectors and plays important role for classifier. The distance between the two lines on the class points that are closest to one another is known as a margin. This approach is well suited for linear separable classes only and not for complex data.

4.3 Principal Component Analysis

PCA is process of image reduction of the dimension on larger dataset which compressed in to smaller one without losing the clarity of an image. Reduction on number of variables for the training samples needs a compromise on the model accuracy, but PCA gives good accuracy also preserve data as much as possible.

5. Conclusion

Extracting necessary information or knowledge from given images always remains a biggest task for many researchers. There are many advantages in using image processing technologies in the field of agriculture for finding the solutions to many problems. Tomato crop is very useful crop in day to day life. Almost every parts of the tomato fruit are edible and can be used in many agro based products. The bacterial and viral attracts found on the tomato crop remains a massive thread in attaining economical standard for the yielding crop. The digital image processing can be a solution to solve many issues related to tomato crop infection by giving necessary remedies and suggestions to farmers. This paper is useful in explaining the preprocessing technique followed for the collected data set from Planet village database. The paper examines the necessary classification techniques and preprocessing methodologies implemented. It gives an idea about color saturation and intensity of color variation in affected parts of the crop. Further the examined images are taken for feature extraction process, which is considered to be the second step of the image processing.

References

1. Amara, J., Bouaziz, B., &Algergawy, A. (2017). A Deep Learning-based Approach for Banana Leaf Diseases Classification. (págs. 79–88). Stuttgart: BTW workshop.
2. M. E. Chowdhury, A. Khandakar, S. Ahmed, F. Al-Khuzaei, J. Hamdalla, F. Haque, et al., (2020) "Design, construction and testing of iot based automated indoor vertical hydroponics farming test-bed in qatar," Sensors, vol. 20, p. 5637.
3. R. N. Strange and P. R. Scott, (2005)"Plant disease: a threat to global food security," Annual review of phytopathology, vol. 43.
4. E. Oerke, "Crop losses to pests,(2006)" The Journal of Agricultural Science, vol. 144, p. 31.
5. F. Touati, A. Khandakar, M. E. Chowdhury, S. Antonio Jr, C. K. Sorino, and K. Benhmed, (2020) "Photo-Voltaic (PV) Monitoring System, Performance Analysis and Power Prediction Models in Doha, Qatar," in Renewable Energy, ed: IntechOpen,

Note: All figures in this chapter were captured/created by Author using Python spyder's software.

Human Machine Interaction in the Digital Era – Prof. J. Dhilipan et al. (eds)
© 2024 Taylor & Francis Group, London, ISBN 978-1-032-54998-9

Operating Wind Energy Turbine Generator as Smart Factory Using Management Information System Equipped with Industry 4.0 Standards

38

E. Kirubakaran[1]
Research Scholar, Department of Management Studies,
Karunya Institute of Technology and Sciences Coimbator, India

K. Karthikeyan[2]
Assistant Professor, Department of Management Studies,
Karunya Institute of Technology and Sciences Coimbatore, India

Abstract Smart factory concept is based on the principle of convergence of Operational Technology (OT) and Information Technology (IT) resulting in a flexible system that can self-optimize performance, self-adapt to learn from new conditions in real time and autonomously run the various functional activities. The basic characteristics of smart factory are connection, optimization, transparency, proactivity and agility. These requirement can be met by the various tools of Industry 4.0 and by practicing management concepts like JIT and LEAN. The smart factory concept can be applied to Wind Turbine Generator (WTG), which basically act as a factory producing electricity using the kinetic energy of the wind as the input. The basic operations of sensing and actuation are taken care by IoT and IIoT; connectivity is established using various types of networking features of Industry 4.0; AI and Machine learning are used to study, analyse and predict the health conditions of the components, which helps to take proactive action and avoid unexpected failure. Since all the components are made to communicate with each other, the entire process is agile and smart. The overall benefits achieved are reduction of cost, increase in efficiency, improvement quality, improved predictability and improved safety and increase in customer satisfaction.

Keywords OT, IT, WTG, IoT, IIoT, AI, ML, Connectivity, Transparency, Agility, Savings of cost

1. Introduction

The Wind Turbine Generator is a device that coverts kinetic energy of the wind to electric energy. The generating capacity of most of the wind turbine generators ranges from 250 KW to 2500 KW and the height from 75 to 150 meters. Even though the expected age of the WTG is about 20 years, there are a good number of WTGs under operation for more than 30 years. These wind turbines are installed in the good wind belt areas and they may be installed either on shore or off shore. Since the manual accessibility to the wind turbines is difficult (because of the remote area of installation and also of the height), a smart way of managing the operations of these wind turbines is very much desirable. The tools and techniques of Industry 4.0 Standards facilitates to operate the wind turbine generators as 'smart factory' efficiently with minimum human intervention. The smart factory style of managing facilitates ease of operation and also result in higher productivity, increase in transparency, reduction in cost, increase in profit and better safety.

[1]ekirubakaran@gmail.com, [2]karthikeyan@karunya.edu

DOI: 10.1201/9781003428466-38

2. Wind Turbine Energy Generator

2.1 Basic operations of Wind Turbine Generator

Wind Turbines Generator converts the kinetic energy of the wind into mechanical energy (by the rotation of shaft and gear) and then the mechanical energy to electric energy using electric generator (Fig. 38.1 and 38.2). A group of WTGs form a Wind Energy Farm. The individual WTGs are connected to the electric grid. Inside each WTG, there is an unit called Supervisor Control and Monitoring System (SCADA) which is basically a computer based control system developed by the original Equipment Manufacturer (OEM) of the wind turbine. This control system receives data from the various sensors of WTG and based on this data, it automatically control the operations (like starting or stopping of the blade movements, yawing of nacelle, pitching of blade etc). Apart from this there are other manual operations like forced shutdown for inspection, preventive maintenance and break down maintenance.

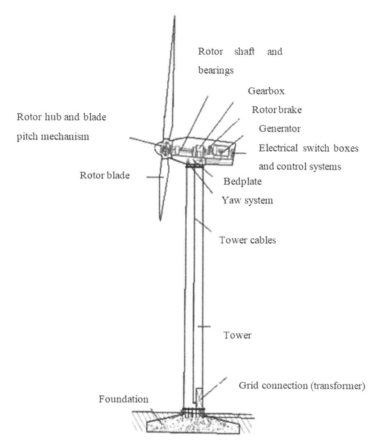

Fig. 38.1 Wind turbine generator

Source: Ref: https://www.hsa.ie/eng/Your_Industry/Renewable_Energy/Micro-Wind-Turbines.pdf

2.2 Basic activities at Wind Turbine Generator

The activities of the Wind Turbine Generator can be broadly classified in the following three categories:

(a) Automated activities handled by SCADA and Control System—Operations of the various components like the Blade, Gear, and generators.

(b) Manual activities—Periodic maintenance, Breakdown maintenance, Preventive maintenance, manual inspection done by the WTG maintenance team in co-ordination with the functional areas (like Material management, Human Resource management, Finance, Service and Inspection)

(c) Information system activities—data acquisition, processing, analysis and Management information System

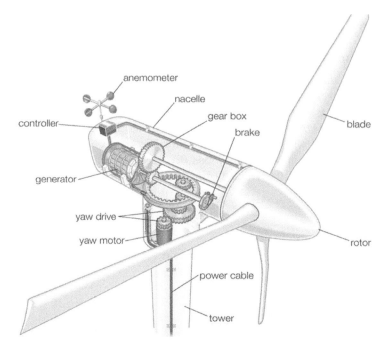

Fig. 38.2 Main components inside the nacelle of wind turbine generator

Source: Review of small wind turbine construction instructions and specifically for structural supports and foundations Prepared by Helacol Services Ltd for the Health and Safety Executive 2016

2.3 Requirement for making Wind Turbine Generators as Smart Factory

"The smart factory is a flexible system that can self-optimize performance across a broader network, self-adapt and learn from new conditions in real or near-real time, and autonomously run entire production processes. (*Source:* 'The smart factory', Deloitte).

Looking at the nature of the operations of Wind Turbine Generators, in order to transform it as Smart Factory, automation are needed need:

- *Sensing the physical parameters*—wind speed, temperature at various locations
- *Actuation based on sensed data*—movement of yaw, pitching, starting and stopping of blade rotation
- *Storing and analysing various parameters*—Wind speed, power generation
- *Communication (Human Machine Interface)*—to the various internal and external functional areas
- *Corrective action*—warning, alert, stopping/stating operations, condition based maintenance

3. Industry 4.0 as Enabler for Making WTG as Smart Factory

The basic tools of Industry 4.0 relevant to Wind Turbine Generators are : IoT & IIoT based sensors, Artificial Intelligence and Data Analytics, Networking, Virtual reality and Augmented reality. This enables to create a Cyber Physical World for the Wind Turbine Generators.

3.1 Internet of Things (IoT) and Industrial IoT (IIoT)

IoT and IIoT refers to interconnected sensors, instruments, and other devices networked together with the concerned industrial applications. This connectivity allows for data collection, exchange, and analysis. The role of IIoT in the Industrial process using Industry 4.0 is shown in Fig. 38.3. It facilitates the various process in the following ways:

Technical assistance by empowering—'Smart objects to reduce human intervention' (automatic starting/stopping the WTG, making required rotation, giving warning signals for safety)

Interoperability—Facilitating communication between heterogeneous objects-(communication within the WTG as well as communication to the exterior world like grid and statutory authorities)

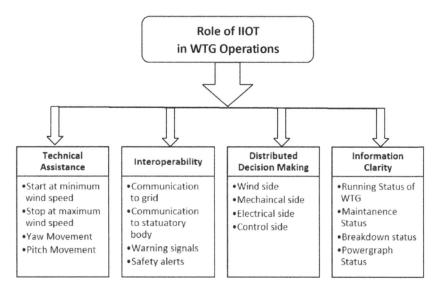

Fig. 38.3 Role of IIOT in industrial process for wind turbine operations

Source: Author's proposed architecture

Distributed decision making – enabling smart objects to take independent autonomous decisions- (the wind side, mechanical side).

Information clarity - pictorial representation of error conditions with different colours.

3.2 Artificial Intelligence and Machine Learning

Artificial Intelligence and Machine Learning are two important technologies used in Data Science to help in the decision making processes. Machine learning develops algorithms to analyse data to learn to predict trends. AI uses this data and predictions for decision-making.

The basic steps in machine Learning are: Data acquisition and preprocessing, Feature selection and extraction, Model selection and validation. The following are the main outcomes of Data Analytic:

Descriptive analysis—answers the question, "What happened?"

Diagnostic analysis—"Why it happened?" Diagnostic data analytics gives analysts a clearer understanding of why something happened and how it happened.

Predictive analysis—"What will happen?" by combining the results of the above and predict the likely happening.

Prescriptive analysis—"What is the optimal course of action?" Prescriptive analytics is done by using advanced processes and tools to analyze data and content. Based on this recommendation is suggested for the optimal course of action or strategy.

3.3 Virtual Reality and Augmented Reality

Virtual Reality and Augmented reality helps to visualize the information in a manner that is easily comprehendible by using pictorial representation, three dimensional view of the objects under study, two dimensional and three dimensional charts etc. In addition, the conditions of the components that is of concern (under high stress/ high temperature/high pressure) can be suitably shown with appropriate colours. The nature of signals (warning, safety hazards, fire alertness) can be expressed in a manner that is comprehendible. The current status of the WTG (the speed and angle of the blade, the tilt of the nacelle position, the vibration of the tower, sound of vibration) can be taken expressed in a pictorial manner, and made accessible to the user through internet.

3.4 Networking Layers with IIoT base

The various layers of networking with IIoT as the base is shown in Fig. 38.4.

The bottom most layer comprises of the physical devices and they are connected with IoT / IIoT devices which takes care of sensing and actuation. These devices communicate with each other and share the information that is needed among these devices. The information is also passed on to the Edge and Gateway. Devices connected to this Edge/Gateway can share the

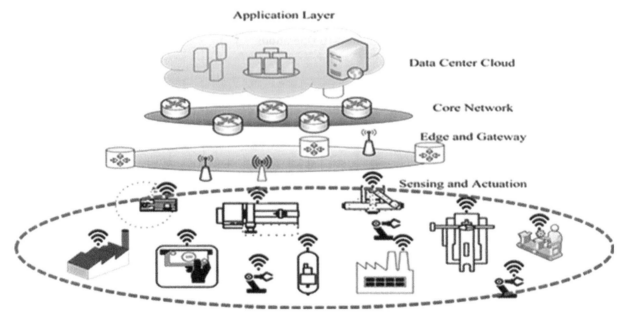

Fig. 38.4 Layer wise architecture of IIoT and networking

Source: "Securing the Internet of Things: A Proposed Framework", Cisco, Online: https://www.cisco.com/c/en/us/about/security-center/secure-iot-proposed-framework.html

required information. Further the Edge/Gateways is connected to the Cloud based Data center through the Core network. It may be noted that all data/information need not go to the apex level and what is required alone is passed on. The integrated information is available in the Data Centre for various applications.

3.5 Cyber Physical System of Wind Turbine Genertors

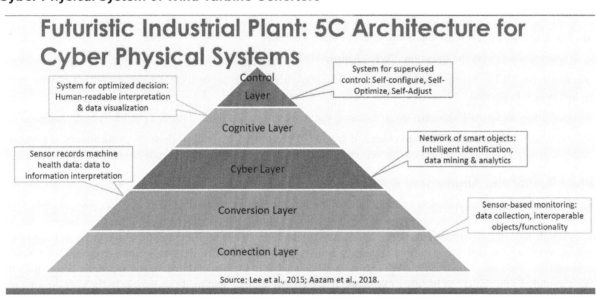

Fig. 38.5 Cyber physical system [1]

The Cyber Physical System (Fig. 38.5) comprises of the following five layers.

1. *Connection Layer*—Sensors collect various parameters (Eg. temperature, pressure, speed, movement)
2. *Conversion Layer*—converts the data collected (above data health of the WTG is arrived)

3 *Cyber Layer*—with the relevant objects and intelligently analyze the operations (need for maintenance or not)

4. *Cognitive Layer*—convert to human readable interpretation- (any severity of the condition to run the WTG or not)

5. *Control Layer*—self optimize and adjust – Eg. Adjusting the SCADA parameters for various operations of WTG

4. Hybrid Management Information System

The real time operational data from the SCADA of Original Equipment Manufacturer (OEM) and also data from the additional IOT devices (after suitable filtering and pre processing) are stored in a data base. The data arising out of the various functional areas (like Human resource, Material management, Operation, Planning, Production, Maintenance and Finance) are also taken and a Hybrid Information System is formed and hosted in the Cloud. Access for the various stake holders like the Wind Turbine Owners, Maintenance Team, Regulatory bodies etc (with respect to their relevant information) is provided with suitable Human Machine Interface. Information Security is of vital importance and an appropriate Information Security Management System (ISMS) policy, which takes care of data security through appropriate policies and procedures is also established. Thus a Hyhrid Management information System is implemented (Fig. 38.6) for providing the right information to the right agencies (human and functional areas) at the right time to take the right decision. Necessary external data like the seasonal wind speed, turbulence, temperature of the geographical location of the WTG is also taken. This MIS enables the Wind Turbine Generator to act as a the Smart factory The Wind turbine with the assistance of this Hybrid MIS is able to function as a smart factory exhibiting the characteristics of agility, higher availability, transparency of operations, better resource utilization, improved planning, long range decision making.

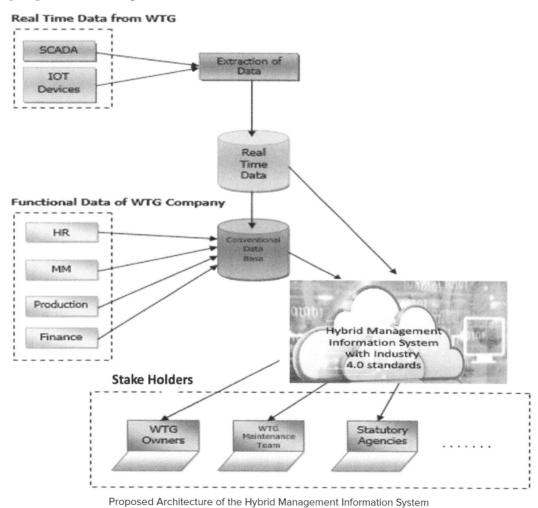

Proposed Architecture of the Hybrid Management Information System

Fig. 38.6 Hybrid management information system

5. Conclusion

It could be seen that while the WTG is made to perform like a Smart machine, the following benefits are achieved.

- Agility,
- Higher availability (Breakdown maintenance is minimized and converted as preventive maintenance),
- Better prediction of performance with AI and Machine Learning of Data Science in trend analysis.
- Transparency of operations
- Better resource utilization (of manpower, equipment and material),
- Improved planning (scheduling maintenance and inspection)
- Long range decision making (relocation or renovation of WTG)
- Better Customer satisfaction

The management concepts like Just in Time (Inventory) and LEAN management for elimination of waste and Customer Satisfaction are achieved by using this Management Information System. However while enhancing the Information System with more sophisticated tools of Industry 4.0, the cost- benefit and the residual life expectancy also has to be considered.

References

1. Lee, Jay, Behrad Bagheri, and Hung-An Kao. "A cyber-physical systems architecture for industry 4.0-based manufacturing systems." Manufacturing Letters 3 (2015): 18–23.
2. Cisco, "Securing the Internet of Things: A Proposed Framework", Online: https:/ /www.cisco.com/c/en/us/about/security-center/secure-iot-proposed-framework.html
3. Deloitte Univeqrsity Press. The smart Factory. Online. URL: https:// www2.deloitte.com/content/dam/insights/us/articles/40 51_The-smart-factory/DUP_The-smart-factory.pdf
4. Imre Delgado and Muhammad Fahim "Wind Turbine Data Analysis and LSTM-Based Prediction SCADA System" (2020). https://dx.doi.org/10.3390/en14010125
5. Yongchao Zhu, Caichao Zhu, Chaosheng Song "Improvement of reliability and wind power generation based on wind turbine realtime condition assessment" International Journal of Electrical Power & Energy SystemsVolume 113, December 2019, Pages 344–354.
6. I. P. Okokpujie, U.C.Okonkwo, C.A.Bolu, O.S.Ohunakin, M.G.Agboola, A.A.Atayero "Implementation of multi-criteria decision method for selection of suitable material for the development of horizontal wind turbine blade for sustainable energy generation"(2020) https://doi.org/10.1016/j.heliyon.2019.e03142
7. Huan Long, Long Wang, Zijun Zhang, Zhe Song, and Jia Xu "Data-Driven Wind Turbine Power Generation Performance Monitoring" IEEE TRANSACTIONS ON INDUSTRIAL ELECTRONICS 2015
8. Colombo, Armando & Karnouskos, Stamatis & Bangemann, Thomas. (2014). Towards the Next Generation of Industrial Cyber-Physical Systems. Industrial Cloud-Based Cyber-Physical Systems: The IMC-AESOP Approach.
9. Khairy Sayed, Ahmed G. Abo-Khalil & Ali M. Eltamaly 6Wind Power Plants Control Systems Based on SCADA System(2021)., https://link.springer.com/chapter/10.1007/978-3-030-64336-2
10. Ahlam Althobaiti, Anish Jindal, Angelos K. Marnerides "SCADAagnostic Power Modeling for Distributed enewable Energy Sources" (2020) IEEE 21st International Symposium
11. I Ishaq, et al., "IETF standardization in the field of the internet of things (IoT): a survey", J. of Sens. and Act. Netw. 2, vol. 2 (2013): 235–287.
12. Ustundag, A., & Cevikcan, E (2018). Industry 4.0:Managing The Digital Transformation. Springer.
13. Chakravarti, S., & Jain, A. (2018). Why Your Product Must be Smart and Connected. Online. URL: http://sites.tcs.com/insights/perspectives/why-your-products- must-be-smart-and-connected.

Human Machine Interaction in the Digital Era – Prof. J. Dhilipan et al. (eds)
© 2024 Taylor & Francis Group, London, ISBN 978-1-032-54998-9

A Review on Machine Learning Techniques to Identify Plant Diseases

39

Sona Maria Sebastian[1]

P.hD Research Scholar, Department of Computer Applications,
SRM IST, Ramapuram Campus, Chennai, India

J. Jebamalar Tamilselvi[2]

Associate Professor, Department of Computer Applications,
SRM IST, Ramapuram Campus, Chennai, India

Rubin Thottupurath Jose[3]

Associate Professor & Head, Department of Computer Applications,
Amal Jyothi College of Engineering, Kanjirappally,India

Abstract An ecosystem that is balanced depends heavily on the agriculture sector. Growing and producing plants and food becomes crucial as a result. The production of plants and crops requires careful attention from farmers. Today, a number of ailments affecting plants have been discovered. Plant disease outbreaks may have a significant impact on crop production, which would slow the nation's economic growth rate. Early detection and treatment are possible for the plant disease. The early identification and classification of diseases may be greatly aided by machine learning (ML), computer vision-based techniques and deep learning (DL). Many researchers have successfully developed various plant disease recognition algorithms and models in the past. A few machine learning-based approaches and techniques that are currently in use to detect diseases in various plant species were looked at and assessed in this study. Additionally, the challenges and limitations of various strategies were analysed and compared based on the success rate of each.

Keywords Deep learning, Convolutional neural networks, K-nearest neighbour, Support vector machine, Classification algorithm, Plant disease detection

1. Introduction

The economic structure in developing nations is supposed to be supported by agriculture, which is a crucial part of the economy. Food crops that are necessary for human consumption are produced through agriculture. Agriculture is crucial to human life since it provides us with food and other necessities for a healthy lifestyle. This encourages researchers to look for new, accurate, and efficient technologies for high production. Furthermore, agriculture offers a variety of advantages and plays a significant role in our daily lives.

Farmers must pay close attention while plants and crops are produced. Many illnesses that harm plants have been found in the modern era. The output of crops may be significantly impacted by plant disease outbreaks, which would reduce the rate

[1]ss5261@srmist.edu.in, [2]jebamalj@srmist.edu.in, [3]rubinthottupuram@amaljyothi.ac.in

DOI: 10.1201/9781003428466-39

of economic growth in the country. It is feasible to identify and treat the plant disease early. The key to a productive farming system is the early detection of plant diseases. A farmer can typically identify disease symptoms in plants using simple eye-to-eye inspections, although this requires constant observation. With large plantations, this procedure is more expensive and less precise. Now a days there are several studies undergoing in this area. The early identification and classification of diseases may be greatly aided by machine learning (ML), computer vision-based techniques and deep learning (DL).

In addition to providing a thorough study of several machine learning approaches used to diagnose plant diseases, this paper provides an illustration of the traditional plant disease detection system.

2. Traditional Plant Disease Detection System

Following are the phases of the conventional plant disease detection method.

2.1 Image Acquisition

This is the process of retrieving images from external resources. The first stage in the detection of plant diseases is image collection. High quality images from external sources can be acquired for further.

2.2 Image Pre-processing

Image pre-processing refers to a series of operations or techniques applied to images before they are fed into a machine learning or computer vision model. It involves applying various techniques to enhance, clean, and prepare the raw image data before feeding it into a machine learning model or performing further analysis. With this, sick areas in leaves or stems can be extracted from the background.

2.3 Segmentation

Image segmentation entails breaking up a picture into groups of pixels that can be represented by masks or labelled images.

2.4 Feature Extraction

This method converts raw data into manageable numerical features while preserving the information from the original data set. In this stage, it is possible to extract the shape, texture, and colour properties of the afflicted plant sections

2.5 Classification

This idea of supervised learning primarily classifies a group of data into groups. Eventually, the numerous plant diseases can be classified using any of the classification methods.

3. Related Work

A thorough comparison of disease detection in plants and classification methods in machine learning has been done. We evaluated how well the Support Vector Machine (SVM), K-Nearest Neighbour Classification Method, and Convolutional Neural Network Classification Approaches performed to identify diseases in plants.

3.1 SVM

One of the most widely used supervised learning techniques, Support Vector Machine (SVM), is utilised for both classification and regression issues The following authors employed SVM to recognise plant diseases. Using a multiclass SVM classifier, a study was conducted to quantify the severity of the illness on soybean leaf. [2]. To compose a feature database this research employs the Grey Level Co-occurrence Matrix (GLCM). In this study reported 87.3% accuracy rate. The size of the database affects how well this research performs. In order to identify tomato leaf disease, a modal implemented based on computer vision and machine learning algorithms [4]. The experimental result shows an accuracy rate of 88%. In a study done for identifying diseases affected in the leaves of Ixora coccinea, Nerium oleander and sweet-scented geranium[11],the experimental result shows accuracy rate of 92.08% with SVM classifier

3.2 KNN

One of the most fundamental supervised learning-based machine learning algorithms is the K-Nearest Neighbour algorithm(K-NN). Although regression and classification problems can be addressed using the K-NN approach, it is frequently used for classification issues. For plant diseases identification the following authors used KNN. A model developed for detecting tomato leaf disease combining computer vision and machine learning techniques [4], the result shows the accuracy rate of 97%. An investigation using GLCM and KNN-based algorithms for the recognition of plant illnesses [6], achieved the accuracy rate between 80-90%. To determine the extent of and find the Grey Mildew illness in cotton plants [7] using KNN method examined 40 images and achieved the accuracy rate of 82.5%. A KNN-based algorithm was created and suggested for recognising the presence of Leaf Scorch Disease in sugarcane leaves. It had an accuracy rate of 95%. [8].

3.3 CNN

A deep learning architecture that directly learns from data is the convolutional neural network, or CNN. For plant diseases identification the following authors used CNN Methods. A method for disease identification in tomato leaves utilising computer vision and machine learning algorithms was put forth in 2022 [4]. The experimental finding indicates a 99.6% accuracy rate. A CNN based modal was developed to identify diseases affected in grapes and tomatoes [9] achieved the overall accuracy rate of 98. 40% and 95.71 respectively. A study done for identifying diseases affected in 14 different plant species using EfficientNetB0 CNN architectural model [10]. An accuracy percentage of 99.56% was attained by the suggested approach for classifying diseases. Utilizing Deep Convolutional Neural Network Computational

Techniques for IoT and Future Wireless Applications to Detect Plant Disease. [12] The proposed method is developed for identifying diseases affected multiple plant species and acquired disease classification accuracy rating of 99.96% using CNN architectural model.

Table 39.1 provides a summary of the contrast of several Machine Learning algorithms used for identifying plant diseases based on the accuracy of the results.

Table 39.1 Comparison of different ML techniques used by various authors [2], [4], [6], [7], [8], [9], [10], [11], [12]

ML Techniques	Plant Species	Result (Accuracy in %)
SVM	Soybean[2]	87.3%
	Tomato[4]	88%
	Nerium oleander, Ixora coccinea sweetscented geranium [11]	92.08%
KNN	Multiple crops[6]	80-90%
	Tomato[4]	97%
	Cotton[7]	82.5%
	Sugarcane[8]	95%
CNN	Tomato[4]	99%
	Grapes, tomatoes [9]	98.40%, 95.71%
	14 different plant species [10]	99.56%
	Multiple plant species [12]	99.96%

4. Conclusion

In this review, a comparison of three machine learning algorithms is made in order to identify plant diseases. The outcome demonstrates that approaches based on CNN accurately identify more diseases. In future, disease detection in plants and autonomous disease diagnosis for farmers may be made possible with the integration of other advanced technologies with CNN.

References

1. Anshul Bhatia, Anuradha Chug, Amit Prakash Singh, Ravinder Pal Singh, Dinesh Singh (2022). A machine learning-based spray prediction model for tomato powdery mildew disease. Indian Phytopathology volume 75, pages 225–230, 2022.
2. Sachin B. Jadhav, Vishwanath R. Udupi,Sanjay B. Patil(2019).Soybean leaf disease detection and severity measurement using multiclass SVM and KNN classifier. International Journal of Electrical and Computer Engineering (IJECE), Vol. 9, No. 5
3. Usama Mokhtar, Mona A. S. Alit, Aboul Ella Hassenian, Hesham Hefny. Tomato leaves diseases detection approach based on support vector machines.11th International Computer Engineering Conference (ICENCO).

4. Sunil S. Harakannanavar, Jayashri M. Rudagi, Veena I Puranikmath, Ayesha Siddiqua, R Pramodhini (2022). Plant leaf disease detection using computer vision and machine learning algorithms. International Conference on Intelligent Engineering Approach (ICIEA).

5. Meghana Govardhan, Veena M B (2019). Diagnosis of Tomato Plant Diseases using Random Forest. 2019 Global Conference for Advancement in Technology (GCAT) Bangalore, India. 978-1-7281-3694-3/19/

6. Gautam Kaushal, Rajni Bala (2017). GLCM and KNN based Algorithm for Plant Disease Detection. International Journal of Advanced Research in Electrical,Electronics and Instrumentation Engineering,Vol. 6, Issue 7.

7. Aditya Parikh, Mehul S. Raval, Chandrasinh Parmar, Sanjay Chaudhary (2016). Disease Detection and Severity Estimation in Cotton Plant from Unconstrained Images. IEEE International Conference on Data Science and Advanced Analytics

8. Umapathy Eaganathan, Jothi Sophia, Vinukumar Lackose, Feroze Jacob Benjamin (2014). Identification of Sugarcane Leaf Scorch Disease using K-means Clustering Segmentation and KNN based Classification. International Journal of Advances in Computer Science and Technology (IJACST), Vol. 3, No. 12, Special Issue of ICCEeT, Dubai, pp. 11–16.

9. Ananda S. Paymode,Vandana B. Malode (2022).Transfer Learning for Multi-Crop Leaf Disease Image Classification using Convolutional Neural Network VGG. Artificial Intelligence in Agriculture Volume 6, Pages 23–33.

10. Sk Mahmudul Hassan, Arnab Kumar Maji, Michał Jasinski, Zbigniew Leonowicz, Elzbieta Jasinsk (2021). Identification of Plant-Leaf Diseases Using CNN and Transfer-Learning Approach. New Technological Advancements and Applications of Deep Learning

11. Paramasivam Alagumariappan, Najumnissa Jamal Dewan, Gughan Narasimhan Muthukrishnan, Bhaskar K. Bojji Raju, Ramzan Ali Arshad Bilal,Vijayalakshmi Sankaran (2020). Intelligent Plant Disease Identification System Using Machine Learning. Eng. Proc. 2020, 2, 49.

12. J. Arun Pandian, V. Dhilip Kumar, Oana Geman, Mihaela Hnatiuc, Muhammad Arif, K. Kanchanadevi (2022). Plant Disease Detection Using Deep Convolutional Neural Network.Computational Methods for Next Generation Wireless and IoT Applications.

13. Bhatia A, Chug A, Singh AP (2020). Plant disease detection for high dimensional imbalanced dataset using an enhanced decision tree approach. Int J Future Gener Commun Netw 13: 71–78.

14. Bhatia A, Chug A, Singh AP (2020). Hybrid SVM-LR classifier for powdery mildew disease prediction in tomato plant. 7th International conference on signal processing and integrated networks (SPIN), pp 218–223.

15. Aanis Ahmad, Dharmendra Saraswat, Aly El Gamal (2023). A survey on using deep learning techniques for plant disease diagnosis and recommendations for development of appropriate tools. Smart Agricultural Technology,100083 [16] Amreen Abbas, Sweta Jain, Mahesh Gour , Swetha Vankudothu (2021). Tomato plant disease detection using transfer learning with C-GAN synthetic images. Computers and Electronics in Agriculture Volume 187.

16. Aanis Ahmad, Dharmendra Saraswat, Varun Aggarwal, Aaron Etienne, Benjamin Hancock (2021). Performance of deep learning models for classifying and detecting common weeds in corn and soybean production systems. Computers and Electronics in Agriculture Volume 187, 106279.

17. Y. Tian et al (2019). Apple detection during different growth stages in orchards using the improved yolo-v3 model. Comput. Electron. Agric.

18. M. Agarwal et al. Toled(2020).Tomato leaf disease detection using convolutional neural network. Procedia Computer Science

19. M. Akila et al(2020).Detection and classification of plant leaf diseases by using deep learning algorithm.Int. J. Eng. Res. Technol

Human Machine Interaction in the Digital Era – Prof. J. Dhilipan et al. (eds)
© 2024 Taylor & Francis Group, London, ISBN 978-1-032-54998-9

SPK-A Traffic Violation Detection System

40

V. Dinesh[1]

Assistant professor,
Easwari Engineering College (Anna university) Chennai India

S. Phvan[2] and P. K. Raghul[3]

Students,
Easwari Engineering College (Anna University) Chennai India

Abstract With the growing population and increasing congestion of transport networks, traffic violations are on the rise and therefore it has become critical to implement systems to detect traffic violations and penalties for the violators. Existing systems like speed cameras do help but, their implementation is limited due to various factors like implementation cost and lighting conditions. This creates the need for a system with ease of implementation in a feasible manner. The proposed system takes these factors into consideration. It consists of a primary unit that can be installed into existing vehicles or can be built-in into future versions of the vehicle. The primary unit works by collecting the onboard diagnostic data (OBD data) which provides various parameters of the vehicle such as speed, acceleration, and braking. This data can be combined with the data from various secondary units which can be developed in accordance to help detect specific traffic violations. The data from these secondary units, for example, a secondary unit attached to a light pole in highways that transmits the region's speed limit can be used by the vehicle for comparison of the vehicle's speed. If the speed of the vehicle in the primary unit is higher than the speed value provided by the secondary unit, then, the primary unit is capable of generating a message along with the location where the violation occurred and sends this message to servers through the onboard GSM module. The standard operating procedure is followed for the e-Challan generation which is then sent to the vehicle owner or to the registered contact number. This system uses a completely automated approach which makes the system highly accurate in the detection of traffic violations.

Keywords Arduino, GPS module, GSM module, Traffic violation detection system, Traffic violation, Vehicle telemetry data

1. Introduction

Traffic rule violations have become one of the biggest problems in this emerging world. With more than 1.5 billion vehicles on the road, managing traffic and proper implementation of traffic rules has become critical tasks in many nations. In India, which has the third widest road network in the world there are more than 300 million active vehicles and there is a critical deficit in manpower for managing traffic and implementing traffic rules. This has led to more than 4 lakh accidents and more than 1.5 lakh people were deceased on average in a year. With the growing vehicle demands and increasing traffic, this issue has started to worsen, and many nations have started to develop and implement various systems to tackle the growing demand. Speed cameras are one of the current systems that are under development and are being preferred by most nations. But this system has

[1]dinesh.v@eec.srmrmp.edu.in, [2]phvan2018@gmail.com, [3]raghul78343@gmail.com

DOI: 10.1201/9781003428466-40

some major drawbacks, and its implementation is limited to only a few use case scenarios. This creates the need for a system that is capable of monitoring all vehicles in real-time to detect traffic violations. The proposed system gathers data directly from the vehicle to determine whether a violation has occurred or not. This makes the system very effective as the drivers have to be conscious throughout their journey to obey traffic laws as their vehicle is directly monitoring them. The Primary objective of this project is to make the process of traffic violation detection automated and efficient.

2. Literature Review

Roopa Ravish, et al. propose a system that aims to ease the movement of traffic to help prevent accidents or crashes using vehicle density data. Dr. Yeresime Suresh, et al. propose a system for the management of traffic using image processing technologies for detecting major violations like speeding. Zhao, J., et al. developed a system that uses a GRU model for traffic speed prediction under nonrecurrent congestion, it uses a GPS map matching algorithm to obtain GPS data from trucks. R. Shreyas, et al. propose a system that uses automatic number plate recognition using image processing technology for monitoring road traffic. T. T. Dandala, et al. discusses the concept of Internet of Vehicles(IOV) that proposes a connection between CCTV cameras, traffic lights, etc,

3. Existing System

Speed cameras work by recording vehicles' speed using radar technology and taking photographs of cars that exceed threshold limits. If a speeding offense is detected in an area where a camera is installed, it will record a digital photograph of the vehicle. The digital photos capture important information for enforcement, such as the Date and time of the offense, The vehicle's direction of travel, and Details of the location where the offense occurred, including the posted speed limit, and the speed at which the vehicle was being operated.

Often, these cameras shown in Fig. 40.1 also use detectors installed on road surfaces alongside radar technology to monitor multiple lanes. This way, vehicles cannot avoid detection by straddling lanes. Additionally, speeding vehicles can still be detected even when driving in heavy traffic. Typically, these are set at an angle that allows them to take images of individual vehicles even if other cars are close by.

Fig. 40.1 Speed camera

Source: https://metro.co.uk/2016/05/31/average-speed-cameras-are-monitoring-drivers-on-more-than-250-miles-of-british-roads-5914406/

4. Proposed System

Proposed system presents a new way to automate Traffic Violations created by vehicles using the On-board Diagnostic (OBD) data. This system consists of two major units.

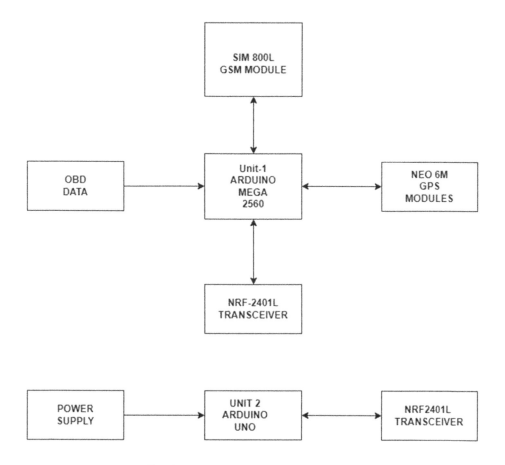

Fig. 40.2   Proposed system architecture

Source: Made by Author

The First unit which is installed in the vehicle gathers the OBD data and analyzes parameters such as speed, acceleration, etc. The system additionally consists of a GPS module and a GSM module. The module used here is SIM800l(GSM) module and NEO 6M(GPS) module. The system also consists of an RF transceiver (NRF2401L) for communicating with unit 2.

The second unit consists of a controller board (Arduino Uno) and a transceiver (NRF2401L). This unit is installed on locations where a specific speed limit has to be set and can be set on existing light poles, sign boards, etc. The speed limit and transmission range is set into the controller which then broadcasts this around the entire region.

When the vehicle is in motion and reaches a region where a specific speed limit has been set, the transceiver installed in the vehicle picks up the specified speed limit set on unit 2 which may be installed on a pole. Once this specified speed is received, the unit in the vehicle starts to analyze the OBD data and compares the vehicle's speed to the specified speed. If the vehicle's speed is found to be over the specified speed. The controller records the vehicle's speed and triggers the GPS module and gathers the location coordinates where the violation occurred. The controller then sends this data to the control room through the GSM module along with the vehicle ID and vehicle registration number. Then the standard operating procedure for e-challan generation will be followed and will be sent to the vehicle owner or the registered contact number.

This system poses various advantages compared to the existing system such as: Can be easily adopted into newer vehicles or easily integrated into existing vehicles. The cost of implementation is lower compared to existing systems. The system fully automates the task of monitoring and e-challan generation for violations. The system ensures stricter surveillance and proper implementation of traffic rules.

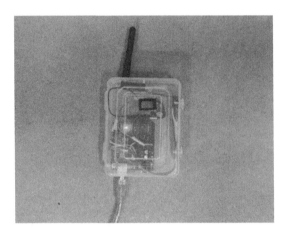

Fig. 40.3 Pole unit and transmitter log

Source: Made by Author

5. Results and Discussion

The simulation's outcome of our system are shown in Fig. 40.3. The secondary units of the system are tested separately for proper functionality. The entire system's functionality is also tested with various speed limit values. The efficiency of the unit broadcasting the specified speed is also tested at various distances. The transmitting unit is tested across various distances and through obstacles to replicate real-time scenarios in highways and freeways. Its transmission efficiency was tested across various distances and the output was verified.

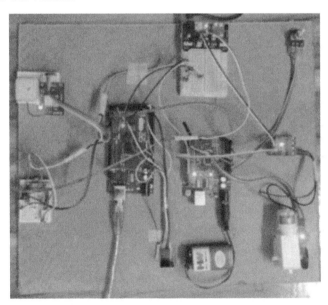

Fig. 40.4 Vehicle unit

Source: Made by Author

Various parameters of this unit were tested such as GPS signal reception in various climatic conditions, reception through obstacles, the accuracy of the data obtained, etc. The GSM module was also tested for signal reception through obstacles and proper transmission of data. Speed data transfer from controller board 1 (Arduino UNO) to controller board 2(Arduino Mega) using I2C protocol was also tested for various values and verified.

Fig. 40.5 GSM and GPS functionality logs

Source: Made by Author

6. Output

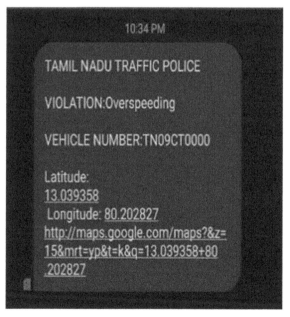

Fig. 40.6 Output

Source: Made by Author

7. Conclusion and Future Work

The proposed system to detect traffic violations completely automates identifying and fining traffic rule violators, eliminating the need for manpower. This proposed system can be better implemented and serves as an economical and feasible solution compared to existing systems such as speed cameras. The existing system (speed cameras) generates many false-positive results, requiring further monitoring by a human operator. This issue is completely mitigated in this journal proposed system as it uses the data directly gathered from the vehicle to identify violations.

This system can be further developed in the future to support various other violations through the development of violation-specific secondary units such as no parking zones, insurance validity, headlight and taillight functionality, seatbelt detection, lane speed detection, etc.

Reference

1. Ravish, Roopa & Shenoy, Datthesh & Rangaswamy, Shanta. (2020). Sensor-Based Traffic Control System. 10.1007/978-981-15-2188-1_17.
2. Pan, Fuquan & Yang, Yongzheng & Zhang, Lixia & Ma, Changxi & Yang, Jinshun & Zhang, Xilong. (2020). Analysis of the Impact of Traffic Violation Monitoring on the Vehicle Speeds of Urban Main Road: Taking China as an Example. Journal of Advanced Transportation. 2020. 1–11. 10.1155/2020/6304651.
3. Yeresime, Suresh & Anusha, Chillara. (2022). TRAFFIC RULES VIOLATION DETECTION SYSTEM. 10.13140/RG.2.2.10612.42884.
4. J. Zhao et al. "Truck Traffic Speed Prediction Under Non-Recurrent Congestion: Based on Optimized Deep Learning Algorithms and GPS Data," in IEEE Access, vol. 7, pp. 9116–9127, 2019, doi: 10.1109/ACCESS.2018.2890414.
5. Shreyas, R. & Kumar, B. & Adithya, H. & Padmaja, Bhimireddy & M P, Sunil. (2017). Dynamic traffic rule violation monitoring system using automatic number plate recognition with SMS feedback. 1–5. 10.1109/TEL-NET.2017.8343528.
6. T. T. Dandala, V. Krishnamurthy and R. Alwan, "Internet of Vehicles (IoV) for traffic management," 2017 International Conference on Computer, Communication and Signal Processing (ICCCSP), 2017, pp. 1–4, doi: 10.1109/ICCCSP.2017.7944096.
7. Jain, Neeraj & Saini, R. & Mittal, Preeti. (2019). A Review on Traffic Monitoring System Techniques: Proceedings of SoCTA 2017. 10.1007/978-981-13-0589-4_53.
8. Nellore, Kapileswar & Hancke, Gerhard. (2016). A Survey on Urban Traffic Management System Using Wireless Sensor Networks. Sensors. 16. 157. 10.3390/s16020157.
9. Ameen, Hussein & Mahamad, Abd & Saon, Sharifah & Bin Ahmadon, Mohd Anuaruddin & Yamaguchi, Shingo. (2021). Driving Behaviour Identification based on OBD Speed and GPS Data Analysis. Advances in Science Technology and Engineering Systems Journal. 6. 550–569. 10.25046/aj060160.
10. B. Malarvizhi, V. Dinesh, M. Janani, R. Gunaseeli, B. Abarna, "An IoT based Staple Food Endowment System and Waste Management System for Foster Care using Arduino and Blockchain," 2020 International Conference on Recent Trends in Computer Science & Information technology" ICRCSIT - 20, ISBN No.978-93-80831-66-4
11. Rachana K P, Aravind R, Ranjitha M, Spoorthi Jwanita, Soumya K, 2021, IOT Based Smart Traffic Management System, INTERNATIONAL JOURNAL OF ENGINEERING RESEARCH & TECHNOLOGY (IJERT) NCCDS – 2021 (Volume 09 – Issue 12).
12. B. Malarvizhi V. Dinesh M.Janani, R.Gunaseeli, B.Abarna, "IoT based Staple Food Endowment System and Waste Management System for Foster Care," 2020 Sambodhi- UGC Care Journal, ISSN:2249-6661, Volume-43, UGC Care
13. S. Misbahuddin, J. A. Zubairi, A. Saggaf, J. Basuni, S. A-Wadany and A. Al-Sofi, "IoT based dynamic road traffic management for smart cities," 2015 12th International Conference on High-capacity Optical Networks and Enabling/Emerging Technologies (HONET), 2015, pp. 1–5, doi: 10.1109/HONET.2015.7395434.
14. Hirawan, Dedeng & Hadiana, A & Abdurakhim, A. (2019). The Prototype of Traffic Violation Detection System Based on Internet of Things. IOP Conference Series: Materials Science and Engineering. 662. 022084. 10.1088/1757-899X/662/2/022084.

Human Machine Interaction in the Digital Era – Prof. J. Dhilipan et al. (eds)
© 2024 Taylor & Francis Group, London, ISBN 978-1-032-54998-9

Hybrid Leach Protocol and Genetic Algorithm to Obtain Route Node in Under Water Sensor Network

41

K. Prakash[1]

Research Scholar, Department of Computer Science,
VELS Institute of Science, Technology & Advanced Studies (VISTAS), Chennai, India and
Assistant Professor in Sri Sarada Mahavidyalayam Arts and Science College for Women's, Ulundurpet

S. Sathya[2]

Assistant Professor, Department of Information Technology VELS Institute of Science,
Technology & Advanced Studies (VISTAS), Chennai, India

Abstract In underwater wireless sensor network (UWSN) reliable data transmission is a major challenging task for transmitting the data to destination. Extension of lifetime of network also a risky process for researchers working in wireless sensor network. To solve this error, the proposed research is implemented using Genetic Algorithm which is used to solve the complex problem by optimization. The proposed research is compared with existing model where it was proposed by LEACH protocol and it performs the task to transmit the data using energy of every node. In this research paper, the GA algorithm is proposed along with LEACH protocol to reduce the transmission time and increase the data transmission from the source to the base station (BS) through cluster head (CH) nodes. The data transmission is executed with higher efficiency in less time successfully.

Keywords Under water sensor networks (UWSN), LEACH protocol, Genetic algorithm, Signal transmission, Clustering, Optimization

1. Introduction

Now a days WSN is widely used field by built with sensors and nodes which will used for providing the physical phenomenon of surroundings such as pressure, heat and lights with sensors [1]. The structure of the system will be improved their capacity and accuracy of the system by implementing the WSN which is most efficient model to develop information and system for communication. Wireless sensor network is considered as very simple and easiest approach for elaborating to achieve the device flexibility by comparing with wired devices. Sensors will be used for developing faster to believe the technologies using sensors and WSN will be the leading technology using internet of things by designing some protocols with serious issues to reach the highest performance in energy with sensor nodes for providing the small and limited power battery points for spreading the signals manually. The cluster methods are required for regulating the scalable and active works using wireless networks [2] [3]. The gathering of data can be performed by cluster head and it helps to assemble the data through the nodes of cluster and it will combine the data by compressing before begin the transmission to the base stations (BS) and the results looks like cluster heads for consuming the more energy. LEACH algorithm is most proposed algorithm, it has some set of rules for clustering to address the role about cluster head (CH) as minor issues for cluster energy. This algorithm will turn into head of the cluster in

---
[1]prakash.staff@gmail.com, [2]ssathya.scs@velsuniv.ac.in

DOI: 10.1201/9781003428466-41

smart way for better improvement of network performance. US-NOAA has undertaken a survey, stated that oceans occupied 97% of surface of the earth which will increase the underwater of environment for different purposes such as climate change, oil platform, surveillance, etc., and all the applications used to involve the communication medium on two different side for research in underwater wireless communication with more benefits for better communications.

Communication devices based on acoustic waves, high frequency waves, and optical waves are used to introduce underwater wireless communications today. Since they can communicate over long distances, underwater acoustic wireless networks have become one of the most commonly used underwater wireless communications systems. In UWASN which are made up of several autonomous and self-organizing sensor nodes. To collect exact information from deep water, these nodes are manually dispersed in various depths in underwater environments [4]. Collecting data to a sink on the water's surface through acoustic waves.

The normal sensor nodes are present in these networks, along with an acoustic modem for connectivity. Sinks are fitted with both acoustic and radio modems, allowing them to receive data from underwater nodes via acoustic waves and transmit it via radio waves to the base station. The traditional method of disaster control is to deploy an underwater sensor and manually collect data. The disadvantage of the traditional approach of not being able to provide timely information inspired a wave of interest in the concept of an underwater sensor network. To deal with these disasters using an underwater sensor network, a reliable communication medium to exchange data is needed. Electromagnetic waves are initially effective for short-distance communication, but due to high attenuation in the sea, they are insufficient for long-distance communications, while optical waves are easily absorbed and dispersed in the water, allowing for very short-distance communication.

Several quality-of-service aware routing protocols, such as vector-based forwarding (VBF) and depth-based forwarding (DBR), have been proposed for UWASN in recent years. In terrestrial wireless sensor networks, a wide number of network applications have been proposed for determining a route from source to sink. These protocols are built on an end to-end approach, which is ineffective in high-dynamic topology networks with long propagation delays. Since UWASN is such a new topic in this field, most researchers are focusing on the physical layer, connection layer, and localization, while network layer analysis is still in its infancy. As a result, only a few routing protocols for UWASN have been established. Optimization technique to increase the energy efficiency of data transmission in routing Path from a source node to a destination node in UWASN is the most promising approach in an underwater environment.

2. Literature Review

A. Wahid, et al. propose a functional node cooperation (NC) protocol to improve collection efficiency by taking advantage of the fact that underwater nodes can hear other nodes transmissions. The underwater data collection area is divided into several sub-zones to minimise the source level of underwater nodes, but in each sub-zone, the mobile surface node using the NC protocol could move adaptively between selective relay cooperation (SRC) and dynamic network coded cooperation (DNC) [5].

X. Zhuo et al. purpose of this report is on the use of underwater acoustic sensor networks (UASNs), which are distinguished by their large scale, dispersed distribution, and differing traffic loads. The strategy to access the popular communication medium is needed to improve the efficiency of UASNs because the underwater sensor channel is known for its restricted bandwidth, time variation, and high propagation delays [6].

M. Molins et al. proposed a new protocol is known as slotted FAMA because it employs time slotting. Time slotting reduces the need for unnecessarily long control packets, resulting in energy savings. Via simulation of a mobile ad hoc underwater network, protocol efficiency in terms of throughput and delay is evaluated, demonstrating the presence of an acceptable rated power to be used for a given user density [7].

Hongyu Cui et al. present the platforms built by our lab, which are based on the well-known ns2 and ns-miracle simulators and feature a layered structure that supports cross layer signalling. The simulation platform runs on a machine with the Linux operating system installed. The platform's architecture is divided into four layers: an application layer, a routing layer, a medium access layer, and a data access layer [8].

Jun-Hong et al. present a novel networking model for exploring aqueous environments is the large-scale mobile underwater wireless sensor network (UWSN). Remote UWSNs, on the other hand, vary greatly from ground-based wireless sensor networks in terms of communication bandwidth, propagation delay, moving node mobility, and failure probability [9].

Z. Wu et al. presented a new underwater acoustic network localization algorithm is implemented. To improve the sensitivity of random noise, the algorithm uses the raw data before calculating the localization. We use a more reliable set of coordinates and a modified calculation technique to reduce the consistency of the calculation results. The new algorithm is better suited to the underwater acoustic sensor network [10].

Xie, et al. a network simulator for underwater sensor networks. Aqua-Sim is based on NS-2, among the most commonly used network simulators. It is constructed in an object-oriented manner, with all network entities configured as classes. Aqua-Sim accurately simulates underwater acoustic channel attenuation and impact behaviours in long delay acoustic networks [11].

T. Hu, et al. proposed on standard MAC protocols and aims to increase the lifespan of networks by uniformly distributing residual energy from sensor nodes. All throughout routing process, the energy consumption of each node, and also the power generation among a group, is taken into account to measure the reward function, which aids in the selection of appropriate packet forwarders [12].

Yi Cui, et al. maximum lifetime routing problem is given a novel utility-based nonlinear formulation. A completely distributed localised routing algorithm based on this concept is also introduced, and it is shown to converge at the optimum stage, where the network lifetime is maximised. To validate the proposed solution, proper mathematical analysis and simulator results are presented [13].

A. Sankar et al. propose a distributed routing scheme that, in a linearly low percent error, reaches the optimal solution. Our method is formulated on the basis of a multi-commodity flow, which allows various power consumption models and bandwidth constraints to be considered. It works with both static and dynamic networks that shift slowly [14].

V. D. Park et al. presented for remote, multi-hop wireless networks, we present a new distributed routing mechanism. The protocol is part of a group of algorithms known as" connection reversal" algorithms. The protocol's response is organised as a time-ordered series of dispersing calculations, each of which is made up of a set of guided connection reversals [15].

Abdellah, and Hssane, implemented enhanced LEACH by extension of the model by improving the stable area of clustering hierarchy to decrease the proability about failure nodes with some parameters in networks. Some higher nodes called CAG become cluster head for collecting the data about the cluster members to transfer the data to sink for minimizing the consumption of energy of cluster head, which will employ the information to route from CH to sink which allows to reduce the failure of CH to increase the network lifetime [16][17].

Farooq et al., implemented multi- hop routing (MHR) integrated with LEACH partition of networks with various layers of clusters. Each layer of cluster will have CH will be collaborated with adjacent layers for transmitting the data from sensors with BS. Proposing the multi- hop LEACH adopts with optimal paths between CH and BS [18][19].

Liao, and Zhu, proposed an energy balanced algorithm for clustering based on LEACH protocol which depends on residual energy with some distance factors to improve the electing the cluster head (CH) and the strategy for non – CH node selection will be the optimal CH [20].

3. Methodology

In this section, the existing and proposed research is discussed in detail with work flow and comparison tables as follows. The proposed research was implemented with genetic algorithm which was randomly implemented for processing the images to fuse it together. Genetic algorithm was proposed and enhance some methods for better transmission of signals from the underwater environment by carrying the data. The following methods are discussed in detail,

1. Low energy adaptive clustering hierarchy protocol
2. Genetic algorithm (GA)

3.1 LEACH Protocol

Low energy adaptive clustering hierarchy protocol is a hierarchical routing protocol implemented in wireless sensor network for better expansion of network signal lifetime. Sensors are arranged automatically in cluster nodes by themselves and one node specifically will be performed as cluster head node.

LEACH protocol was the widely used routing protocol for fusing the data with clustering strategies to solve security issues and challenges. The process of protocol was divided into different stages to represent the routing protocol which is organized and self-adaptive to decrease the unwanted costs of energy. Cluster node which became head cluster will send signal to member nodes to notify about the process based on the formula,

$$T(n) = P/1 - Px \, (r \bmod P^{-1}) \tag{1}$$

$$T(n) = 0$$

Were,

n denotes random number 0 to1.

P denotes cluster head.

G denotes set of nodes grouped together. To balance the consumption of energy cluster node cannot be again become as head until all the member nodes act as head cluster once in a round. LEACH is a protocol for processing the transmission with efficient low energy method and find the shortest path by generating the form of groups with cluster which is called as NTP method and the proposed research is upgraded with hybrid genetic algorithms (GA).

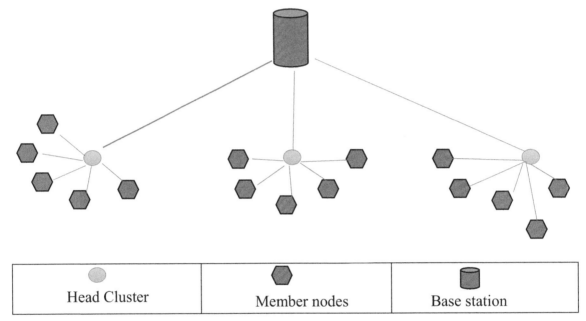

Fig. 41.1 LEACH protocol for WSN

Source: Created by Authors

Merits:

1. Control information not required
2. Save energy
3. Distributed method.

Demerits:

1. Cluster becomes useless, if cluster head dies.
2. Randomly clusters are divided

3.2 Genetic Algorithm (GA)

GA is used for solving risky problems with high quality solutions to optimize for searching the problems. It was developed by holland to give solution for complex optimization problems and to simulate the process of natural environment species. Genetics represents the code of individual to set the data which is known as chromosome for identifying the individual with their entire entity. It also used to identify the local optimal solutions for complex problems which require lifetime for solve the issues. GA will solve the problem using ML with clear execution of result.

The below figure explained that, in the begin stage population of n chromosome will generated randomly and the initial population stage will begin with group of nodes known as population. Each node has a solution for the issues and it will collect some values from the parameters. In fitness stage, the node is calculated for their fitness and state the strength of the fitness on individual nodes using fitness score which will help to identify the node suited for better transmission.

In selection level, two different nodes are predicted among the population and select some particular tasks to allow the node to transfer the data. Crossover is the major focused level which will select the node randomly with pair of head cluster nodes to speed up the process of data transmission. Termination is the process of reintroducing the node from the population to end the task and it has a series of approaches to the proposed model.

This genetic algorithm is proposed in our research to transmit the data process with some consistency and synchronized models. The GA will have a stable clock with same values which will neglect the dead nodes and lacking of energy consumption. GA will combine the existing algorithms and protocols to create a new node path for better transmission of data and finds the best path among other nodes in a cluster group which will help to perform the transmission in efficient time with less error occurrence. All these data are kept recorded for creating the threshold to hold the best root node which is called as malicious node.

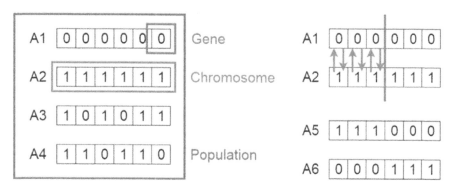

Fig. 41.2   Genetic algorithm (GA) overview

Source: Wiki Media Commons

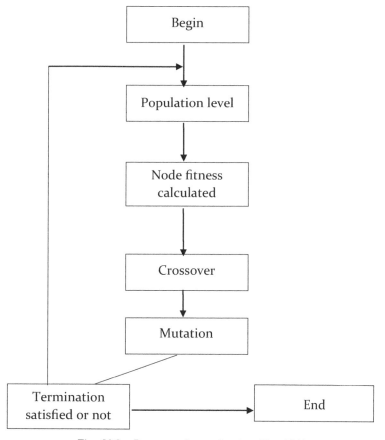

Fig. 41.3   Process of genetic algorithm (GA)

Source: Created by Authors

Clustering:

Clustering is one of the major considered methods in wireless sensor network (WSN) which will increase the lifetime of network and it will involve the sensor nodes together to form a clusters and head of the clusters for all the member nodes. The clusters are formed based on how all the nodes will transmit the packet of data directly to the sink node continuously until the packet recached the proper node successfully. Sink node will be located outside the area which is supervised and the nodes are transmitted with higher energy to reach the destination (sink) directly. Clustering of nodes have a huge benefits like energy efficiency, good communication through network, reduced delays, etc.,

The below Fig. 41.4 will describe the overall process of our proposed research in detail, the input will be considered in terms of parameters such as number of populations, number of nodes, number of mutation and number of standards. All these parameters are compared with both existing research and proposed research for better understanding of the proposed up gradation. The values of all parameters are changed dynamically to execute different results and the outcome of proposed genetic algorithm is highly efficient that existing research.

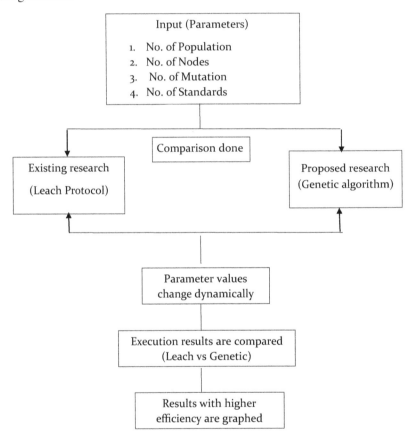

Fig. 41.4 Flow chart of proposed methodology

Source: Created by Authors

Table 41.1 Comparison of LEACH protocol and genetic algorithm

S. No	LEACH Protocol	Genetic Algorithm
1.	Speed varies based on system	No speed changes
2.	Sequence is different in transmission	Transmission sequence is parallel
3.	Dead nodes presence	No dead nodes
4.	Difference in clock cycle process	Clock cycle process will remain same
5.	Accuracy predicted based on some specific terms	Accuracy based on population
6.	No specific history set for transmission track	History for every transmission recorded
7.	LEACH protocol was not synchronized	Genetic algorithm set synchronized in process.
8.	Transmission based on energy	Transmission based on time, speed.

Source: Created by Authors

4. Results and Discussion

In this section, the proposed research with GA algorithm for efficient and highly executed results. Results from LEACH protocol are discussed from Fig. 41.5, that indicates the ratio level of every performing node.

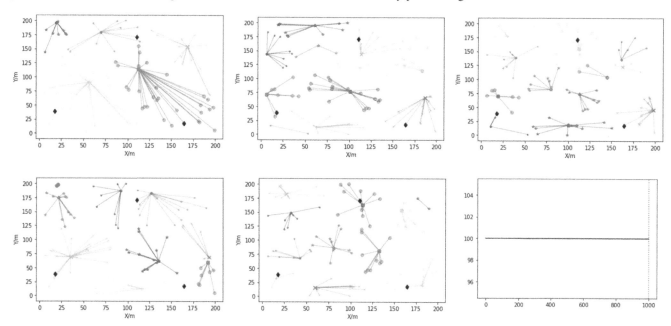

Fig. 41.5   Existing research (LEACH Protocol) outcomes

Source: Created by Authors

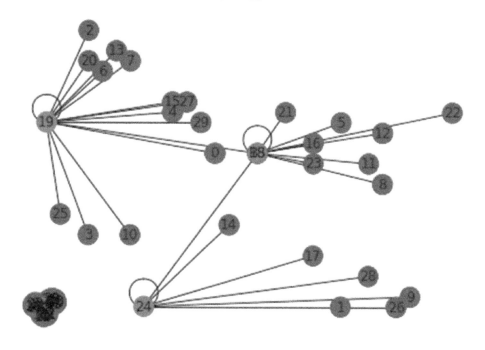

Fig. 41.6   GA implemented for analysing NT on round 10

Source: Created by Authors

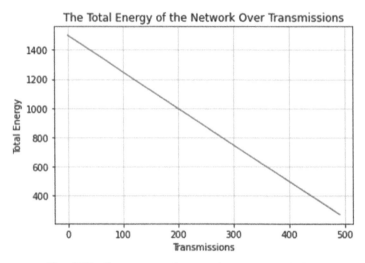

Fig. 41.7 Energy execution graph over transmission

Source: Created by Authors

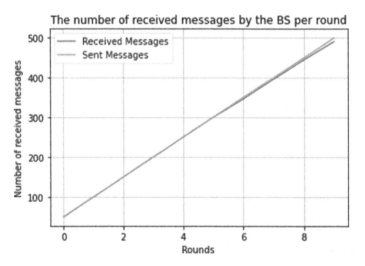

Fig. 41.8 Messages received per round at base station (BS)

Source: Created by Authors

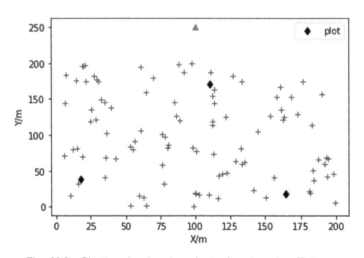

Fig. 41.9 Plotting dead nodes, cluster head, node efficiency

Source: Created by Authors

LEACH protocol is compared with proposed algorithm named as genetic algorithm (GA) where the dead nodes are occurred due to improper transmission of data with higher energy. But in the proposed algorithms, the genetic algorithm is performed with avoiding the nodes to die and transmit the signals efficiently to the base station at destination. In Fig. 41.9, dead nodes are indicated in red color, cluster head indicated in black color and efficiency of the nodes are indicated in blue color. Minimum of 10 dead nodes are identified during transmission out of 200 nodes atleast. The percentage of node performance and its ratio are calculated with N number of iteration which will take some time to process.

Table 41.2 Consumption of energy

Methods	Time (1ms)						Energy remaining
	0	100	200	300	400	500	
LEACH	29.2%	37.4%	45.5%	52.3%	59%	66.1%	34.8%
LEACH + GENETIC	70%	60%	50%	40%	30%	20%	75.6%

Source: Created by Authors

Consumption of energy is calculated for LEACH protocol along with Genetic algorithm using time taken in 1 milli seconds. Based on the calculation of transmission and energy assumed remaining energy are calculated. The below mathematical formula is used to calculate the consumption of energy,

$$Energy\ Consumption = \frac{Total\ Energy}{Sum\ of\ Jules} \times 100 \tag{2}$$

The total joules of 2000 are taken for calculating consumption of energy by multiplying with 100 percentage upto least energy of 300. By comparing with previous leach protocol, it consumes upto 34.8% where the proposed model Leach along with genetic consumed 75.6% of remaining energy to prove the uniqueness of proposed model.

N number of nodes are grouped together to form a cluster with the values of $x = 200$ and $y = 200$, which mentioned about the nodes. The GA is used to find the path which is near to each other and transmit the data by neglecting the higher energy by performing each group with individual process. The proposed research will track the transmission of data to which it will be transmitted. The proposed research stated that, the consumption of power energy may be avail between 12 v to 24 v, where the transmission is transferred by speed and time. The proposed research implementation will be created an individual history of every transmission to easily process the task next time by reducing time. For example, data with 10 mb, 20 mb, 35 mb size is transferred over nodes by finding the shortest path and reach the destination. When the process is initiated for next time with 20mb of data size, the history previous stored was referred and easily transmitted the data to destination by reducing the time.

5. Conclusion

In this research paper, the underwater wireless sensor network (UWSN) is proposed using genetic algorithm by comparing with existing research known as LEACH. The proposed research was implemented with new method by finding the shortest path to send the data to destination in a short time. The energy of every node will be used to speed up the transmission process in existing research, and the proposed research can create an individual history as backup for every transmission and it will be used for later transmission without any time delay. The result executed in proposed research is highly efficient for data transmission by avoiding node dead and delayed transmission of data. Any one particular parameter is selected for boosting the particular method as future enhancement.

References

1. Zhang, et. al. "UASN based simulation and emulation platforms", IEEE China OA, pp. 1–5, 2016.
2. Wu, et. al. "Enhanced UAN with localization algorithms", Journal of communication, volume 12, issue 3, pp. 77–83, 2015.
3. Yang, et.al., "Awareness of Delay and queue scheduling methods based on MAC protocol for UASN", IEEE access, volume 7, pp. 56263–56275, 2019.
4. Jiang, et. al. "Self-deployment of node based on cluster uneven model by correcting the radius of underwater sensors", Sensors, Volume 16, issue 1, 2016.
5. Long, et. al. "Data collection methods with high availability based on multiple data for UWSN", IEEE transactions on mobile computing, volume 19, issue 5, pp. 1010–1022, 2022.

6. Rani and Malhotra, "Routing protocols based on energy efficient chain-based model", Journal of network and computer applications, volume 92, pp. 42–50, 2017.

7. Jiang, et. al. "Efficiency of energy routing model for UWSN", International journal of communication system, volume 30, issue 15, pp. 3303–3309, 2017.

8. Mridula, et. al. "Localization of anchor node for acoustic sensor network under water model", International journal of communication system, volume 31, issue 2, 2018.

9. Liao, et. al. "Balanced energy clustering method using LEACH algorithm", 2nd International conference on system engineering and modelling, volume 5, issue 3, pp. 90–95, 2013.

10. Alghamdi, et. al. "Awareness of proactive routing approach of energy with path efficiency in WSN", IEEE access, volume 7, pp. 140703–140722, 2019.

11. Goyal, Verma, et. al. "Secured authentication for data aggregation scheme to enhance QoS for UWSN", Wireless personal communication, volume 113, issue 1, pp. 1–15, 2020.

12. Sandhu, et. al. "Data agglomeration using ML for UWSN", International journal of management and technology in engineering, volume 9, issue 6, pp. 240–245, 2019.

13. Farsad, et. al. "Survey on advancement of molecular communication", IEEE communication surveys, volume 18, issue 3, pp. 1887–1919, 2016.

14. Verma, et. al. "Cluster method to detect fault and recover techniques in UWSN", International journal of communication system, volume 31, issue 4, 2018.

15. Rahman, et. al. "Optimal energy protocol for efficient relay selection on UWSN", IEEE access, volume 6, 2018.

16. Ahmed & team, "Tolerance of delay in underwater WSN", Journal of mobile informative system, 2016.

17. Buckingham, et. al. "Characteristics of underwater environment", Journal of applied acoustics, Elsevier, 2017.

18. Cai, et. al. "Collection of data in UWSN based on mobile computing", IEEE access, volume 7, 2019.

19. Wang, et. al. "Location point covering data collection algorithm to avoid 3D UAWSN", IEEE access, volume 5, 2017.

20. Vieira, et. al. "Path joining model for suitable USN", IEEE transaction & sustainable computation, volume 4, issue 4, 2019.

Human Machine Interaction in the Digital Era – Prof. J. Dhilipan et al. (eds)
© 2024 Taylor & Francis Group, London, ISBN 978-1-032-54998-9

A Drastic Review on Self-Efficacy of Hard of Hearing and Mutant Students Using Assistive Technologies in Higher Education

42

Agusthiyar R.[1]
Prof & Head, Department of Computer Applications,
SRM Institute of Science and Technology, Ramapuram, Chennai

A. Chitra[2]
Research Scholar-Part time (External), Department of Computer Science,
SRM Institute of Science and Technology, Ramapuram, Chennai

Abstract A human's ability to envision, listen, talk, and respond in accordance with circumstances is one of their most priceless gifts. Despite the fact that their vocal chords may not be affected, people who are born deaf or who lose their hearing at a young age are unable to communicate. Due to their inability to imitate spoken sounds due to hearing loss, they become deaf and dumb. Peer communication between hearing-impaired and hearing-normal peers has never been easy. To aid in their learning, deaf or hearing-impaired students may have a note-taker or a sign interpreter present in the classroom. "Education of special needs children" In India, 18 million people who are deaf are illiterate or partially illiterate. Blind persons can easily communicate using conventional language; however, hearing-impaired people use their own manual-visual language called "sign language." Thus, there is an obstacle to communication among these two communities as a result of this. Many makers throughout the world have created numerous systems of sign language; however, the end users find them neither versatile nor cost-effective. In this work, the numerous teaching strategies that are employed to instruct and educate the pupils will be discussed. For those who have hearing loss, using captioning services, sign language interpretation, warning devices, telecommunication devices, frequency modulation, as well as loop systems, and hearing assistive equipment can help them communicate and learn more effectively. There are various educational technologies that can promote the efficient use of assistive technology aids to support in-class instruction. The foundation of an e-learning system is formal education, but it is supported by electronic resources that are not found in a regular classroom. With a large variety of students, including deaf and dumb people, the entire programme is offered online. E-learning, sign language interpreters, sign language gloves, sign language professional academic videos, sign language keyboards, text and video materials, optical character recognition (OCR), tele-typewriting (remote keystroke), and other learning strategies for disabled and hard-of-hearing students in higher education will be discussed.

Keywords OCR, Sign language, E-learning, Capturing devices, Frequency modulation

1. Introduction

People who are profoundly deaf are referred to as "deaf," but those who are mildly to moderately deaf are stated to as "hard of hearing." Hearing impaired people have either pre- or post-lingual deafness. Individuals who are pre-lingually deaf have impaired hearing before learning how to speak. Those with post-lingual deafness lost their hearing after learning to speak.

[1]hod.dca.rmp@srmist.edu.in, [2]chitramscmphilcs@gmail.com

DOI: 10.1201/9781003428466-42

Each group will experience hearing loss and the degree of deafness differently. Hearing-impaired students could come across as isolated in the classroom. If students don't have many chances to meet and talk with other students, it could affect how well they learn. When transitioning to the post-secondary educational environment, some hearing-impaired students who have only experienced the school system may need some time to acclimatise because they are used to a structured learning environment [1]. Anxiety over performance in front of others may result from communication issues and changes. This might make it more difficult for students to participate in sessions, especially if their hearing loss has affected how well they can speak.

2. Literature Review

Aderonke Kofo Soetan, et al. [2], proposes the integration of technology has the possible to be a significant instrument for providing effective learning materials. Numerous studies in the area of education and in the field of education in general have demonstrated that assistive technology are instruments that, when used effectively, would aid in the advancement of learning amongst students with disabilities. Idowu Olatunji [3], et al. establishes the students were aware of and generally used assistive technology for learning, according to this study's findings. It is suggested that the government integrate access to assistive devices in educational plans and programmes and that seminars and workshops on the use of cutting-edge assistive technologies be regularly offered for students and lecturers. Tatiany X. de Godoi, et al. [4], proposes the analysis used a qualitative research approach, and information was gathered by surveying participants and watching them use assistive technology. From the standpoints of usability, user experience, and accessibility, challenges and barriers could be identified using the grounded theory methodology.

Sarah Ezekiel Kisanga, et al. [5], establishes the thematic analysis found that the lack of hearing aids, their poor quality, the lack of sign language interpreters, and unsupportive classroom environments are all resource barriers that make it hard to learn. Dhruv Jain, et al. [6], presents during a three-week implementation in four DHH homes, participants' awareness of their homes and of themselves improved. However, the absence of sight lines and difficulties with sound classification were also revealed. Oliver Alonzo, et al. [7], proposes certain text difficulty levels were discriminated against among participants who had lowering literacy levels through subjective evaluations and comprehension tests with simple linguistic structures. Only individual differences in text complexity levels among people with higher levels of literacy could be distinguished. Matthew Seita, et al. [8], establishes previous research has found that Americans with DHH identities have distinct literacy profiles and, on average, lower reading literacy levels than their hearing counterparts.

3. Materials and Methods

Cochlear implants, hearing aids, and certain other devices can all be used to help people with hearing loss, but they can't fully replace the clarity and sharpness of natural hearing. However, there are other approaches that can improve a student's understanding and participation in the classroom. This option enables learners who are deaf or hard of hearing to take part more fully in classroom instruction because it is portable, lightweight, and unobtrusive. There are many non-invasive assistive hearing technologies available today those teachers can incorporate into their curricula. E-learning, sign language gloves, sign language interpreters, infrared, IP and video relay services, frequency modulation, a sign language keyboard, text and video resources, tele-typewriting (remote keystroke), as well as other learning techniques.

3.1 E-learning

By utilising new technology, e-learning is a great strategy to raise the proportion of educated Deaf individuals. Many modern systems have been created to make it easier for Deaf persons to learn and train online. Deaf persons need the e-learning system to acquire knowledge. The following are required to implement e-learning: virtually equipped classroom, Sign language translator, hand gesture graphics, sign language related educational video, sign language teaching video, text and video resources, sign language dictionary, computer and communication tools.

3.2 Sign Language Interpreters

A Deaf or Hard-of-Hearing individual with particular training in interpreting and the court system is known as an intermediary interpreter. The intermediary interpreter, often known as a Deaf interpreter, is a professional who is fluent in both visual-gestural language elements and American Sign Language (ASL) shown in figure 1. When communication is difficult and/or dangerous, Deaf interpreters ensure that almost all deaf people have equivalent and full right to use to the legal system. In classroom and lecture halls, the translator provides services to deaf and hard-of-hearing learners utilizing American Sign Language or another manual sign method. The cultural identity of deaf and hard of hearing staff and students as well as lectures, discussions, announcements, chats, conferences, events, and other spoken word scenarios [9].

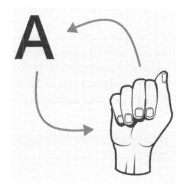

Fig. 42.1 Sign language interpretation

Source: Adapted from https://apps.apple.com/us/app/sign-languagetranslator/id1458992650

3.3 Sign language Gloves

A glove interprets and speaks the user's signs, providing assistance to those who have hearing loss shown in Fig. 42.2. A right-handed glove translates sign language into spoken English using a machine learning (ML) algorithm. The MPU-6050 (a three-axis accelerometer and gyroscope) is used in our gadget to identify the orientation and rotational movement of the hand. Five-Spectra Symbol Flex- Sensors are used to measure how far each finger is bent. An ATmega1284p microcontroller is used to read the sensors, average them, and group them into packets. Then, a Python script is delivered along with these packets in serial form to the user's PC. The user creates data sets from the glove for each motion that should eventually be interpreted, and the system learns from these datasets to predict what a user is signing in real time [10].

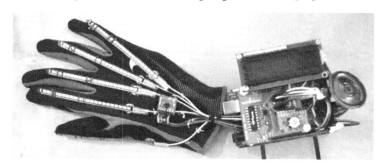

Fig. 42.2 Sign language gloves

Source: Adapted from https://hackaday.com/2010/10/04/from-sign-language-to-spoken- language/

3.4 Sign Language Keyboards

ASLized, a nonprofit company that creates and advertises educational resources in American Sign Language, created the sign language keyboard application called Signily (ASL). Emoji-like hands depicting precise marks and shapes are used on the keyboard shown in Fig. 42.3. Even though many Deaf people use email, text, and handwriting in addition to sign language, the ability to use sign language is no different from using one's phone in their native language, even if they understand English, because sign language differs from location to location and doesn't always translate flawlessly to English. Additionally, the skin tone of the keyboard's hands can be changed.

Fig. 42.3 Sign language keyboard

Source: Adapted from https://www.etsy.com/listing/597105843/82-asl-stickers-aslkeyboard-stickers

3.5 Tele-typewriting

A digital communication channel is built into a teletypewriter, also known as a teletype. A teleprinter, also known as a teletypewriter, teletype, or TTY for Teletype or Teletypewriter, is a now mostly extinct electro-mechanical typewriter that can be used to transmit typed messages over a basic electrical communications channel, which is frequently only a pair of wires shown in Fig. 42.4. The most advanced version of these gadgets is all electronic and uses a screen in place of a printer. TDDs (Telecommunications Equipment for the Deaf) or TTYs are still used by the deaf for telephone communications that require typing.

Fig. 42.4 Teletypewriter

Source: Adapted from https://www.sutori.com/en/item/tty-teletypewriter-412d

4. Conclusion

According to the study, using assistive technology helps students feel more capable. This means that effectively involving children who have hearing impairments in classroom activities requires creating a sense of belonging for them. In order to effectively deploy assistive devices and tools to enhance classroom instruction, it was advised that schools hire educational technologists. Assistive technology serves as one of the crucial elements for promoting the inclusion of individuals with hearing loss, along with other supports including sign language interpreters, personal assistance, and barrier reduction. In order for many people to receive and benefit from education, it is essential that they have access to auxiliary technology and devices for those with impairments.

References

1. Ali Mohammed Ridha, et al. "Assistive Technology for Hearing-Impaired and Deaf Students Utilizing Augmented Reality", Conference Paper, September 2021.
2. Idowu Olatunji, et al. "Undergraduates with Special Needs' Awareness Towards the Use of Assistive Technology for Learning at The University Level", ASEAN Journal of Community and Special Needs Education 1(1) (2022) 45-54.
3. Aderonke Kofo Soetan, et al. "Attitude of Hearing Impaired Students Towards Assistive Technology Utilization in Oyo State Adopting the Survey Method", Indonesian Journal of Community and Special Needs Education 1 (2) (2021) 103-118.
4. Tatiany X. de Godoi, et al. "A Case Study about Usability, User Experience and Accessibility Problems of Deaf Users with Assistive Technologies", Research Gate, Chapter-July 2020, DOI: 10.1007/978-3-030-49108-6_6.
5. Sarah Ezekiel Kisanga, et al. "Barriers to Learning Faced by Students Who are Deaf and Hard of Hearing in Higher Education Institutions in Tanzania", PED NO. 37, VOL. 2, 2019.
6. Dhruv Jain, et al. "HomeSound: An Iterative Field Deployment of an In-Home Sound Awareness System for Deaf or Hard of Hearing Users" CHI '20, April 25–30, 2020, Honolulu, HI, USA.
7. Oliver Alonzo, et al. "Comparison of Methods for Evaluating Complexity of Simplified Texts among Deaf and Hard-of-Hearing Adults at Different Literacy Levels", CHI '21, May 8–13, 2021, Yokohama, Japan.
8. Matthew Seita, et al. "Automatic Text Simplification Tools for Deaf and Hard of Hearing Adults: Benefits of Lexical Simplification and Providing Users with Autonomy", CHI 2020, April 25–30, 2020, Honolulu, HI, USA.
9. Mr. S. Shiva Prakash, et al. "Educating and communicating with deaf learner's using CNN based Sign Language Prediction System", International Journal of Early Childhood Special Education (INT-JECSE), Vol 14, Issue 02, 2022.
10. Steven M. Goodman, et al. "Toward User-Driven Sound Recognizer Personalization with People Who Are d/Deaf or Hard of Hearing", Proc. ACM Interact. Mob. Wearable Ubiquitous Technol., Vol. 5, No. 2, Article 63. Publication date: June 2021.

Human Machine Interaction in the Digital Era – Prof. J. Dhilipan et al. (eds)
© 2024 Taylor & Francis Group, London, ISBN 978-1-032-54998-9

Uncovering the Secrets of Stegware: An In-Depth Analysis of Steganography and its Evolving Threat Landscape

43

M. Anitha[1]

Research scholar, Department of Computer Science and Engineering,
SRM Institute of Science and Technology, Ramapuram, Chennai

M. Azhagiri[2]

Assistant Professor, Department of Computer Science and Engineering,
SRM Institute of Science and Technology, Ramapuram, Chennai

Abstract Steganography is the technique that hides the information from the plain text and conceals the secret text. Stegwares are masked into the digital files which are not visible to the naked eye and cause a serious security threat. These files are of low security risks and are a gateway for cyber attackers to conceal malicious content. Amidst the growth of security, malware are growing at a very rapid rate on each day making the researchers and security analyst difficult to analyze and detect the malware. A stegware is a stealthy malware that is increasing rapidly. Stegware authors and cyber security professionals tend to increase the devising techniques and reinvent the existing ones to hunt the stegware. Recent years, the growth in stegware has increased which makes the security analysts and researchers discover the techniques. While, the attackers and hackers master the stegware and their applications. The growth in stegware causes a great threat to the social and economic sector. These causes are non-negligible as the attackers have evolved over time and exploit the victims to the maximum. Similar to stegware, phenomena like mobile malware, fireless malware are also widespread and are a serious threat to digital content on the internet. Thus the development of tools or models to identify the stegware has spurred in recent days. The rise in stegware has resulted in several manipulations in the confidential data. This study dissects the stegware and analyzes its behavior through which a detailed report on the stegware family can be studied. Further, the history of stegware, comparison between stegware and steganalysis. The detailed reports on various tools to detect stegware are also discussed along with the detection process.

Keywords Stegware, Mobile malware, Stealthy malware, Fireless malware

1. Introduction

The method of concealing the restricted information in a standard document or message that avoids the detection is known as Steganography. Then, at that point the mysterious content is separated at the objective. Steganography along with encryption is used to protect the data as an additional step. Steganos is derived from the Greek word which refers to the concealed meaning or covered. Various forms of steganography have evolved in the past centuries, where a secret text is hidden with any available technique [1]. The hidden text is concealed and is encrypted with a secret key and the secret text is concealed within any type of data of a digital content. Encryption is done to expand the trouble of recognizing the mysterious content. In general, steganography is used to transfer a secret code. With the alarming increase of malware threats, steganography is used to hide

[1]anithamuthulingam.26@gmail.com, [2]azhagiri1687@gmail.com

DOI: 10.1201/9781003428466-43

the malicious code into the text. Thus the fraudulent actions caused by the hackers can be prevented and can make a copyright protection. The process of analyzing and extracting the data is steganalysis, which is the rear side of steganography. It detects the hidden information and either extracts or destroys the hidden data that is found on the stego image. Stegware is a type of malicious software that is designed to hide its presence by using steganography, which is the technique of concealing data within another type of data. Stegware can be used for a variety of malicious purposes, such as stealing sensitive information, initiating attacks on other systems, or simply remaining undetected on a compromised system. Detection of stegware can be challenging, as traditional antivirus software may not detect it, and manual analysis may be required to uncover its presence.

2. Related Work

"Stegomalware: A Systematic Survey of Malware Hiding and Detection in Images, Machine Learning Models and Research Challenges"[2] by Chaganti, R et al. is a research paper that provides a comprehensive overview of the use of steganography in malware, as well as the techniques and challenges associated with detecting such malware. The author discusses the different types of steganographic techniques used in malware, including LSB embedding, discrete cosine transform (DCT), and spread spectrum. It also examines the use of machine learning algorithms for detecting steganographic malware; including supervised and unsupervised learning techniques, such as support vector machines (SVMs), deep learning, and clustering. The authors provide an overview of the current state of the art in steganographic malware detection, including various tools and techniques used to identify malware hidden in images.

The ensemble-based stegware detection system [3] proposed by Monika A and Eswari R is designed to detect information hiding malware attacks that use steganography to evade detection. Steganography is a technique that involves hiding information within other information, such as hiding a message within an image or audio file. The proposed system uses multiple machine learning algorithms to analyze various features of the files on a system. The system then combines the outputs of these algorithms to improve detection accuracy and reduce the risk of false positives or false negatives. The algorithms used in the paper include the decision tree algorithm, the random forest algorithm, and the support vector machine (SVM) algorithm.

"Comparison of the State-of-the-Art Reviews in Multimedia Steganography and Steganalysis"[4] by Doaa et. al. is a research paper that compares and contrasts the state-of-the-art reviews in multimedia steganography and steganalysis. The paper provides an overview of the latest techniques and methods used in both fields and discusses the strengths and weaknesses of each. The paper first provides a brief introduction to multimedia steganography and steganalysis, describing the basic concepts and techniques used in each field. It then provides an overview of the latest research and developments in multimedia steganography, including various embedding techniques, such as least significant bit (LSB) embedding, and the machine learning algorithms for steganography detection.

The MalJPEG paper, by A.Cohen et al. presents a machine learning-based solution for detecting malicious JPEG images [5]. The authors propose a deep learning model that can identify malicious JPEG images by analyzing their features and comparing them to a database of known malicious images. The model is trained on a large dataset of benign and malicious images and achieves high accuracy in detecting malicious images. The authors demonstrate the effectiveness of their approach by testing it on a dataset of over 100,000 JPEG images, including both benign and malicious images. They show that their model outperforms existing methods for detecting malicious images and can accurately identify a wide range of image-based attacks, such as steganography and image-based malware

The author Jessica Fridrich et al. on her work "Practical Steganalysis of Digital Images – State of the Art"[6] focuses on the state of the art in practical steganalysis of digital images is constantly evolving, as new techniques and algorithms are developed. One approach is the use of statistical analysis to identify anomalies in the image data. For example, hidden information may alter the distribution of pixel values in an image, which can be detected through statistical analysis. Another effective approach is to analyze the image at multiple scales, using techniques such as wavelet analysis or multiresolution analysis. This can help to detect hidden information that has been embedded at different levels of resolution, as well as to identify areas of the image that have been specifically targeted for embedding.

In the paper "Never Mind the Malware, Here's the Stegomalware [7]", the author Luca et al. Focus is on the Stegomalware, which is typically distributed via email or social media, disguised as a harmless file or image. Once the file is opened, the stegomalware is activated, allowing the attacker to gain access to the victim's computer and steal sensitive information. Stegware is particularly dangerous because it can be used to bypass traditional cybersecurity measures such as firewalls and antivirus software. The author Luca Caviglione tries to combat the threat of stegomalware, cybersecurity experts are developing new techniques for detecting and removing hidden code from files and images.

Thus, the rise of stegomalware highlights the need for continued innovation in the field of cybersecurity. As attackers continue to develop new and more sophisticated techniques for hiding their malicious code, cybersecurity experts must remain vigilant and adapt their strategies accordingly. The paper "An emerging threat Fileless malware: a survey and research challenges [8]" by S. Kumar provides an overview of fileless malware, a type of malware that operates in memory and does not leave any traces on the victim's hard drive. The author discusses the various techniques used by fileless malware to evade detection, such as injecting malicious code into legitimate processes, using PowerShell scripts, and exploiting vulnerabilities in software.

3. Proposed Work

With the traditional antimalware systems in practice, the users protect their data confidentiality for a very long time span with their signatures. Since, there is evolution of newer stegware and its associated security threats are in a considerable manner to the user and society ML and AI are used intensively to detect and work with the most innovative stegware. The adversaries are increasing at an alarming rate each day coming up with newer stegware. In recent days, there have been reports of stegomalware being used in various cyber attacks. Stegomalware is a particularly insidious type of malware because it can be difficult to detect. Traditional antivirus software may not be able to detect the hidden code within the image or file, and the malware can remain undetected for long periods of time, allowing it to steal data or carry out other malicious activities. With the several improved factors in detecting the stegware like antivirus, certain times false negatives appear which is unavoidable. This happens while detecting the stegware in multimedia files like image, audio, video files. This study focuses on analyzing the stegware, stegware types, history, detection tools and methods, along with a comparative study. This work may stand as a base to deal with the stegware and as a reference on Machine learning models for steganalysis for images. This paper's motivation is to give a detailed summary of stegware and its analysis with varied examples and comparisons.

4. The History of Stegware

Steganography has a long history that dates back to ancient times when people used to write secret messages on the wax surface of tablets before covering them with a layer of plain wax.Advanced Persistent Threat (APT) malware is a type of malware that is specifically designed to remain undetected within a system for a long period of time. Some notable examples include: Operation Aurora, Stuxnet, Duqu, Flame.Zeus Trojan, also known as Zbot, is a type of malware that was first discovered in 2007. It is a sophisticated form of banking malware that is designed to steal login credentials and other sensitive information from victims, including banking credentials, credit card numbers, and personal identification numbers is a type of Trojan malware. Some of the most notable Zeus variants include: GameOver Zeus, Ice IX, and Citadel. Stegosploit is a type of malware that uses steganography to hide its malicious code in an image or advertisement that is then distributed through a legitimate website. The malware was first discovered in 2015 and was notable for its ability to infect victims' computers simply by visiting a compromised website, without the need for the victim to download or execute any files. Stegosploit works by embedding its malicious code in an image or advertisement that is displayed on a website.

5. Historical Malware Variants for Employing Steganography

Malware has been using steganography [16] for exploitation for quite some time, and there have been several historical malware variants that have employed steganography for their malicious activities. The Table 43.1 gives a brief summary of the Malware variant and its purpose along with its year of identification. Just a few examples of historical malware variants that employed steganography for exploitation.

Table 43.1 The history of stegware variants and their purpose

Malware Variant	Year	Steganography Technique	Purpose
JFIF Injector	2001	LSB	Hide malicious code in JFIF (JPEG File Interchange Format) files
MP3Stego	2001	LSB	Hide malicious code in MP3 files
Outguess	2001	F5 algorithm	Hide malicious code in image
S-Tools	2001	LSB	Hide malicious code in image files
Steganos	2001	LSB	Hide malicious code in image files
StegFS	2002	File system steganography	Hide malicious files within a file system
Steganography Tool	2002	LSB	Hide malicious code in image files
StegoMagic	2002	LSB	Hide malicious code in image files

Malware Variant	Year	Steganography Technique	Purpose
MP3Stego	2003	LSB	Hide malicious code in MP3 files
SNS-Steg	2003	LSB	Hide malicious code in image files
SteganoNet	2003	LSB	Hide malicious code in network traffic
Trojan.Stegoloader	2012	Multiple techniques	Hide malicious code in image files, HTML files, and other file types
Backdoor. Vestibular	2015	LSB	Hide malicious code in image files
Stegano	2017	Multiple techniques	Hide malicious code in image files and other file types

Source: Made from various blogs and sites

6. Steganography and Stegware

Steganography is the practice of hiding a message or information within another object, such as an image, audio file, or video, in such a way that it is not easily detectable. The goal of steganography is to conceal the existence of the message, as opposed to cryptography, which aims to make the message unreadable to unauthorized parties. In digital steganography, the message is hidden within the bits of the cover object, which can be any digital file format.

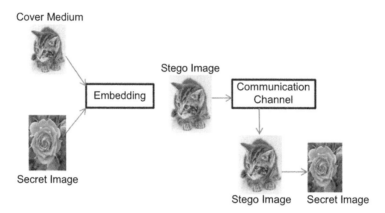

Fig. 43.1 Steganography process

Source: Made by Author

Stegware [9] can be used for both legitimate and malicious purposes. Stegware can come in many forms, including standalone applications, plugins for popular software programs, or scripts that can be executed on a command line. Detecting stegware often requires specialized tools and techniques, such as steganalysis algorithms, that are designed to detect and analyze steganographic payloads in digital media. Some examples of steganography techniques include least significant bit (LSB)[11] insertion, phase encoding, spread spectrum, and transform domain techniques. Table 43.2 gives an overall idea of Multimedia Steganography and steganalysis in various aspects.

Table 43.2 Comparison multimedia steganography and steganalysis

Aspect	Multimedia Steganography	Steganalysis
Definition	The process of hiding secret information within multimedia files such as images, audio, or video.	The process of detecting the presence of hidden information within multimedia files.
Objective	To conceal the existence of the secret information by making it indistinguishable from the original file.	To detect the presence of hidden information, determine the type of information, and recover the original data if possible.
Techniques	LSB, DCT, Spread Spectrum (SS)	Statistical analysis, visual analysis, bit pattern analysis, etc.
Tools	Steganography tools such as OpenStego, Steghide, S-Tools, etc.	Steganalysis tools such as Stegdetect, OutGuess-Extractor, StegExpose,
Applications	Confidential communication, copyright protection, data hiding.	Digital forensics, cybercrime investigations, information security, etc.
Challenges	Maintaining the visual and/or auditory quality of the original file while embedding the secret data.	Dealing with the high computational complexity and diversity of steganography techniques.
Counter measures	Detection tools, encryption of the secret data, hiding the metadata of the multimedia file.	Using steganography-aware tools, analyzing the file format and structure, analyzing the statistical properties of the multimedia file.

Source: Made from various blogs and sites

6.1 Study of Stegware Creation Tools

The creation of stegware [12], which is malware that employs steganography to conceal its malicious payload, requires specialized tools and techniques. Here are some of the common tools used for creating stegware: Steganography software, Malware construction kits, Code obfuscation tools, Virtual private networks (VPNs), and Exploit kits.

Table 43.3 Historical malware variants employing steganography

Stegware variant	Year	Steganography technique	Exploitation stage	File type	Purpose
MP3Stego	2001	LSB	Delivery	MP3 files	Hide malicious code in MP3 files
Outguess	2001	F5 algorithm	Delivery	Image files	Hide malicious code in image files
S-Tools	2001	LSB	Delivery	Image files	Hide malicious code in image files
Steganos	2001	LSB	Delivery	Image files	Hide malicious code in image files
StegFS	2002	File system steganography	Execution	File systems	Hide malicious files within a file system
Stegano Tool	2002	LSB	Delivery	Image files	Hide malicious code in image files
StegoMagic	2002	LSB	Delivery	Image files	Hide malicious code in image files
MP3Stego	2003	LSB	Delivery	MP3 files	Hide malicious code in MP3 files
SNS-Steg	2003	LSB	Delivery	Image files	Hide malicious code in image files
SteganoNet	2003	LSB	C&C	N/W traffic	Hide malicious code in network traffic
Trojan. Stegoloader	2012	Multiple techniques	Delivery	Image,HTML files	Hide malicious code in various file types

Source: Made from various blogs and sites

The table shows the different stegware and the year they were discovered along with their techniques used to exploit the different file types. This also covers the purpose of their exploitation.

7. Steganalysis

Steganalysis is the process of detecting the presence of hidden information within a digital image, audio file, or video file. The goal of steganalysis [11] is to identify the existence of a covert message or payload that has been embedded within a cover medium using steganography. There are different techniques and methods used in steganalysis, and they can be classified into two main categories: Statistical and Heuristic.

7.1 Feature Extraction

In steganalysis, features are extracted from the stego object and used to identify whether a message has been embedded within it. The process of feature extraction [14] involves analyzing specific characteristics of the stego object and extracting numerical features that can be used to differentiate between stego objects and cover objects. The choice of features to extract will depend on the specific steganography technique used, the type of digital object being analyzed, and the characteristics of the embedding process. There are various methods and techniques used in feature extraction for steganalysis, and they include Statistical features, Structural features, Model-based features, Machine learning-based features.

7.2 Stegware Detection Frameworks

A stegware detection framework [13] is a set of methods and tools designed to identify the presence of steganographic techniques in digital files. The framework can be used to detect stegware in a variety of contexts, such as in network traffic, email attachments. The framework typically consists of components such as Steganalysis algorithms, Signature databases, Network traffic monitoring tools, File integrity checkers, System logs and event monitoring.

7.3 Stegware Creation Process

The process of creating stegware [15] involves the following steps: Selecting a cover medium, Selecting a steganography technique, Preparing the payload, Embedding the payload, Testing the stegware, Distributing the stegware.

7.4 Stegware Analysis Framework

A stegware analysis framework is a set of methods and tools designed to analyze and detect steganography techniques in digital files. The framework typically consists of several components, such as Steganography analysis algorithms, Network traffic monitoring tools, File integrity checkers, System logs and event monitoring, Reverse engineering tools. Thus, a stegware analysis framework is a set of tools and methods designed to detect the presence of steganography in digital files and to analyze and identify the steganography technique used, the payload being hidden, and the purpose of the stegware.

8. Conclusion

After conducting extensive research on stegware, it is evident that this type of malware is becoming increasingly sophisticated and challenging to detect. The use of steganography techniques to hide malicious code within seemingly innocuous files presents a significant threat to individuals and organizations alike. This work highlights the detailed history of various steganography, steganalysis and its methods, as well as the various tools and methods used by attackers to create and distribute stegware. The importance of keeping software and operating systems up-to-date, implementing robust security measures, and educating employees to recognize potential threats cannot be overstated. In conclusion, stegomalware is a complex and rapidly evolving threat that requires constant attention and effort to combat. By staying informed, implementing best practices, and remaining vigilant, individuals and organizations can reduce the risk of falling victim to stegomalware attacks. This work may serve as a base for those who work on stegware and its associated processes.

References

1. Oleg Evsutin, Anna Melman, and Roman Meshcheryakov. Digital steganography and watermarking for digital images: A review of current research directions. IEEE Access, 8:166589–166611, 2020.
2. Chaganti, R., Ravi, V., Alazab, M., & Pham, T. D. (2021). Stegomalware: A Systematic Survey of Malware Hiding and Detection in Images, Machine Learning Models and Research Challenges. arXiv preprint arXiv:2110.02504v1
3. Monika, A., & Eswari, R. (2021). An Ensemble-based Stegware Detection System for Information Hiding Malware Attacks. National Institute of Technology Tiruchirappalli.
4. Shehab, D.A. & Alhaddad, M.J. (2022). Comprehensive Survey of Multimedia Steganalysis: Techniques, Evaluations, and Trends in Future Research. Symmetry, 14(1), 117. https://doi.org/10.3390/sym14010117
5. Cohen, A., Nissim, N., & Elovici, Y. (2020). MalJPEG: Machine Learning Based Solution for Detection of Malicious JPEG Images. Expert Systems with Applications, 136, 112895. https://doi.org/10.1016/j.eswa.2019.112895
6. Fridrich, J., & Goljan, M. (2012). Practical Steganalysis of Digital Images – State of the Art. Proceedings of the SPIE, 8303, 83030L. https://doi.org/10.1117/12.912528
7. Caviglione, L., & Mazurczyk, W. (2018). Never Mind the Malware, Here's the Stegomalware. IEEE Security & Privacy, 16(6), 56-63. https://doi.org/10.1109/MSP.2018.2701046
8. Kumar, S. (2020). An emerging threat Fileless malware: a survey and research challenges. Cybersecurity, 3(1), 6–15. https://doi.org/10.1186/s42400-019-0016-7
9. Vaidya, N., & Rughani, P. (2019). An Efficient Technique to Detect Stegosploit Generated Images on Windows and Linux Subsystem on Windows. International Journal of Computer Sciences and Engineering, 7(12), 21. Doi: 10.26438/ijcse/v7i12.2126.
10. Caviglione, L., & Choras, M. (2020). Tight Arms Race: Overview of Current Malware Threats and Trends in Their Detection. IEEE Access, 8, 218515-218539. https://doi.org/10.1109/ACCESS.2020.3048319.
11. Çataltaş, Ö., & Tutuncu, K. (2017). Improvement of LSB based image steganography. International Journal of Electrical, Electronics and Data Communication, 5(9), 1-5. ISSN: 2320–2084.
12. Roseline, A., & Geetha, S. (2021). A comprehensive survey of tools and techniques mitigating computer and mobile malware attacks. Journal of Computers & Electrical Engineering, 92, 107143. https://doi.org/10.1016/j.compeleceng.2021.107143
13. Choudhary, Kaustubh. (2020). IMPLEMENTATION, DETECTION AND PREVENTION OF STEGOMALWARE. Doi :10.13140/RG.2.2.19870.97609.
14. Qian, Y., Dong, J., Wang, W., & Tan, T. (2018). Feature learning for steganalysis using convolutional neural networks. Multimedia Tools and Applications, 77(15), 19633–19657. Doi: 10.1007/s11042-017-5326-1
15. Fridrich, J. (2005). Feature-Based Steganalysis for JPEG Images and Its Implications for Future Design of Steganographic Schemes. In Proceedings of the Information Hiding Conference (pp. 67-81). Springer. Doi: 10.1007/978-3-540-30574-3_5
16. P.M. Kavitha and B. Muruganantham, "A study on deep learning approaches over Malware detection," 2020 IEEE International Conference on Advances and Developments in Electrical and Electronics Engineering (ICADEE), Coimbatore, India, 2020, pp. 1–5, doi: 10.1109/ICADEE51157.2020.9368924.

Human Machine Interaction in the Digital Era – Prof. J. Dhilipan et al. (eds)
© 2024 Taylor & Francis Group, London, ISBN 978-1-032-54998-9

Wheelchairs—A Boon in the Rehabilitative Era

44

P. Raja Rajeswari Chandni[1]

Assistant professor (O.G), department of Biomedical engineering,
Sri Ramakrishna Engineering College.

Shiddharth S.[2], V. S. R. Kamaleshwaran[3], B. Aarthi[4], M. Jeyaganesh[5]

Department of Biomedical Engineering,
Hindusthan College of Engineering and Technology, Coimbatore, Tamil Nadu, India

Abstract Mobility impairment is one among the distressful disabilities, which almost ceases the ability to move around freely. Wheelchairs are considered among the better assistive devices that aid people with physical and mental disorders. This mobility aid minimizes the efforts of any individual to push the wheelchair, with the adaptation of robotic mechanization brought to use in the field of rehabilitative medicine. According to the developing technology, wheelchairs are developed from manual to automated and infused with monitoring sensors, IoT, comforts, and varied mobility controls. This paper reviews all the possible tech features upgraded so far and that are yet to be included shortly.

Keywords Arduino UNO, Buzzer, Bluetooth, Pulse sensor, DC Motors, Fall detector sensor, Mobile

1. Introduction

The wheelchair is the most commonly used assistive devices that promote mobility in enhancing the quality of life for the people who have difficulties in selfmobility. In addition to mobility, a suitable wheelchair benefits the physical health and quality of life by reducing common problems such as pressure sores, and progression of deformation, and improving respiration and digestion. Numerous upgrades are updating each year as technology progresses.

Table 44.1 Percentage of disabled to total population India, 2011

Sr. No.	Residence	Male	Female	Persons
1	Rural	2.43	2.03	2.24
2	Urban	2.34	1.98	2.17
	Total	2.41	2.1	2.21

Table 44.2 The proportion of disabled population by type of disability India, 2011

Sr. No.	Types of Disabilities	Male	Female	Total
1	In Seeing	17.6	20.2	18.8
2	In Hearing	17.9	20.2	18.9
3	In Speech	7.6	7.4	7.5
4	In Movement	22.5	17.5	20.3
5	Mental Retardation	5.8	5.4	5.6
6	Mental Illness	2.8	2.6	2.7
7	Any Other	18.2	18.6	18.4
8	Multiple Disability	7.8	8.1	7.9
	Total	100	100	100

Source: World Health organization as of data collected with the year 2011.

[1]chandnipasupathy@gmail.com, [2]shidsun2001@gmail.com, [3]vsrkamal3011@gmail.com, [4]aarthibharathi2002@gmailcom, [5]jeyag45@gmail.com

DOI: 10.1201/9781003428466-44

2. Mobility Control

The wheelchair can be run either by hand or automatically, the manual mode includes a joystick, mouse, and keypad. Likewise the automatic mode includes head movements, eye tracking and most of the manual designs need the seated person to be able to control the wheelchair, wheelchair movements by hand or finger. For the most acute patient with SCI, the primary means of locomotion is by wheelchair.[17] Initial examination of wheelchair mobility includes absorbing the individuals, ability to manage, wheelchair part example wheel lock, footrest, and so on. The mobility controls have three functional areas:

- *Bed mobility:* the ability of the patient to move around in bed, including moving from lying to sitting and vice-versa.
- *Transferring:* the action of a patient moving from one surface to another.
- *Ambulation:* the ability to walk. This includes assistance from another person or an assistive device such as a cane, walker, or crutches.[18]

In cases with nervous and muscular disorders due to which lack of mobility signal supply from the brain. Such patients could not be supported on a joystick controlled system, so automatic wheelchairs that work with the signal generated from the brain can be used to control the wheelchairs. By the use of this method traditional manual wheelchairs can be eliminated. [19]

Figure 44.1. [33] Is a sample of mobility controlled automated model

Automated wheelchair models were designed relying on the information that disabled patients generate and repeat from the brain. This type can also be used for people with non-paralytic face muscles. Head gestures and eye iris position tracking system to control the wheelchair movement using a USB camera mounted in front of the face to track the users eye movement towards the left or right. A video Charge Coupled Device (CDD) camera and a frame grabber were used to analyze a series of human images to track the eye movement of the user to control the wheelchair. [7] Another model proposed by [8] uses lead control to wheelchairs especially for handicapped people with severe disabilities. It has a 2D color face

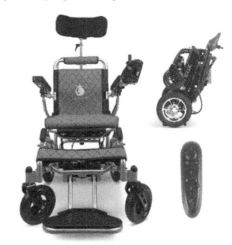

Fig. 44.1 Proposed a sample of mobility controlled automated model [7]

tracker and a detector to detect the face movement of the user and generate some comments to drive the wheelchair. The face inclination and mouth shape information to determine the direction of the wheelchair and captured the facial expressions using a digital camera the collected images are pre-processed and interpreted by an application running on a laptop computer on the chair. Eye blinks, eye movements, and head movements generate a command sent to a robotic wheelchair.

3. Safety Features

Safety is one of the imminent feature for all devices. For wheelchairs it includes fall detection, obstacle finding, seat belt, etc., are some notable features implemented in wheelchairs. A gyroscope, GPS module, FSR pressure sensor, or microcontroller are implemented into the system for fall detection. The gyroscope is used to detect the position of the wheelchair while the FSR pressure sensor which is placed on the sit pad of the wheelchair will be used to detect and recognize the user's gesture. [9]

Another important feature is the obstacle avoidance, which can be defined as the ability to react when unscheduled events occur. Figure 44.2 [34] Classical obstacle avoidance algorithms are based on a behavioral approach, consisting in converting sensor reading to motion instructions through state machines (actions). Several techniques can be applied for obstacle avoidance such as fuzzy logic, neural networks, and so on. Naturally, these methods can be neither exhaustively tested nor proved effective in all cases. [10]

The wheelchairs with seat belt features are also a safety feature to prevent falling patient who is completely paralyzed, weak, or having any disability which establishes the automatic fall of the subjects. Wearing the seat belt when operating a wheelchair outdoors, utilize the seat belt on the wheelchair for safety. [20]

4. Convertible Features

To give new hope to the physically disabled person, they developed a convertible wheelchair into a stretcher. This will help the attendee of a patient, or disabled person to avoid heavy lifting and save energy as well as time. This is a friendly assistive

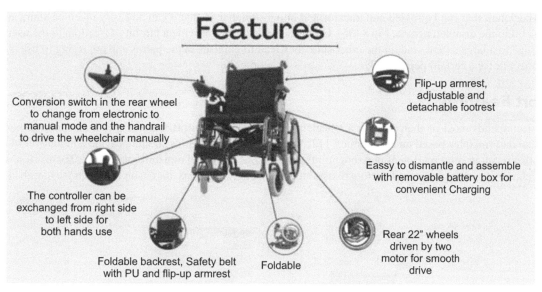

Fig. 44.2 Features of the wheelchair [1]

Fig. 44.3 Convertible features [1]

device for the physically challenged person who cannot move independently. [21] It is a device with a simple mechanical and pneumatic control mechanism is does not skilled persons to operate. Since shifting a patient from a wheelchair to a stretcher is reduced, it will prevent damage to a patient and increase comfort. [22]

A modified wheelchair that can be folded and transformed into a stretcher so that it can be easily taken on stairs and anywhere else is very useful to the disabled person. Fig 8 [35]. On one side stretchers provide a fair bit of freedom to the users, limit their upper body usage to a large extent, and on the other side, they lead to calluses in the palms and are trying to use after traveling a certain distance, or for a certain period. [23]

5. Comfort Features

Since the subject spends most of their time in a wheelchair needs some comfort features like a cushion, suspension, shock absorber, and so on principles based on ergonomics. [12] Wheelchair seat cushions should provide a high level of comfort and prevent sores that result in compression of pressure points to keep cool and avoid moisture stagnation. There are several types of seat cushions chosen according to your detectors like foam cushions, gel cushions, air cushions, and honeycomb cushions. [13]

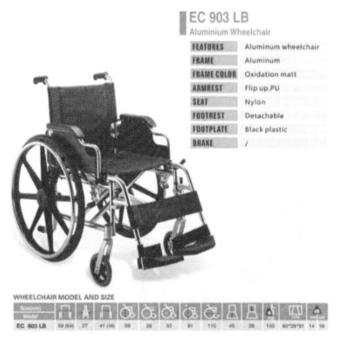

Fig. 44.4 Ergonomic features of the wheelchairs [12]

Every comfort feature intimate the ergonomic principle for the subject's comfort. The principle of ergonomics applied in wheelchairs like adjustable arm support, seat depth, and seat surface height gives each user a custom fit as shown in

Figure 44.9 [36]. Wheelchair seatbelts are used to protect the occupant of the wheelchair to prevent them from falling out of the wheelchair and the reason to use the wheelchair is to position the occupant perfectly with the wheelchair comfortably. There are several seatbelts available in the market, some of them are as follows: Padded pelvic wheelchair stabilizer belt, which attaches to even the tightest of spaces via grommets for compatibility with wheelchairs.[2]

- 2-point and 4-point hip wheelchair belts, connect with 2-points/4-point for quick and easy installation.
- Stayflex wheelchair harnesses have dual stretch zone technology to prevent riding up and reduce strangulation risk.[3]
- 2-piece auto buckle positioning belt, designed to provide safe and comfortable seat positioning for the user, and to help a light restraining user from falling or slipping out of the wheelchair.
- De Royal Torse support wheelchair positioning harness provides postural support around the chest and shoulders.
- The wheelchair chest belt has an infinitely adjustable hook-and-loop closure for a personalized fit and so on. [24]

The shock absorber allows the occupant of the wheel confidently moves across the roughest of terrains without damaging themselves or the wheelchair. Wheelchair suspension provides comfort stability and traction. Comfort reduces the force that impacts the wheelchair user and contributes to loss of position discomfort or fatigue. Stability to help a subject feel safe and confident driving the wheelchair over a variety of terrain and surface. Traction allows the wheelchair to be driven over a wide variety of surfaces including grip, cuts, and ramps reducing the risk of high cantering or getting stuck which promotes greater wheelchair user independence in foam and community mobility. [25]

6. Accessories

Initially, wheelchairs are inverted to carry individuals with a disability to move around. It has traveled so far from simple wheelbarrows to fully automated wheelchairs that can make a disabled person climb stairs. Also accessories like seat cushions, back cushions, wheelchair safety belts, wheelchair gloves, wheelchair backpack, wheelchair table tray, portable ramp, transfer board, mobility manual hand cycle, mobility electric hand cycle, bag holder, wheelchair cargo, shelf, front armrest bag, cup holder, beverage holder, can oxygen attachment, wheelchair safety, fall detection sensor, protective helmet, etc. [26]

7. IoT Application

Traditional simple wheelchairs failed to detect the sense of external environment during use and have a single method. There is a need for IoT wheelchairs with three functions namely:

- Occupant Wheelchair Environment Multimode Sensing to recognize gestures information, to sense positioning, speed, and posture information.
- The fusion control scheme, a mobile control scheme based on rocker and gesture recognition as well as a backrest and footrest, lifting lowering, and movement control scheme based on tensed could and mobile app.
- Human Machine interaction, the wheelchair is locked to tense 107 explorers that esp8266 Wi-Fi module using MQTT protocol is used to upload sensor data, while the wheelchair stretch can be viewed and controlled on the app.
- Smart wheelchair, in which cameras ladder ECG sensors are installed to build sensor networks. Provide users with a full range of sensing humanmachine interaction remote monitoring and mobility control functions.[14]

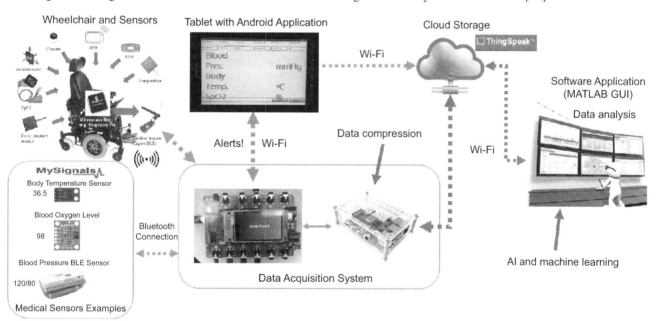

Fig. 44.5 Applications of IoT based wheelchair model incorporated with smart features [14]

8. Conclusion

Of course, wheelchairs also come with some limitations and disadvantages, they can result in repetitive strain issues and are not ideal for going long distances or up and they need assistance to get around a part that the benefit from using wheelchairs or more than their limitations with the innovations of accessories and modification of simple wheelchairs to smart automated wheelchairs

References

1. Kedar Sukerkar, Darshitkumar Suratwala, Anil Saravade, Jairaj Patil, Rovina D'britto. Smart Wheelchair: A Literature Review. International Journal of Informatics and Communication Technology (IJ-ICT). 2018.

2. Wieczorek, B.; Warguła, Ł.; Rybarczyk, D. Impact of a hybrid assisted wheelchair propulsion system on motion kinematics during climbing up a slope. Appl. Sci. 2020.

3. Wieczorek, B.; Kukla, M. Effects of the performance parameters of a wheelchair on the changes in the position of the center of gravity of the human body in dynamic conditions. PLoS ONE 2019

4. Elliott, M.A.; Malvar, H.; Maassel, L.L.; Campbell, J.; Kulkarni, H.; Spiridonova, I.; Sophy, N.; Beavers, J.; Paradiso, A.; Needham, C.; et al. Eye-controlled, power wheelchair performs well for ALS patients. Muscle Nerve 2019

5. Meena, Y.K.; Cecotti, H.; Wong-Lin, K.; Prasad, G. A multimodal interface to resolve the Midas-Touch problem in gaze controlled wheelchair. In Proceedings of the 2017 39th Annual International Conference of the IEEE Engineering in Medicine and Biology Society (EMBC), Jeju Island, Korea, 11–15 July 2017.

6. Fornaser, A.; De Cecco, M.; Leuci, M.; Conci, N.; Daldoss, M.; Armanini, A.; Maule, L.; De Natale, F.; Da Lio, M. Eye tracker uncertainty analysis and modeling in real-time. J. Phys. Conf. Ser. 2017.

7. De Cecco, M.; Zanetti, M.; Fornaser, A.; Leuci, M.; Conci, N. Inter-eye: Interactive error compensation for eye-tracking devices. In AIP Conference Proceedings; AIP Publishing: Ancona, Italy, 2016.

8. Zanetti, M.; De Cecco, M.; Fornaser, A.; Leuci, M.; Conci, N. The use of INTEREYE for 3D eye-tracking systematic error compensation. In Proceedings of the ELMAR, 2016 International Symposium, Zadar, Croatia, 12–14 September 2016.

9. Wästlund, E.; Sponseller, K.; Pettersson, O.; Bared, A. Evaluating gaze-driven power wheelchair with navigation support for persons with disabilities. J. Rehabil. Res. Dev. 2015.

10. Maule, L.; Fornaser, A.; Leuci, M.; Conci, N.; Da Lio, M.; De Cecco, M. Development of innovative HMI strategies for eye controlled wheelchairs in virtual reality. In Proceedings of the International Conference on Augmented Reality, Virtual Reality and Computer Graphics, Lecce, Italy, 15–18 June 2016.

11. Singer, C.C.; Hartmann, B. See-Thru: Towards Minimally Obstructive EyeControlled Wheelchair Interfaces. In Proceedings of the 21st International ACM SIGACCESS Conference on Computers and Accessibility, Pittsburgh, PA, USA, 28–30 October 2019.

12. Subramanian, M.; Songur, N.; Adjei, D.; Orlov, P.; Faisal, A.A. A.Eye Drive: Gazebased semi-autonomous wheelchair interface. In Proceedings of the 2019 41st Annual International Conference of the IEEE Engineering in Medicine and Biology Society (EMBC), Berlin, Germany, 23–27 July 2019.

13. Jafar, F.; Fatima, S.F.; Mushtaq, H.R.; Khan, S.; Rasheed, A.; Sadaf, M. Eye Controlled Wheelchair Using Transfer Learning. In Proceedings of the 2019 International Symposium on Recent Advances in Electrical Engineering (RAEE), Islamabad, Pakistan, 28–29 August 2019.

14. Peleshko, D.; Ivanov, Y.; Sharov, B.; Izonin, I.; Borzov, Y. Design and implementation of visitors queue density analysis and registration method for retail video surveillance purposes. In Proceedings of the 2016 IEEE First International Conference on Data Stream Mining & Processing (DSMP), Lviv, Ukraine, 23–27 August 2016.

15. Ivanov, Y.; Peleshko, D.; Makoveychuk, O.; Izonin, I.; Malets, I.; Lotoshunska, N.; Batyuk, D. Adaptive moving object segmentation algorithms in cluttered environments. In Proceedings of the Experience of Designing and Application of CAD Systems in Microelectronics, Lviv, Ukraine, 24–27 February 2015.

16. Frego, M.; Bertolazzi, E.; Biral, F.; Fontanelli, D.; Palopoli, L. Semi-analytical minimum time solutions with velocity constraints for trajectory following of vehicles. Automatica 2017.

17. Bertolazzi, E.; Frego, M. Interpolating clothoid splines with curvature continuity. Math. Methods Appl. Sci. 2018.

18. Li, Z.; Xiong, Y.; Zhou, L. ROS-Based Indoor Autonomous Exploration and Navigation Wheelchair. In Proceedings of the 2017 10th International Symposium on Computational Intelligence and Design (ISCID), Hangzhou, China, 9–10 December 2017.

19. Romero-Ramirez, F.J.; Muñoz-Salinas, R.; Medina-Carnicer, R. Speeded up Detection of Squared Fiducial Markers. Image Vis. Comput. 2018.

20. M. H. Alsibai, H. Manap, and A. A. Abdullah, "Enhanced face recognition method performance on android vs windows platform", Proceedings of International Conference on Electrical, Control and Computer Engineering, 2015.

Human Machine Interaction in the Digital Era – Prof. J. Dhilipan et al. (eds)
© 2024 Taylor & Francis Group, London, ISBN 978-1-032-54998-9

CoV-3DM: COVID-19 Disease Detection Based on Ensemble SMOTE Technique with Routine Blood Test Parameters

45

L. William Mary[1]

Research Scholar,
SRM Institute of Science and Technology, Kattankulathur, Chennai

S. Albert Antony Raj[2]

Deputy Dean, Head & Professor,
SRM Institute of Science and Technology, Kattankulathur, Chennai

Abstract The deadly disease SARS-CoV-2 caused an outbreak of the emerging infectious ailment of COVID-19. The different pathogenies have been categorized as variants of concern. The Omicron variant had a higher affinity and caused more asymptomatic infections than the other. Identifying and preventing COVID-19 in addition to combating the fast spreading of the disease at an early stage is imperative. This paper proposes a CoV-3D model for deadly diseases (COVID-19). It uses the ensemble machine learning (SMOTE) technique with laboratory blood test parameters. The model has been developed to predict the deadly disease from standard blood parameters: neutrophil count, complete lymphocyte count, and hematocrit. The proposed classical model has been created, tested, and evaluated using an open-source blood test data set. The first stage classifier used a state-of-the-art supervised learning algorithm. The subsequent phase used K-Nearest Neighbors, Support Vector Machine, Naive Bayes, Decision Tree, Random Forest, and Logistic Regression. The proposed model employs a synthetic minority oversampling technique to balance a data distribution. The proposed method attains an excellent prediction accuracy of 0.80%, the AUC (0.84%), and F1 Score (0.80%).

Keywords COVID-19, Blood test, Diagnostic model, Ensemble learning, Machine learning

1. Introduction

The most widespread outbreak in December 2019 was caused by the Infectious disease COVID-19, and the entire world is engaged in a battle to contain this contagious disease. The virus was designed to be eradicated by vaccines, but it quickly evolves as it spreads. The global pandemic is wreaking havoc in some regions. Early disease detection and prediction will limit its spread and lessen its social burden. RT-PCR (Fang Y et al. 2020), Blood Tests, X-rays, and CT-Scan (Zhao D et al. 2020) based on infected individuals' respiratory and other laboratory samples can be used to identify the lethal COVID-19 disease. Although this test is the most accurate way to diagnose COVID-19, it is technically difficult and takes 48 hours (Long DR et al. 2021). The RT-PCR tests are also taken in addition to the RAT (Rapid Antigen Tests) (Corman VM et al. 2020).

The popular concept in the current era is Artificial Intelligence (AI). The most significant component of AI is machine learning (ML), which incorporates the ideas of model development, statistics, algorithms, and simulations and aids in clinical research (Browning L et al. 2021). In the blood test samples from COVID-19-infected and non-infected patients, ML approaches have been used to predict the novel coronavirus in this article.

[1]wl6649@srmist.edu.in, [2]hod.dca.ktr@srmist.edu.in

DOI: 10.1201/9781003428466-45

CT scan exposure can cause radiation. CT scanners are also cost-effective but unavailable everywhere (Coppock H et al. 2021), (Tena A et al. 2022). According to various medical research studies, blood test parameters and laboratory indicators are included in the primary screening process (Akhtar A et al. 2021), (Ferrari D et al. 2020), (Alballa N et al. 2021), (Chadaga K et al. 2021), (AlJame M et al. 2020). Blood tests used by machine learning algorithms to detect diseases are less expensive, simpler to use, and take less time than expensive testing. In developing nations that lack the equipment and laboratory supplies, testing based on blood samples can be used. Testing can be expedited using this low-cost method (Schwab P, et al. 2020). The key conclusions and contributions are:

1. Review and thoroughly examine various machine learning programs that use numerous blood test parameters to identify and forecast COVID-19 infectious illnesses.
2. Several cutting-edge machine learning techniques that correctly forecast COVID-19 cases.
3. The observed platelets, leukocytes (WBC), and lymphocytes were the most significant markers of the occurrence of coronavirus for the considered data, which was done using the Boruta technique for feature selection.
4. Various blood parameters are also considered in diagnosing the COVID-19 virus.
5. For handling imbalanced data SMOTE technique was used.

2. Materials and Methods

2.1 Dataset Description

The main objective is to diagnose COVID-19 disease using blood test data. The open-source blood test dataset was collected for this investigation. According to the test ranges and results, some samples tested positively and others negatively. The data were collected with a mean value of zero to produce an accurate normal distribution. The label has 511 instances of positive and negative values with 26 variables. The dataset has inadequate positive and negative cases, making it imbalanced. Analysis of the study that displays the proportion of positive and negative cases. Positive cases are denoted by the number 1, whereas negative cases are denoted by the number 0. In this analysis, 319 COVID-19-infected and 232 uninfected data points were considered for the investigation.

2.2 Data Preprocessing

Data pre-processing is converting raw (unstructured) data into a usable format. After normalization, the collected data is incomplete, inconsistent, redundant, and noisy. Transforming raw data into processed data requires several processes. The following Figure illustrates the various preparation phases.

Data Cleaning

Data cleaning improves the accuracy of outlier removals and imputations for missing values. Two-dimensional data tables are used to hold the operational values. This procedure identifies erroneous records and missing data in the data set. The performance of classification algorithms is adversely affected by missing values (Zhang S et al. 2010). The most common value, the mean value, and regression-based techniques are used, among other categories, to replace missing values (Kotsiantis SB, et al. 2007).

Data Reduction

Digital data, including audio, video, and image files, is simplified through data reduction. This method aids in breaking down large amounts of data into smaller pieces.

Data Transformation

The four data transformation steps are normalization, standardization, aggregation, and generalization. There are two distinct categories: normalization and standardization. The model training process is aided by normalization. The two most widely used normalization techniques are standardization scaling and min-max scaling. The difference between the lowest and highest values in each column is used by min-max scaling to split the range. The column values are calculated with 0 and 1 as the lowest and highest, respectively. Standardization involves setting a variable's center value to zero and its variance value to one. These methods are used, although the distribution of the data is unknown.

2.3 Co-relation Analysis and Feature Importance

A measure of an association between variables is a correlation. A change in one variable's magnitude is also correlated when two variables are related. There is an association between positive correlation in one direction and negative correlation in the other direction. The linear association amongst two continuous variables is measured by using the correlation method, and to calculate we use the Pearson correlation coefficient method. The restricted confidence intervals function best when the correlation coefficient is 0 or nearly zero. Pearson Coefficient and Spearman's rank are the two most essential correlation coefficients. The Pearson Coefficient ϱ denoted as calculating population parameter, and r is denoted by sample statistic. The sample calculation formula is as follows:

$$r = \frac{\sum_{i=1}^{n}(x_i - x)(y_i - y)}{\sqrt{\left[\sum_{i=1}^{n}(x_i - \overline{x})^2\right]\left[\sum_{i=1}^{n}(y_i - \overline{y})^2\right]}}$$

2.3 Methodology for Feature Selection

Boruta feature selection techniques

A feature selection algorithm is called Boruta. It functions as a wrapper algorithm for Random Forest, and the package's name is derived from Slavic mythology, which was in pine trees. Predictive modeling requires feature selection when the data set contains multiple variables. For the feature selection in the scenario below, the Boruta technique is used to pick features. Seven COVID-19 features from dataset 19 have been verified, compared to 16 that have not.

2.4 Methodology for Handling Data Imbalances

SMOTE

(Marcio Dorn et al. 2021) Chawla developed the technique SMOTE (Synthetic Minority Over Sampling Technique) as the boosted algorithm for imbalanced data. A small number of samples are randomly interpolated between samples. New models are obtained with the SMOTE algorithm. With fewer data, the ratio of data imbalance increases. It enhances the imbalanced data set's impact on classification. Overfitting is a prevalent issue that is frequently caused by data imbalance. Synthetic minority oversampling is a valuable technique for ML to address dataset imbalance (SMOTE). To modify the initial training dataset, SMOTE uses an oversampling method. When the data are balanced, the models are trustworthy. The dataset is shown in the graphic before and after SMOTE.

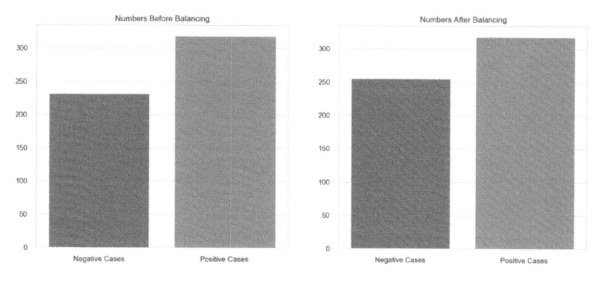

Fig. 45.1 Before balancing and after balancing

Source: Worked coding from the online dataset Zerrin Gamsızkan (2021)

Process of SMOTE Techniques

Procedure 1: Determine each minority sample's k nearest neighbors by calculating the distance between it and the other instances, where x_i ($i = 1, 2..., n$).

Procedure 2: Each sample x_i's random m nearest neighbors are chosen randomly and referred to x_{ij} ($j = 1, 2..., m$) by means of a subset of the k-nearest collection, following which a synthetic minority sample p_{ij} is calculated $p_{ij} = x_i = \text{rand}(0, 1) \times (x_{ij} - x_i)$)

rand value (0,1) generated a random number that is evenly distributed between [0,1] (Wang S et al. 2021).

2.5 Classification Methodologies of Machine Learning

The five different algorithms are used for the prediction:

- *Logistic Regression:* Making use of a supervised learning model, data is predicted to have a probability that ranges between 0 and 1 (Eunnuri Cho et al. 2022).
- *k-Nearest Neighbors:* It is used the nearest data points value to predict new data (Muller AC, et al. 2016).
- *Decision Trees:* The variable is divided into two for each branch in this supervised learning model, which classifies input based on predetermined standards. There is no requirement for Linearity, normality, and equal variance. Continuous variables are considered discontinued values. Furthermore, it breaks the search area into smaller chunks and searches for each one using a series of yes/no questions. The combination of decision trees is the Random Forest algorithm (Goodfellow I, et al. 2015).
- *Support Vector Machine:* After being proposed, this approach has been in use frequently since the late 1990s (Cho, E, et al. 2022). To classify outputs, the support vector classifier locates the hyperplane that distinguishes between classes (Belavagi MC, et al. 2016).

3. Experimental Studies

3.1 Experimental Implementation of Models

The experiments on open-source datasets that are conducted to show the effectiveness of the algorithms are described in this section. The COVID-19 blood test data set, which after pre-processing has 511 instances, 26 numerical attributes, and no missing values in the dataset, is used to assess the prediction accuracy. COVID-19 results by class are both positive and negative. In the tests, the instance was pre-processed to have a variance of 1.0, and each attribute was given a range of values between 0 and 1. An age-wise analysis was conducted to determine which age group the acquired dataset had the most impact on. The age range taken into consideration for this analysis was 18 to 93. The data shows that people aged 30 to 44, 69 to 79, and 83 to 84 are the least affected. The virus impacted people in the 70s, 71s, and 72s, the most extensive age range.

3.2 Analysis of Machine Learning Algorithms

Logistic Regression

A complete blood test data set is employed in the following analysis to determine the prediction accuracy, confusion matrix, and classification report. One of the statistical analysis techniques that forecast binary outcomes is logistic regression. This model takes various inputs into account. It also indicates the dependent data variable from the datasets. The result of the Logistic regression F1 score is 77% for the given blood test dataset. The classification report has shown the predicted result of precision, recall, and f1-score values. The Area Under the Curve is used to find the classification threshold probability of the COVID-19 disease, whether the patient has an infected coronavirus or not, based on the target value. The AUC () function calculates the score values. The chart below shows the statistics for the true positive class value and false positive class value (0.814).

k-Nearest Neighbors

This algorithm predicts an accuracy of 72.17% for the dataset. The confusion matrix shows the actual predicted values. The classification reports the precision value for the positive cases (0.69%), recall (0.83%), f1-score (0.75%), and accuracy (0.72%). The AUC measure displays the true positive class value and false positive (AUC = 0.774).

Decision Trees

Decision Trees predict an accuracy of 67.83% for the dataset. The confusion matrix shows the actual predicted values. The classification reports the precision value for the positive cases (0.67%), recall (0.73%), f1-score (0.70%), and accuracy (0.68%). Values for both the true and false positive rates (AUC = 0.742).

Support Vector Machine

The prediction accuracy of this algorithm is 80% for the processed dataset. The confusion matrix shows the actual predicted values. The classification provides the precision value (0.79%), recall (0.83%), f1-score (0.81%), and accuracy (0.80%) for the positive examples. The AUC (AUC=0.846) displays the data for the true positive and false positive values.

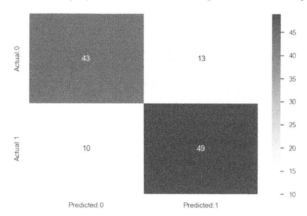

Fig. 45.2   Confusion matrix for support vector machine

Source: Worked coding from the online dataset Zerrin Gamsızkan (2021)

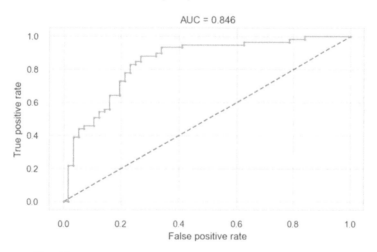

Fig. 45.3   AUC value for true positive and false positive rate

Source: Worked coding from the online dataset Zerrin Gamsızkan (2021)

4. Results

4.1 Comparative Analysis

With each method's accuracy, AUC, and F1 score values, a comparative examination of multiple state-of-the-art algorithms is displayed. Decision trees (0.67% accuracy), K-nearest Neighbors (0.72% accuracy), Logistic Regression (0.76% accuracy), and Support Vector Machine (0.80% accuracy).

Table 45.1   Model Comparison

ML Algorithm	Accuracy	AUC	F1 Score
Logistic Regression	0.765217	0.813559	0.765217
K-Nearest Neighbors	0.721739	0.773608	0.753846
Decision Trees	0.678261	0.742282	0.699187
Support Vector Machine	0.800000	0.845642	0.809917

Source: Worked coding from the online dataset Zerrin Gamsızkan (2021)

5. Conclusion

Boruta feature selection approach is used to select blood test features and automatically choose the best feature that uses the SMOTE technique to balance the imbalance dataset. To determine the best accuracy, we applied cutting-edge algorithms. The Support Vector Machine (SVM) accuracy for the selected samples from the 511 blood test datasets obtained using the suggested technique shows 80%. We examine the static model using the most real-time datasets for forthcoming work.

References

1. Fang Y, Zhang H, Xie J, Lin M, Ying L, and Pang P. (2020). Sensitivity of chest CT for COVID-19: comparison to RT-PCR. Radiology. 296: E115–7. doi: 10.1148/radiol.2020200432.
2. Zhao D, Yao F, Wang L, Zheng L, and Gao Y, Ye J. (2020). A comparative study on the clinical features of coronavirus 2019 (COVID-19) pneumonia with other pneumonias. 71: 756–61. doi: 10.1093/cid/ciaa247.
3. Long DR, Gombar S, Hogan CA, Greninger AL, O'reilly-Shah V, BrysonCahn C. (2021). Occurrence and timing of subsequent severe acute respiratory syndrome coronavirus 2 reverse-transcription polymerase chain reaction positivity among initially negative patients. Clin Infect Dis. 72: 323–326.
4. Corman VM, Landt O, Kaiser M, Molenkamp R, Meijer A, Chu DK, Bleicker T, Brünink S, Schneider J, Schmidt ML, and Mulders DG. (2020). Detection of 2019 novel coronavirus (2019-nCoV) by real-time RT-PCR. doi.org/ 10.2807/1560-7917.es.2020.25.3.2000045.
5. Browning L, Colling R, and Rakha E, et al. (2021). Digital pathology and artificial intelligence will be key to supporting clinical and academic cellular pathology through COVID-19 and future crises: the Path LAKE consortium perspective. J Clin Pathol 74(7): 443–447. https://doi.org/10.1136/jclinpath-2020-206854.
6. Coppock H, Gaskell A T, and zirakis P et al. (2021). End-to-end convolutional neural network enables COVID-19 detection from breath and cough audio: a pilot study. BMJ Innov 7: 356–362. https://doi.org/10.1136/bmjinnov-2021-000668.
7. Tena A, Clarià F, and Solsona F. (2022). Automated detection of COVID-19 cough. Biomed Signal Process Control 71:103175. https://doi.org/10.1016/j.bspc.2021.103175.
8. Coppock H, Jones L, Kiskin I, and Schuller B. (2021). COVID-19 detection from audio: seven grains of salt. Lancet Digit Health 3(9):e537–e538. https://doi.org/10.1016/s2589-7500(21)00141-2.
9. Akhtar A, Akhtar S, Bakhtawar B, Kashif AA, Aziz N, and Javeid MS. (2021). COVID-19 detection from CBC using machine learning techniques. 1(2): 65–78. https://doi. org/10.54489/ijtim.v1i2.22.
10. Ferrari D, Motta A, Strollo M, Banf G, and Locatelli M. (2020). Routine blood tests as a potential diagnostic tool for COVID-19. Clin Chem Lab Med 58(7): 1095–1099. https://doi.org/10.1515/ cclm-2020-0398.
11. Alballa N, and Al-Turaiki I. (2021). Machine learning approaches in COVID-19 diagnosis, mortality, and severity risk prediction: a review. Inform Med Unlocked 3:100564. https://doi.org/10.1016/j. imu.2021.100564.
12. Chadaga K, Prabhu S, and Vivekananda BK et al. (2021). Battling COVID-19 using machine learning: a review. Cogent Eng 8(1):1958666. https://doi.org/10.1080/23311916. 2021.1958666.
13. AlJame M, Ahmad I, Imtiaz A, and Mohammed A. (2020). Ensemble learning model for diagnosing COVID-19 from routine blood tests. Inform Med Unlocked 21: 100449. https://doi.org/10.1016/j. imu.2020.100449.
14. Soares F. A. (2020). Novel specific artificial intelligence-based method to identify COVID-19 cases using simple blood exams. MedRxiv, doi: https://doi.org/10.1101/2020.04.10.20061036.
15. Schwab P, Schütte AD, Dietz B, and Bauer S. (2020). Clinical predictive models for COVID-19: a systematic study. J Med Internet Res 22(10):e21439. https://doi.org/10.2196/preprints.21439.
16. Zhang S, Wu X, and Zhu M. (2010). Efficient missing data imputation for supervised learning. In: 9th IEEE international conference on cognitive informatics (ICCI'10). IEEE, pp 672–679.
17. Kotsiantis SB, Kanellopoulos D, and Pintelas PE. (2007). Data preprocessing for supervised learning. Int J Comput Inf Eng 1: 4104–4109.
18. Patrick Schober, Christa Boer, and Lothar A Schwarte. (2018). Correlation Coefficients: Appropriate Use and Interpretation, PMID: 29481436;126(5): 1763–1768. doi: 10.1213/ANE.0000000000002864.
19. L M Rubenstein, and C S Davis (1999), Estimation of the average correlation coefficient for stratified bivariate data, Stat Med. 15; 18(5): 567–80. doi: 10.1002/(sici)10970258(19990315)18:5<567::aid-sim52>3.0.co;2-f
20. Swinscow TDV. (1997). In: Statistics at square one. Nineth Edition. Campbell M J, editor. University of Southampton; Copyright BMJ Publishing Group.
21. Nitesh, V. C., Kevin, W. B. and Lawrence, O. H. (2002). SMOTE: Synthetic minority over-sampling technique. J. Artif. Intell. Res. 16(1), 321–357.
22. Mi, Y. (2013). Imbalanced classification based on active learning SMOTE. Res. J. Appl. Sci Eng. Technol. 5(3), 944–949.
23. Chawla, N.V. Bowyer, K.W. and Hall, L.O. et al. (2002). SMOTE: Synthetic minority over-sampling technique. J. Artif. Intell. Res. 16, 321–357

24. Han, H.; Wang, W.Y.; Mao, B.H. (2005). Borderline-smote: A new over-sampling method in imbalanced data sets learning. Adv. Intell. Comput, 3644, 878–887

25. Wang, S., Dai, Y., and Shen, J. (2021). Research on expansion and classification of imbalanced data based on SMOTE algorithm. Sci Rep 11, 24039. https://doi.org/10.1038/s41598-021-03430-5

26. Eunnuri Cho, (2022). Data Preprocessing Combination to Improve the Performance of Quality Classification in the Manufacturing Process, https://doi.org/10.3390/electronics11030477.

27. Muller AC, and Guido S. (2016). Introduction to machine learning with python. O'Reilly Media Inc, California

29. Goodfellow I, Bengio Y, and Courville A. (2015). Deep learning. MIT Press, Cambridge.

30. Cho, E. Chang, T.W. and Hwang, G. (2022). Data Preprocessing Combination to Improve the Performance of Quality Classification in the Manufacturing Process. *Electronics*, *11*, 477.

31. https://doi.org/10.3390/electronics11030477

32. Belavagi MC, and Muniyal B (2016) Performance evaluation of supervised machine learning algorithms for intrusion detection. Procedia Comput Sci 89: 117–123

Human Machine Interaction in the Digital Era – Prof. J. Dhilipan et al. (eds)
© 2024 Taylor & Francis Group, London, ISBN 978-1-032-54998-9

A Research on Patient Engagement Using Predictive Analytics

46

N. Revathi[1] and P. Roshni Mol[2]

Assistant Professor, Department of Computer Applications,
SRMIST College of Science and Humanities, Ramapuram Campus,
Chennai, Tamil Nadu, India

J. Thimmia Raja[3]

Dr. Mahalingam College of Engineering and Technology,
Pollachi, Tamil Nadu, India

C. Immaculate Mary[4]

Associate Professor & Head, PG & Research Department of Computer Science,
Sri Sarada College for Women (Autonomous),
Fairlands, Salem, Tamil Nadu, India

Abstract Predictive analytics reasonable assessment has turned into a fundamentally enchanting issue concerning examination scene as additional affiliations handle that farsighted assessment draws in them to bring down risks, pick careful decisions, and make separated client experiences. The significant objective of this substance is to propose a judicious rendition for sensible execution of practical assessment in affiliations in family engagement. Creating interest for endoscopic methods, gotten on the whole with diminishing security reimbursement, has required improvement in endoscopy unit functional execution measures, for instance, loosening up through put and reducing labour force additional time without an assortment in calm keeping up with up time.

Keywords Predictive analytics, Family engagement, Governance, Patient engagement

1. Introduction

Data is getting speedier than at the contrary time from a collection of assets along with electronic media, cells, and thus the net of things. Also, colossal realities can give other monstrous endowments to dating, for example, interfacing new matters and foundations, supporting with improving social gathering client demands, and managing improvement and evaluation use. Patient commitment inside the designing and execution of evaluation probably will need to in like manner improve its understanding into logical practice (Sullivan P, Goldmann D., 2011) In norm, there's making seeing roughly the major control of patient obligation in research, that can improve the examination of health advantages research. Past productive assessments have depicted unique pieces of the obligation cycle (Nilsen ES et al. 2006) (Mockford C et al. 2012) Along these strains, PCORI charged a conscious overview that wants joining the float evidence about liberal obligation in examination to help experts in engineering and driving basic patient obligation in clinical manual exploration.

[1]revathiphd3@gmail.com, [2]roshnimphil@gmail.com, [3]thimmia@gmail.com, [4]cimmaculatemary@gmail.com

DOI: 10.1201/9781003428466-46

2. Literature Review

2.1 The Developing Interest for Endoscopic Interaction

Colorectal mischief is that the fourth driving legitimization contamination annihilation internal us. The utilization of colonoscopy for screening is filling in go over (Harewood, G.C. and D. Lieberman, 2004) and has end up the decision screening strategy in the us Federal medical care people (Schenck A.P et al. 2009) All the while, monetary necessities are restricting prospering systems' capacity to apply more gathering of laborers to fulfil this comprehensive need.

2.2 The Rise of Big Data

The verbalization "goliath records" transformed into delivered in mid 1990s and is portrayed as measurements that is unreasonably huge, complex, and dynamic, and out manoeuvres the dealing with limitation of general illuminating document styles of a coalition (Weiss, S. M.,Indurkhya, N.,1998). The records are to a great extent irrationally epic and can't be overseen most certainly, its activities with sumptuous speed rambling in and out, making it extreme to notice.

2.3 The Rise of Business Analytics

Evaluation, as business venture realities, is portrayed as a lot of advances, cycles, and gadgets that utilization realities to expect most likely direct by utilizing people, mechanical get on the whole or unique materials. More data should exchange over into bigger potentials for a business right at the off danger that it might find the significance inward it (Minelli, M., Chambers, M., & Dhiraj, A. ,2013)

2.4 Traditional Versus Current Examination

Inside the first 5 years, progression includes leave surges of alterations inside the BI scene. Progresses in information hoarding, computational power, artificial intelligence, and huge research have delayed the distance of BI past figures. Most importantly, the general strategy trusts data to be as a decent estimated improvement in making self-affiliation assessment. Well known BI levels are being connected with every single one of the shorter distributions of activity.

3. Methodology

3.1 Prescient Analytics Process

Sharp Examination licenses dating to engage ahead looking, hanging tight for and choosing caught in to the realities not on doubts. The going with cycles summarizes drives taken to execute Prescient Examination

- *Issue definition:* frame the task impacts and business objections. See the enlightening archives which might be used;
- *Records assortment:* assemble information (created or unstructured data) from various basic concentrations for assessment;
- *Genuine examination:* Utilize elegant quantifiable styles to incorporate and actually look at the speculations and in like manner the speculations;
- *Showing:* Utilize brilliant appearance to characterize exact sharp models or the most extreme clean distributions of activity around fate;
- *Execute moves:* introduce the fast impacts in constantly exciting cycle. Show screen and study the model presentation to make specific the it gives the standard results.

3.2 Many Ways of Predictive Analytics to Increase Patient Engagement

Judicious exploration is by and by being utilized by retail and friendliness organizations to more without inconveniences understand the customer profiles probably going to make a purchase or have the greatest expanded approaching necessity for a first rate or organization.

3.3 Better Patient Engagement through Big Data

Reasonable medical clinic cure promoters by and by have significantly less hard permission to data the investigation and dynamic inclinations for capacity patients because of the improvement of selling programming and on-line measurements clearinghouses that can make enormous educational files like net site guests, looking for affinities, economics, and geo-explicit data.

3.4 Focus on the Right Patient at the Right Time

The prevalent impacts approaching patients have over their investigation degree, insightful advancing test engage clinical administrations exhibiting and exceed organizations to improve changes. Through the utilization of zeroing in on the greatest attracted victims who are prepared to seek after a logical contribution's decision, this furthermore creates impacted individual trade charges.

3.5 Ethical Considerations

Clearly, more critical detectable quality in like manner raises a lot of moral issues. For instance, point gifted cruel complaint several years sooner for sending regular postal mail coupons purposeful for confident ladies to the spot of an extreme staff younger woman's circle of family members, on this way critical the pregnancy.

3.6 The Rise of Predictive Analytics

Perception and records Representation and Dashboards to Distinct Examination in surrender to Prescient Examination. As we move, from Answering to Prescient Examination, the normal, worn out and examination of assessment gets to the next level. Spellbinding assessment gave clarity in regards to where a coalition stands related to portrayed undertaking measures.

3.7 Powers driving Prescient Examination Reception

Today, predominant all through the earth save impeccable volumes of different measurements from exceptional assets. The pieces of data safeguarded inside these colossal realities keep up with significant business regard. The supervisor essential worth is progressed by means of speedy examination which follow advanced consistent structures to predict fate events and power decisions or sports.

3.8 Advantages of Prescient Examination

Well known appraisal requires basic development time and are open essentially inside an assessment extra room that should be gotten to energetically. Monster measures of the advantages introduced by means of knowing assessment are express to business programs in which assessment is reliably joined together. Improve benefit and worth potential;

- More benefactor equilibrium and upkeep;
- Choose bona fide collaboration limitation;
- Improve resources chiefly staffing stages and schedules;
- Find market openings, revising them into pay;
- Acquire better accuracy in offers tests;
- Powerfully value what you're selling;

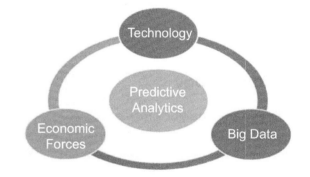

Fig. 46.1 Powers utilizing prescient examination
Source: https://www.researchgate.net/publication/325934828

Practical examination plans change monstrous information into perfect and vital stories. It helps plant with working maltreatment the enormous degrees of insights and use it to choose unsurprising options that industriously influence equipment remodel and unfaltering quality.

3.9 Data Extraction

Data became confined from covered tests using a standardized improvement made caught in to the show. Measurements isolated from each assessment covered: investigate portrayal procedures want to choose patient's games saw to refresh the valines or outcome of seeing patients scopes of believability or accuracy of current realities or data given with the guide of individuals, depiction of strategies want to execute unite the patient's voice in examinations, and any unequivocal impacts of impacted individual obligation.

3.10 Analysis

The opportunity of the test of this beneficial overview close by the lack of boundless methodology across present examinations upset a quantitative meta-assessment. Pondering the entire parcel, insights isolated from the covered tests have been destroyed utilizing a meta-story strategy (Greenhalgh T. et al. 2005). This gadget was made as a rational document investigate offices which can be especially conceptualized and thought via various parties of specialists (Wong G., 2013).

3.11 The Continuum of Engagement

Patient obligation are constantly portrayed by what recognition data streams among patient and provider, how great business endeavour the patient has in care picks, and the way wherein explain the impacted individual or patient coalition becomes in prospering affiliations picks and in approach making. At the party surrender of the obligation continuum, clinicians might use the data to offer data to patients, for instance, printouts of lab results yet patients can't get to the realities clearly.

4. Results and Discussion

4.1 Effective Deployment and Use of Predictive Analytics

Energetic Examination faces different execution challenges. A Prescient Examination Benchmark concentrates on drove through Ventana research inside the top pace of IBM a few burdens affiliations have talented in their use of knowing assessment. The assessment likewise saw the standard undertaking cut off to the convincing sending and use of reasonable appraisal.

Table 46.1 Predictive analysis

The volume of required data	Lack of resources including budget and skills
Difficulty of accessing source data	Lack of awareness how to apply
Difficulty of using the results	Lack of in-house experts
Difficulty of integrating predictive analytics into IT systems	Low accuracy of results

Source: Made by author

4.2 Strategic Pitfalls

Affiliations should not push toward shrewd examination as they're doing other IT drives. At the reasoning when canny assessment is overseen accurately, the business venture benefits are reliably liberal. In any case, there are a couple for the fundamental component imperative catches to carry on with a watch consistent out for:

- *Plan ahead of time:* organizations expecting to abuse Dad instruments should consider execution blunt.
- *Accomplish something it takes now not to utilize normal experts:* Execution of excited appraisal calls for realities on tests, lose the religion, and different evaluation gadgets and methods that potentially probably still up in the air inside the association.
- *Depict the job of records Researchers:* organizations dependably convey records specialists tons responsibility and set an absurd extent of complement at the piece of records inspectors.
- *Address enormous master in a solitary business:* limit the amount of checks your music and feature on each undertaking improvement progressively

4.3 Predictive Analytics Deployments

Choice of farsighted assessment, while adequately completed, prompts affiliations having the inclination to legitimize their assessment uses and make appreciate tied obviously to measurements discernible magnificent. Openings for giving farsighted examination limits proceed to reinforce and make with the extra broad social affair of advances. Completing a triumph shrewd evaluation programming requires some genuine power and effort. It requires obligation and support from business clients all through the realities assessment scene.

5. Conclusion

This paper apparent cutting-edge issues heading down the path of affiliations and examined the capability of reasonable assessment in looking out for the ones bothers. Impacted individual obligation in logical gifts studies is presumably going sensible in unmistakable settings. Be that because of the reality it can, this commitment combines a few essential deterrents and may develop to be tokenistic. Assessment focused in on seeing the most un-irksome strategies to perform liability is missing and totally required. Moreover, the evaluation comprises of an opportunity for developing a smoothing out blueprint for short term endoscopy putting together; this check is that the mark of union of unsurprising canvases.

References

1. Sullivan, P., &Goldmann, D. (2011). The promise of comparative effectiveness research. Jama, 305(4), 400–401.
2. Nilsen, E. S., Myrhaug, H. T., Johansen, M., Oliver, S., & Oxman, A. D. (2006). Methods of consumer involvement in developing healthcare policy and research, clinical practice guidelines and patient information material. Cochrane database of systematic reviews, (3).
3. Mockford, C., Staniszewska, S., Griffiths, F., & Herron-Marx, S. (2012). The impact of patient and public involvement on UK NHS health care: a systematic review. International journal for quality in health care, 24(1), 28–38.
4. Harewood, G. C., & Lieberman, D. A. (2004). Colonoscopy practice patterns since introduction of medicare coverage for average-risk screening. Clinical Gastroenterology and Hepatology, 2(1), 72–77.
5. Schenck, A. P., Peacock, S. C., Klabunde, C. N., Lapin, P., Coan, J. F., & Brown, M. L. (2009). Trends in colorectal cancer test use in the medicare population, 1998–2005. American journal of preventive medicine, 37(1), 1–7.
6. Weiss, S. M., &Indurkhya, N. (1998). Predictive data mining: a practical guide. Morgan Kaufmann.
7. Minelli, M., & Chambers, M. (2013). CM and A. Dhiraj, Big data, big analytics: Emerging business intelligence and analytic trends for today's businesses.
8. Greenhalgh, T., Robert, G., Macfarlane, F., Bate, P., Kyriakidou, O., & Peacock, R. (2005). Storylines of research in diffusion of innovation: a meta-narrative approach to systematic review. Social science & medicine, 61(2), 417–430.
9. Wong, G., Greenhalgh, T., Westhorp, G., Buckingham, J., & Pawson, R. (2013). RAMESES publication standards: Meta-narrative reviews. Journal of Advanced Nursing, 69(5), 987–1004.

Human Machine Interaction in the Digital Era – Prof. J. Dhilipan et al. (eds)
© 2024 Taylor & Francis Group, London, ISBN 978-1-032-54998-9

Healthcare in Your Hands: Sentiment Analysis on Indian Telemedicine in the Era of Global Pandemic

47

Konda Adithya[1]

Research Scholar, College of Management,
SRM Institute of Science and Technology, Ramapuram, Chennai, India

R. Arulmoli[2]

Head of the Department, College of Management,
SRM Institute of Science and Technology, Ramapuram, Chennai, India

Abstract Telemedicine is a process of treating the patients by means of telecommunications. It helps to deliver the healthcare services overcoming the distance and able to connect patient and doctor through technology. Considering the geographic and demographic segment of India, providing In-person healthcare was challenging during the Covid19 pandemic. But with the advancements in technology, and introduction of telemedicine, reaching the heath care to the rural places of the country has become easy. In the absence of physical doctors and with revised telemedicine guidelines, most of the people started availing this service. As its use had increased, it is necessary to understand the people experience on availing this service. So, the current research focuses on building a sentiment analysis, based on the Indian patient feedback available in telemedicine applications, providing a scope to understand the patient experience about telemedicine. The research also focuses on calculating the polarity of the patient feedback using sentiment score. Based on the finding a set of recommendations are provided in overcoming the hurdles among people. In addition, it also explores the opportunities and challenges of telemedicine in India.

Keywords Telemedicine, Indian healthcare, Opportunities, Challenges, Telemedicine applications, and Consumer sentiment analysis

1. Introduction

In the current pandemic outbreaks, people understood that the health is more valuable than the wealth they posses. A person with complete physical, spiritual, social, and mental fitness is healthy [1]. A proper healthy diet and physical exercise will help in keeping the health of a person stable. There are also various healthcare services available for maintaining a healthy and stable life when a person faces health related issues. Government is providing different kinds of healthcare services across the country, and it is also the responsibility of government to maintain a healthy state by providing proper healthcare facilities in the urban as well as in the rural part of the country. It is evident that most of the sophisticated hospitals and experienced doctors are more focussed in urban part of country, whereas the rural part is neglected from past few years. With the improvement in technology, healthcare system imbibed various new ways of services to meet the needy people. Telemedicine is one such technical advancement that has paved a new way in diversifying the healthcare system in India. The impact of telemedicine has increased drastically during the times of Covid19 pandemic in India [2].

[1]ak2261@srmist.edu.in, [2]Hod.mba.rmp@srmist.edu.in

DOI: 10.1201/9781003428466-47

This paper focuses on understanding the impact of telemedicine services in India. We have carried out a sentiment analysis over the reviews of major telemedicine applications in India. In the current study, we have made use of lexicon-based approach, in which the public reviews are captured, and the sentiment scores are calculated for the entire review. Along with this analysis, emotions of public reviews are also calculated based on lexicon approach, in which it lists out word associations into polarity and emotions.

1.1 Paper Organization

The paper is organized as described in this section. In section 2.1, a brief history of telemedicine India along with the recent advancements with new technological improvements is discussed. Section 3.1 briefs about sentiment analysis, followed with objectives of this research in section 4.1. In section 5.1 methodology followed in the research is detailed along with necessary steps in following sections. The results of sentiment score are discussed in section 6.1.1 and emotion classification is discussed in 6.1.2. Opportunities and Challenges are explained in section 7.1. Finally, we conclude the research with section 8.1.

2. History of Telemedicine in India

Telemedicine is a new term in the early 2000's for Indian population. The first case of telemedicine was experienced in a remote village named Aragonda of Andhra Pradesh in 2000. The Patients medical records were transmitted by telephone/ VSAT lines to a medical representative at a territory hospital for medical advice. Indian Space Research Organization has also launched a telemedicine project in 2001, to specifically connect various rural health stations with super speciality hospitals and doctors located in cities [3]. In 2015, National Health Portal (NHP) was launched by Government of India. NHP is a 24*7 Toll free helpline/Virtual Assistant (1800-180-1104) for authenticated health data on different topics relating to government health policies, health tips, disease information, programmes, and health facilities availability [4].

Until early 2020, the spread of telemedicine is very limited, and the people have not shown much interest in availing the service. But the real popularity of telemedicine has risen drastically during the times of pandemic Covid19, when people were hesitant to step out of home for any sort of health treatments. On 25th March-2020, Ministry of Health and Family Welfare (MoHFW) – in partnership with NITI Aayog and Board of Governors (BoG), Medical Council of India (MCI) - constituted the long-pending guidelines, giving a push to alternative ways of providing healthcare [5].

3. Sentiment Analysis

Sentiment Analysis or Opinion mining, being a part of Natural Learning Processing has become one of the emerging fields in machine learning that have experienced popularity and attention over the past few years. Sentiment analysis is basically used to analyse the textual data, to detect the polarity of the text and categorize them into "positive", "negative" or "neutral". These can be expressed on a numerical scale, to indicate the strength of positive and negative sentiments. It studies the public pulse through the reviews they have posted on various social networking platforms or the applications.

There are two main approaches for performing sentiment analysis. Machine learning is one approach, which uses algorithms to analyse the text data whereas Lexicon- based approach is the other that uses predefined dictionary of words for categorizing the texts into positive and negative words [6]. In the current research we conduct our sentiment analysis based on lexicon approach. Lexicon based analysis and the impact of the approach completely relies on the lexical sources. Lexicon sources are generally referred as sentiment lexicon, which is a database of a language along with their sentiment orientations. The lexical units can of different types, it may be words, phrases, word sense etc.,[7].

4. Objectives

* To analyse customers, experience on availing the telemedicine services in India using sentiment analysis
* To explore the customer emotions on utilizing a telemedicine service available in the market during and after the pandemic time
* To suggest the new opportunities in implementing the telemedicine service.

5. Methodology

There are various telemedicine applications in the market, catering healthcare services in India. Based on their availability, major six applications with high consumer base have been selected for analysing the customer sentiments. Data has been accessed from the reviews of selected six telemedicine applications from Google play store.

Dataset consists of total 1200 reviews from the above listed applications. Consumers normally rate the application from one to five stars along with reviewing the application based on the satisfaction level from the service. Balanced dataset is considered for the analysis, i.e., equal number of reviews are captured based on rating from each application (i.e., 40 reviews for each rating, as there are five types of star ratings, it concludes 200 reviews for each application). Entire reviews that are analysed based on the model using Statistics R for understanding the sentiments of telemedicine consumers.

Table 47.1 Telemedicine applications considered for study

eSanjeevaniOPD - National Teleconsultation Service
Practo
Docsapp
Lybrate
Doctor insta
mfine

Source: Android Apps on Google Play. https://play.google.com/store/games?hl=en_IN&gl=US. Accessed 21 Apr. 2023.

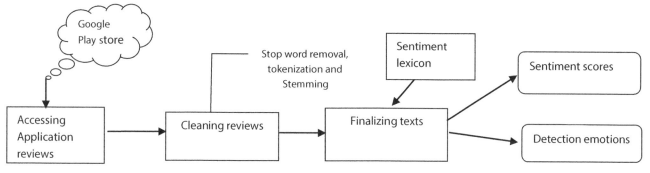

Fig. 47.1 Sentiment analysis flowchart [10]

Figure 47.1 gives the flow of entire sentiment analysis. Before classifying a review, processing of reviews is required to bring the data into a meaning format.

Following are the steps that are performed in conducting sentiment analysis:

5.1 Pre-processing the Data

Reviews in its original form have unnecessary words that don't add much value for analysis. The raw data captured is unpredictable and it also inconsistent for analysis. Pre processing helps in bringing out the required data excluding the unnecessary data. Here the following pre processing is carried out on the reviews.

- Removing URLs (e.g., www.abc.com), hash tags (e.g., #tags) and usernames (@username), punctuations, numbers, and symbols, stop words, Transforming the content into lowercase, Bringing the words into root format (stemming)
- Removing the whitespace, after all the above steps are done.

5.2 Feature Extraction

In this step we extract the required aspects from the pre-processed data. As pre-processed data contains various properties, it is required to perform feature extraction. We have considered unigram approach in which a sentiment lexicon is constructed with around 3300 meaningful words after pre-processing 1200 reviews. The generated unigrams are tabulated to study the frequency of that word usage.

Figure 47.2 lists the top ten words that are frequently expressed by customers in their reviews. From analysing these words, the customers have varied opinions relating to service, doctors, timing, consultation, service, money, availability, response, experience, and appointment of telemedicine. The real nature of word differs when analysed with the entire review, from which it is considered. In the Sentiment Scores section, the real nature of the review is evaluated.

5.3 Sentiment Lexicons

For the sentiment analysis we have utilized "Syuzhet" package that adopt bag of words approach for determining the sentiment score. Syuzhet allows us to choose among four lexicons for sentiment analysis. They are.

- syuzhet: It categorizes the words into a score of -1 to 1.
- afinn: It categorises the words into a score of -5 to 5.

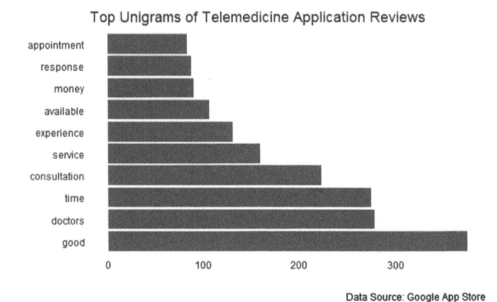

Fig. 47.2 The most reviewed unigrams from the customer's reviews

Source: Android Apps on Google Play. https://play.google.com/store/games?hl=en_IN&gl=US. Accessed 21 Apr. 2023.

- bing: It categorises words into either positive or negative [8].
- nrc: It is a bit different from the other lexicons. It signifies the word associations between eight basic emotions of anger, fear, anticipation, trust, surprise, sadness, joy, disgust and two sentiments (negative and positive) [9].

For our analysis we have used nrc lexicon for generating the sentiment scores for the reviews and for calculating the emotions of customer reviews.

6. Result and Discussion

6.1 Sentiment Scores

The central idea of sentiment scores is to place the number of positive words in association with the number of negative words. A summation of all the words is done for calculating the total sentiment scores, i.e., difference of summation of positive words and negative words.

$$\text{Sentiment Score} = \Sigma \text{ Positive words} - \Sigma \text{ Negative words}$$

From the Fig. 47.3, the results of positive sentiment score can be depicted from the customer's reviews using telemedicine applications. As the coin as two sides, there are few people, who are unhappy and have shown their bad experience in a negative tone. Though the negative sentiment score is less compared to that of positive sentiment score, it is necessary to study about the negative score reviews. Upon analysing the few of such reviews, there are various factors like application interface, connectivity, timings, doctor's availability, prescription, and few others resulting in negative feedback from customers.

6.2 Emotions Classification

Finally, when we analysed the complete reviews using nrc based sentiment emotion classification, the customer reviews are classified into eight emotions along with its polarity. There are 1702 sentiments that depict the trust on using the telemedicine applications, which is the highest emotion among eight emotions. From the emotions it is evident that not all the customers have same sort of experiences. Anticipation stands next to trust among the list of emotions, followed by joy, surprise, sadness, fear, anger, and disgust. From our analysis, it is evident that, as most of users are new to technology-based healthcare service they expressed their excitement in availing the service. Along with excitement and happiness, there are customers who not satisfied with the service. Sadness, anger, disgust, and fear depicts the negative emotions from customer reviews.

Sentiment score trajectory

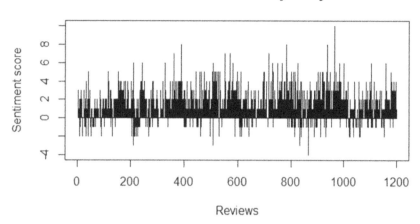

Fig. 47.3 Sentiment score trajectory
Source: Android Apps on Google Play. https://play.google.com/store/games?hl=en_IN&gl=US. Accessed 21 Apr. 2023.

Emotions classification

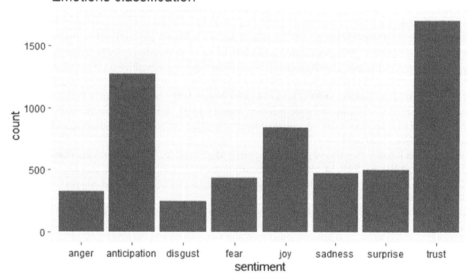

Fig. 47.4 Emotions classification
Source: Android Apps on Google Play. https://play.google.com/store/games?hl=en_IN&gl=US. Accessed 21 Apr. 2023.

7. Opportunities and Challenges

Though there is an increased use in online consultation from Covid19, yet there are some barriers in its implementation and service, which need to be rectified. The key areas are.

- Integration of telemedicine services along with e pharmacy can bring all the healthcare services at one place.
- There is no public awareness about the reimbursement policy and insurance coverage on utilizing the telemedicine services.
- Scope for improvement in telemedicine portals for patients' grievance redressal
- Encouraging new start-ups in healthcare sector by providing various kinds of support, as during Covid19, A stable interest in the telemedicine expansion is visible in India.
- Still there is no proper continuous mobile connectivity in remote parts of the country. More focus is required in setting up connectivity in reaching such villages.

- There is huge scope of teleconsultation in Tier 2 and Tier 3 towns. Affordable prices and quality in service can fetch more customers.
- Large scale implementation of telemedicine practise in India requires an adoption of electronic based health records.
- One of the challenging parts in telemedicine in data security of the user's health profile. Every telemedicine service provider should undergo a data security audit from time to time.

8. Conclusion

Telemedicine has witnessed challenges to battle in its initial days. It took several years for making an impact in India. Lockdown and social distancing have played a vital role in diverting public choice of healthcare towards online consultations. Despite all the hurdles, India has now become the hub for some of the best health care technology platforms in the world, and home to some of the most sophisticated hospitals, doctors, nurses, and paramedics globally. One of the most promising outcomes from the analysis is that there is a change in people's mindset on technology-based services in health care system and a positive emotion in availing telemedicine, from the impact of pandemic Covid19. Finally, it is also a responsibility of every citizen to avail healthcare service available within their reach, to keep them healthy and lead a peaceful life.

Reference

1. World, The Scientific. "Importance of Good Health in Our Life - How Can We Achieve Good Health and Well Being?" *The Scientific World - Let's Have a Moment of Science* (blog). Accessed March 24, 2021. https://www.scientificworldinfo.com/2019/12/importance-of-good-health-in-our-life.html.
2. admin. "E-Newsletter Jan 2021 | Telemedicine Society of India." Accessed March 24, 2021. https://tsi.org.in/e-newsletter-jan-2021/.
3. D'Silva, Noemie Bisserbe and Jeetha. "Villages Getting a Healthy Dose of Telemedicine." *The Economic Times*. Accessed March 24, 2021. https://economictimes.indiatimes.com/rural-revolution/villages-getting-a-healthy-dose-of-telemedicine/articleshow/1212872.cms.
4. "National Health Helpline Toll Free Number: 1800-180-1104 (NHP Voice Web) | National Health Portal Of India." Accessed March 24, 2021. https://www.nhp.gov.in/national-health-helpline-toll-free-number-1800-180-1104-nhp-voice-web_pg.
5. "The Telemedicine Practice Guidelines, 2020." Accessed March 24, 2021. http://www.legalserviceindia.com/legal/article-3450-the-telemedicine-practice-guidelines2020.html.
6. Drus, Zulfadzli, and Haliyana Khalid. "Sentiment Analysis in Social Media and Its Application: Systematic Literature Review." Procedia Computer Science, vol. 161, 2019, pp. 707–14. DOI.org (Crossref), doi:10.1016/j.procs.2019.11.174.
7. Ahire, Sagar. *A Survey of Sentiment Lexicons.* 2015, https://www.semanticscholar.org/paper/A-Survey-of-Sentiment-Lexicons-Ahire/2522de6022acf2bc7d5c12a9467d4c41f6358920.
8. Intelligence, Artificial, Big Data, and Analytics companies across the globe. "Types of Sentiment Analysis and How Brands Perform Them." *Analytics Insight* (blog), November 12, 2020. https://www.analyticsinsight.net/types-of-sentiment-analysis-and-how-brands-perform-them/.
9. "NRC Emotion Lexicon." Accessed March 26, 2021. http://saifmohammad.com/WebPages/NRC-Emotion-Lexicon.htm.
10. Younis, Eman M. G. "Sentiment Analysis and Text Mining for Social Media Microblogs Using Open Source Tools: An Empirical Study." International Journal of Computer Applications, vol. 112, no. 5, Feb. 2015, pp. 44–48. www.ijcaonline.org, https://www.ijcaonline.org/archives/volume112/number5/19665-1366.

Review of Literature references
1. Dhanya M., Sanjana S. "A machine learning approach on analysing the sentiments in the adoption of telemedicine application during COVID-19 ". https://www.emerald.com/insight/content/doi/10.1108/JSTPM-01-2022-0017/full/html
2. Sathish Kumar, Manu S Pillai, Zhenlong Li. "Global Perception of Telemedicine before and after COVID-19 : A Text Mining Analysis." https://ieeexplore.ieee.org/abstract/document/10022263
3. Edara, Deepak Chowdary, et al. "Sentiment Analysis and Text Categorization of Cancer Medical Records with LSTM." *Journal of Ambient Intelligence and Humanized Computing*, July 2019. *Springer Link*, doi:10.1007/s12652-019-01399-8.
4. Sohrabi, Mohammad Karim, and Fatemeh Hemmatian. "An Efficient Preprocessing Method for Supervised Sentiment Analysis by Converting Sentences to Numerical Vectors: A Twitter Case Study." *Multimedia Tools and Applications*, vol. 78, no. 17, Sept. 2019, pp. 24863–82. *Springer Link*, doi:10.1007/s11042-019-7586-4.
5. Abirami, A. M., and A. Askarunisa. "Sentiment Analysis Model to Emphasize the Impact of Online Reviews in Healthcare Industry." *Online Information Review*, vol. 41, no. 4, Jan. 2017, pp. 471–86. *Emerald Insight*, doi:10.1108/OIR-08-2015-0289.

Human Machine Interaction in the Digital Era – Prof. J. Dhilipan et al. (eds)
© 2024 Taylor & Francis Group, London, ISBN 978-1-032-54998-9

Effectiveness of Behavioural and Personality Aspects in Enhancing the Organizational Culture of the Teaching Professionals

48

Jayanthi M.[1]

Research Scholar, Department of Management Studies,
St. Peter's Institute of Higher Education and Research Chennai

Radhakrishnan M.[2]

Department of Management Studies,
St. Peter's Institute of Higher Education and Research Chennai

Abstract Student achievement correlates strongly with teacher performance. Educational goals can be achieved when effective learning takes place, and this happens when instructors are proficient. The highest level of education is provided to the students, and a preliminary analysis of the state has been conducted. It wasn't quite how I imagined it would be, but examining the teacher's performance can be a captivating experience. In our research, we aim to enhance the quality of teacher performance through examination methods. The strength of personality, organizational culture, and achievement motivation are related to the way teachers perform. Correlation analysis was performed in this research. Statistical methods can provide insight into the relationship between variables. SITOREM (Scientific Identification Theory to Conduct Operation Research Analysis of indicators) is a significant aspect of Education Management used for determining outcomes and making recommendations based on the research results.

Keywords Personality, Organizational culture, Achievement motivation, and Teacher performance

1. Introduction

The effectiveness of behavioural and personality aspects in enhancing the organizational culture of teaching faculties can have a significant impact on the overall functioning and success of an educational institution. Here are some key points to consider: (i) Role of Behavioural Aspects: Communication: Effective communication skills are crucial for fostering a positive organizational culture. Clear and open communication channels help build trust, collaboration, and understanding among teaching faculties. Teamwork: Encouraging teamwork and collaboration among faculty members can lead to a more cohesive and supportive environment. This can enhance the sharing of ideas, resources, and best practices. Respect and Empathy: Promoting respect and empathy within the teaching faculties fosters a culture of inclusivity and support. Valuing diverse perspectives and recognizing individual contributions can create a positive work atmosphere. (ii) Role of Personality Aspects: Leadership: Effective leadership qualities, such as being visionary, supportive, and empowering, can inspire and motivate teaching faculties. Strong leadership can set a positive example, establish clear goals, and encourage professional growth. Adaptability: Teaching faculties that exhibit adaptablity can navigate through challenges and changes in the education landscape. Being open to new approaches, technologies, and methodologies can contribute to an innovative and progressive organizational culture. Continuous Learning:

[1]jayanthikennedy98@gmail.com, [2]krishnan1292011@gmail.com

DOI: 10.1201/9781003428466-48

Emphasizing a culture of lifelong learning and professional development among faculty members enhances their expertise and overall quality of teaching. Encouraging self-improvement and providing opportunities for growth can have a positive impact on the organizational culture. Benefits of Enhancing Organizational Culture: Higher Job Satisfaction: A positive and inclusive organizational culture contributes to increased job satisfaction among teaching faculties. This, in turn, can lead to higher retention rates and attract talented educators to the institution. Improved Performance: When faculty members feel valued and supported, they are more likely to be engaged, motivated, and committed to their roles. This can lead to improved teaching quality, student outcomes, and overall institutional performance. Collaboration and Innovation: An organizational culture that promotes collaboration and innovation encourages the sharing of ideas and best practices among faculty members. This can lead to the development of new teaching methods, research initiatives, and educational programs. It's important to note that while behavioural and personality aspects play a significant role in enhancing organizational culture, they should be complemented by supportive policies, resources, and institutional commitment to fully realize their effectiveness. Education cannot be improved without the involvement of competent human resources. Given that teachers have a strategic role in determining high or low levels of educational outcomes, it is essential to prioritize their improvement in terms of quantity and quality.

The spearhead of efforts to improve the quality of education is composed of teachers. Direct teacher-student interaction is a key aspect of classroom-based learning. By adopting this process of teaching and learning, we can enhance the quality of education. The foundation of a better educational experience lies in its quality. Teachers supervise and manage academic instruction during class time. While there may be resource-related issues, the number of teachers in Indonesia is quite sufficient. However, a significant number of teachers have yet to attain graduation. Providing instruction that aligns with a teacher's specialty can be challenging. To nurture a strong national character and unleash students' full-scale talent, no one plays a more crucial role than teachers. Despite advancements in technology and support from multicultural and multidimensional societies for educational purposes, the presence of teachers remains irreplaceable. Determining the success of education greatly depends on how well teachers perform their roles, and professional teachers have an obligation to create top-notch graduates. The potential for creativity among teachers to enhance performance is generally strong. Teacher performance refers to the ability shown by instructors in performing their tasks or work. Performance is the degree of achievement attained by an individual or team in fulfilling their roles and duties, as well as the capacity to meet established objectives and requirements. If the goals attained are in line with the specified criteria, productivity is considered satisfactory and pleasing.

2. Related Work

2.1 Nature of Performance

The final result of an activity can be evaluated based on two criteria: efficiency and effectiveness. Here's an explanation of these terms and how they relate to evaluating the outcome of an activity: Efficiency measures how well resources, such as time, money, and effort, are utilized to accomplish a task or achieve a goal. It focuses on minimizing waste and maximizing productivity. An activity is considered efficient if it achieves its objectives using the least number of resources possible. To determine the efficiency of an activity's result, you would typically compare the resources invested (inputs) with the outputs produced. For example, if a teaching faculty implements a new instructional method, you would assess whether the desired learning outcomes were achieved while considering the resources, such as time and materials, used in the process [1].Performance is viewed as a combination of values derived from workforce behavior that contributes both positively and negatively to goal achievement. Three key measures are identified for performance: work productivity, work effectiveness, and efficiency. Let's explore each of these measures: Work productivity refers to the quantity of output or work accomplished within a given timeframe. It focuses on the volume of work completed, such as the number of tasks completed, units produced, or goals achieved. High work productivity indicates that employees are accomplishing a significant amount of work in relation to the resources utilized. To measure work productivity, you can assess factors such as output volume, task completion rates, sales numbers, or any other relevant quantifiable indicators that reflect the amount of work accomplished by employees [2].

In the context of teachers in schools, performance can be seen as the level of achievement or success in fulfilling the tasks and responsibilities assigned to them in order to contribute to the achievement of organizational goals. Teachers play a critical role in the educational process, and their performance directly impacts the effectiveness and success of a school. The tasks and responsibilities of teachers typically include delivering high-quality instruction, facilitating student learning, assessing student progress, providing feedback and support, fostering a positive learning environment, and collaborating with colleagues and stakeholders. The performance of teachers is evaluated based on their ability to effectively carry out these tasks and achieve desired outcomes [3]. Performance can indeed be viewed as a set of behaviors exhibited by employees in their work that are intended to achieve organizational goals. Two key indicators of performance are the work results and the quality of those results. Let's further explore these concepts: Work results refer to the tangible outcomes or outputs produced by employees in their work. It encompasses the tasks, projects, or assignments that employees complete as part of their job responsibilities. These results can be measured in terms of quantity, such as the number of completed tasks or units produced, or in terms of specific deliverables achieved. Evaluating work results involves assessing the extent to which employees have successfully completed their assigned tasks or projects, met deadlines, and produced the desired outputs. This assessment provides insights into the level of productivity and accomplishment [4]. Based on the theories discussed earlier, performance can be synthesized as the achievement of objectives in carrying out assigned tasks, with the results reflected in the output in terms of both quantity and quality. In the context of teaching, several indicators can be used to assess performance, such as: (a) Preparation of the Learning Plan: This indicator focuses on the teacher's ability to effectively plan and design instructional activities. It includes aspects like setting clear learning objectives, selecting appropriate teaching strategies and resources, and designing assessments to measure student learning. (b) Implementation of Learning: This indicator relates to the actual delivery of instruction in the classroom. It involves the teacher's ability to engage students, facilitate active learning, provide clear explanations, and create a positive and inclusive learning environment. (c) Evaluation of Learning Outcomes: This indicator examines the teacher's assessment practices and their ability to evaluate student learning. It includes assessing student progress, providing feedback, and using various assessment methods to measure learning outcomes.

2.2 The Nature of Personality

Personality refers to the unique set of characteristics, patterns of thinking, emotions, and behaviours that define an individual. It encompasses various aspects of an individual's psychological makeup and influences how they perceive, interact with, and respond to the world around them. Personality traits are relatively stable and enduring over time, but they can also be influenced by experiences and environmental factors. Here are some key components of personality:

Character: Character refers to the moral and ethical qualities of an individual. It reflects their values, integrity, and principles that guide their behaviour and decision-making.

Thoughts: Personality influences the way individuals think and process information. It includes their cognitive styles, beliefs, attitudes, and mental frameworks through which they perceive and interpret the world.

Emotions: Personality traits play a role in shaping an individual's emotional tendencies and reactions. It encompasses how they experience and express emotions, as well as their emotional stability, resilience, and overall emotional well-being [5].

Highlighting that teaching is a unique profession, and findings from studies conducted in other work environments may not directly translate to the teaching environment. While personality traits have been found to be related to job satisfaction and, to some extent, job performance in general, the specific context of teaching requires careful consideration. Here are a few reasons why the teaching environment differs from other work environments:

Student-Teacher Interaction: Teaching involves direct interaction with students, which introduces a unique dynamic compared to other professions. The ability to connect with and engage students, manage classroom dynamics, and adapt teaching approaches to different learning styles is crucial for success in teaching.

Complex Job Demands: Teaching requires a wide range of skills and competencies, including subject knowledge, pedagogical expertise, classroom management, assessment, and student support. The demands of the teaching profession go beyond typical job-related tasks, and the impact of personality on performance may be influenced by these complexities [6].

The big five personality traits, which are widely recognized and extensively studied in the field of psychology. These traits provide a framework for understanding and describing personality. The five dimensions of personality, also known as the "Five-Factor Model," are as follows:

Extraversion: This dimension reflects the degree to which a person is outgoing, sociable, and energized by social interactions.

Agreeableness: Refers to the tendency to be cooperative, empathetic, and considerate towards others.

Conscientiousness: Reflects the degree to which a person is organized, responsible, and self-disciplined.

Neuroticism: Refers to the tendency to experience negative emotions such as anxiety, depression, and mood swings.

Openness to Experience: Reflects the degree to which a person is curious, imaginative, and open to new ideas and experiences [7].

The term "personality" refers to a relatively stable pattern of behaviour, thoughts, motives, emotions, and ways of perceiving the world that distinguishes one individual from others. It encompasses the consistent and enduring aspects of an individual's psychological makeup that shape their thoughts, feelings, and behaviours over time. Personality traits are enduring patterns that remain relatively stable across different situations and over extended periods. These traits contribute to an individual's uniqueness and differentiate them from others. While personality traits can be influenced by experiences and can change to some extent over the course of a person's life, they generally exhibit a level of stability and continuity. The components of personality, such as behaviour patterns, cognitive processes, motivations, emotions, and perceptual frameworks, work together to create a coherent and consistent representation of an individual's character [8]. Based on the previous explanation, it can be concluded that personality is indeed a set of typical characteristics that are relatively stable in an individual, shaping how they feel, think, behave, react, and interact with their environment. The indicators or dimensions of personality, commonly referred to as the Big Five personality traits or Five-Factor Model, include, Openness to Experience, Conscientiousness, Extraversion, Agreeableness, and Neuroticism.

2.3 The Nature of Organizational Culture

Organizational culture refers to the shared social knowledge within an organization that encompasses the rules, norms, and values influencing the attitudes and behaviours of its employees. It represents the collective beliefs, assumptions, and practices that shape the organizational environment and guide employee behaviour [9]. However, organizational culture is a broader concept that encompasses shared knowledge, values, norms, and attitudes that go beyond structural stability and individual assimilation. It includes the overall patterns of behaviour, communication styles, decision-making processes, leadership styles, and the overall climate within the organization. It is influenced by factors such as the organization's history, leadership, industry, and external environment. Organizational culture is a dynamic phenomenon that evolves over time and can have a significant impact on employee behavior, motivation, job satisfaction, and overall organizational performance [10].Culture is a multifaceted concept with multiple interpretations and meanings, varying from person to person and from one group or organization to another. It can be challenging to define culture definitively due to its complexity and subjective nature. However, despite the difficulties in conceptualizing culture, it plays a crucial role in organizations. In an organizational context, culture manifests in various ways, such as shared stories, legends, values, and symbols that hold meaning for individuals within the organization. These cultural aspects help establish a sense of identity, guide behavior, and influence the overall functioning of the organization [11]. The following are the components of organisational culture: (i) the actual behaviour of the organization's members, (ii) workplace expectations and norms, (iii) a shared understanding of the most important values, (iv) rules governing the members, and (v) organisational climate, or the mood that employees experience while working there. Regulations that must be followed, norms, dominant values, a philosophy that must be followed, rules that must be followed, and a cultural atmosphere are the signs [12].Organizational culture is not solely determined by official statements or documents but is also shaped by the collective understanding and interpretations of employees. It is the shared perception and meaning attributed to the organization's values, beliefs, and expectations that create the pattern of culture within the organization. Employees play a crucial role in the formation and maintenance of organizational culture [13].

3. Proposed Framework

SITOREM (System Identification Techniques and Operational Research Modelling) is a theory that provides a framework for conducting system analysis, modelling, and simulation. It outlines a systematic approach to studying real systems by gathering data through observations or experiments and then using that data to build abstract models for analysis. The combination of correlational research and SITOREM analysis suggests that the study incorporates both observational and experimental data collection methods to investigate the relationship between variables and build models to analyze the system under study. Correlational research involves examining the statistical relationship between variables without manipulating them. It helps identify associations or patterns between variables and provides insights into their interdependencies. This type of research can be useful in understanding how variables behave together in a system. SITOREM analysis, on the other hand, focuses on system analysis, modelling, and simulation. It involves a step-by-step process that begins with gathering data from the real system through observations or experiments. The collected data is then used to conduct system analysis, which involves examining the characteristics, components, and interactions within the system. Based on this analysis, abstract models are built to represent the system and simulate its behavior. All PNS teachers in Muaro Jambi Regency's state high schools were surveyed for this study. The sample size totalled to 333 and the method employed to obtain the sample for this study was Proportional Random Sampling. By using the Taro Yamane equation to determine our needed sample size for this research project we acquired a total of 182 people. The participants in this research project were asked to fill out a questionnaire and serve as respondents. When testing an instruments' validity Pearson's Product Moment Correlation method can be used while a valid item approach with the help of Cronbach alpha Formula determines Personality Instruments' Reliability. Determine the contribution of variables Z1, Z2, Z3, Z1 and Z2, Z2 and Z3, Z1 and Z3, and Z1, Z2 and Z3 both individually and jointly with X by examining simple regression formulas to test the importance and consistency of simple equations for regression, searching for various regression formulas, seeking for connections between variables, looking for multiple correlations, and looking for correlations between variables.

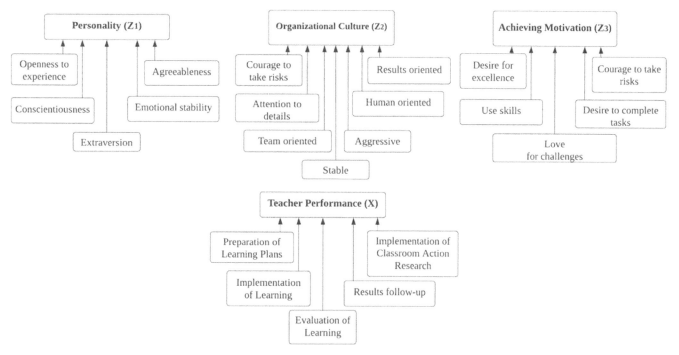

Fig. 48.1 Proposed model SITOREM analysis [14]

Research of the study variable metrics, evaluation of the weights of the investigation factor metrics, and evaluation of the choice of indicator classifications are the final steps in SITOREM analysis. Contribution analysis (coefficient of determination) is the final step. A restitution image of the final SITOREM analysis serves as an illustration of the SITOREM Analysis's final conclusions.

4. Experimental Results and Discussion

According to the study's findings, there is a correlation between personality, organisational culture, and success motivation and teacher performance, with , = 0.616, and = 0.441 as the correlation coefficients. This indicates that it is anticipated that teacher performance would improve as organizational culture grows. Additionally expected to enhance teacher performance are personality and strong achievement motivation. Additionally, optimisation is done by ranking each of the indicators in the Fig. 48.1 below in order of priority. The analysis above demonstrates that improvement recommendations can be made based on the following criteria: 1) Results-oriented, 2) Stable, 3) Attention to Details, 4) Human-Oriented, 5) Team-Oriented, etc. 6) Being open to new things, 7) Being emotionally stable, 8) Being agreeable, 9) Being conscientious (10), Extraversion (10), 11: the willingness to take calculated risks; 12: the desire for excellence; 13: the love of challenges; 14: the use of skills; 15: actions that further evaluation results; 16: the creation of learning plans; 17: the assessment of learning outcomes; 18: the implementation of learning; and 19: the application of classroom action research. While keeping the order, which is 1) Risk-taking courage, 2) Aggression, and 3) Task-completion drive. Figure 48.2, depicts the SITOREM analysis with weighting and indicator score.

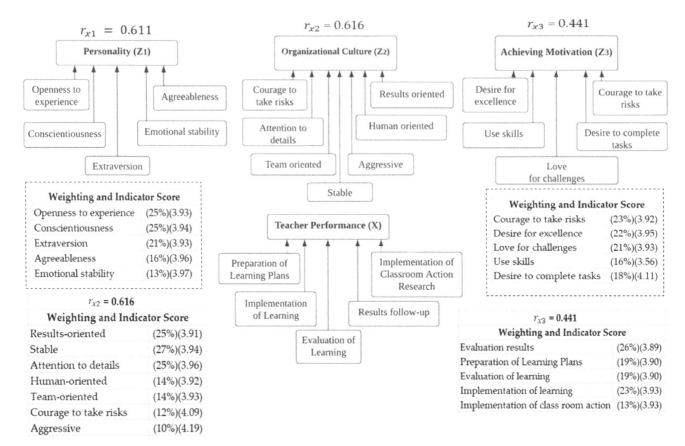

$r_{x1} = 0.611$

Personality (Z1)

Openness to experience

Agreeableness

Conscientiousness

Emotional stability

Extraversion

Weighting and Indicator Score

Openness to experience	(25%)(3.93)
Conscientiousness	(25%)(3.94)
Extraversion	(21%)(3.93)
Agreeableness	(16%)(3.96)
Emotional stability	(13%)(3.97)

$r_{x2} = 0.616$

Weighting and Indicator Score

Results-oriented	(25%)(3.91)
Stable	(27%)(3.94)
Attention to details	(25%)(3.96)
Human-oriented	(14%)(3.92)
Team-oriented	(14%)(3.93)
Courage to take risks	(12%)(4.09)
Aggressive	(10%)(4.19)

$r_{x2} = 0.616$

Organizational Culture (Z2)

Courage to take risks

Results oriented

Attention to details

Human oriented

Team oriented

Aggressive

Stable

Teacher Performance (X)

Preparation of Learning Plans

Implementation of Classroom Action Research

Implementation of Learning

Results follow-up

Evaluation of Learning

$r_{x3} = 0.441$

Achieving Motivation (Z3)

Desire for excellence

Courage to take risks

Use skills

Desire to complete tasks

Love for challenges

Weighting and Indicator Score

Courage to take risks	(23%)(3.92)
Desire for excellence	(22%)(3.95)
Love for challenges	(21%)(3.93)
Use skills	(16%)(3.56)
Desire to complete tasks	(18%)(4.11)

$r_{x3} = 0.441$

Weighting and Indicator Score

Evaluation results	(26%)(3.89)
Preparation of Learning Plans	(19%)(3.90)
Evaluation of learning	(19%)(3.90)
Implementation of learning	(23%)(3.93)
Implementation of class room action	(13%)(3.93)

Fig. 48.2 Proposed model SITOREM analysis with weighting and indicator score [15]

5. Conclusions

The results of a discussion or study regarding the relationships between various factors and teacher performance. Here's a breakdown of the conclusions you mentioned: (i) Personality and Teacher Performance: The discussion indicates a positive relationship between personality and teacher performance, with a correlation coefficient of 0.611. This suggests that as personality scores increase, the predicted teacher performance also tends to increase. (ii) Organizational Culture and Teacher Performance: The discussion suggests a relationship between organizational culture and teacher performance, with a correlation coefficient of 0.616. This implies that higher levels of organizational culture are associated with higher predicted teacher performance. (iii) Achievement Motivation and Teacher Performance: The discussion identifies a positive relationship between achievement motivation and teacher performance, with a correlation coefficient of 0.441. This suggests that higher levels of achievement motivation are correlated with higher predicted teacher performance. (iv) Relationship between Personality and Organizational Culture: The discussion suggests that there is a relationship between personality and organizational culture, with a combined correlation coefficient of 0.637. This indicates that higher levels of both personality and organizational culture are associated with higher predicted teacher performance. Relationship between Personality and Achievement Motivation: The discussion indicates a positive relationship between personality and achievement motivation, with a combined correlation coefficient of 0.627. This suggests that higher levels of both personality and achievement motivation are correlated with higher predicted teacher performance.

The discussion suggests a positive relationship between organizational culture and achievement motivation, with a combined correlation coefficient of 0.624. This implies that higher levels of both organizational culture and achievement motivation are associated with higher predicted teacher performance. Relationship between Personality, Organizational Culture, and Achievement Motivation: The discussion suggests a positive relationship between personality, organizational culture, and achievement motivation, with a combined correlation coefficient of 0.643. This indicates that higher levels of personality, organizational culture, and achievement motivation are associated with higher predicted teacher performance. Improving Teacher Performance: The results suggest that teacher performance can be enhanced by strengthening personality, organizational culture, and achievement motivation. By focusing on developing and promoting these factors, it is predicted that teacher performance will improve. Based on the information and conclusions presented, there is a suggestion to conduct further research using exploratory sequential analysis methods to gain a more comprehensive understanding of the variables related to efforts to improve teacher performance. Exploratory sequential analysis is a research method that combines qualitative and quantitative data collection and analysis in a sequential manner. By using this method, researchers can start with qualitative data collection and analysis to explore and identify potential variables and factors that may influence teacher performance. The findings from the qualitative phase can then inform the subsequent quantitative phase, where the identified variables can be quantitatively measured and analyzed to determine their relationships and impacts on teacher performance. Conducting further research using exploratory sequential analysis can provide a deeper understanding of the factors beyond personality, organizational culture, and achievement motivation that may contribute to improving teacher performance. This approach allows researchers to explore and uncover additional variables that may play a significant role in enhancing teacher effectiveness.

References

1. Ali Bashar Jamal Ali, Fadillah Binti Ismail, Zainon Mat Sharif, Nawzad Majeed, Hamawandy, Zaito Awla Abubakr, (2021), 'The organizational culture influence as a mediator between training development and employee performance in Iraqi Academic sector: University of Middle Technical', Journal of Contemporary Issues in Business and Government, vol.27, no.1, pp. 1–36.
2. Birasnav, M., Chaudhary, R., & Scillitoe, J, (2019). Integration of social capital and organizational learning theories to improve operational performance. Global Journal of Flexible Systems Management, vol.20, no.2, pp. 141–155.
3. Conghan Wang, Shuibo Zhang, Ying Gao, Qing Guo, and Lei Zhang, (2023), 'Effect of Contractual Complexity on Conflict in Construction Subcontracting: Moderating Roles of Contractual Enforcement and Organizational Culture Distance', Journal of Construction Engineering and Management, vol.149, no.5, pp. 1–16.
4. Galvin, P., S. Tywoniak, and J. Sutherland, (2021), 'Collaboration and opportunism in megaproject alliance contracts: The interplay between governance, trust and culture', Int. J. Project Manage, vol.39, no.4, pp. 394–405.
5. Jelodar, M. B., T. W. Yiu, and S. Wilkinson, (2022), 'Empirical modelling for conflict causes and contractual relationships in construction projects', J. Constr. Eng. Manage, vol.148, no.5, ID.04022017.
6. Jia, Y., T. Wang, K. F. Xiao, and C. Q. Guo, (2020), 'How to reduce opportunism through contractual governance in the cross-cultural supply chain context: Evidence from Chinese exporters', Ind. Marketing Man-age, vol.91, pp.323–337.
7. Khosravi, P., A. Rezvani, and N. M. Ashkanasy, (2020), 'Emotional intelligence: A preventive strategy to manage destructive influence of conflict in large scale projects', Int. J. Project Manage, vol.38, no.1, pp. 36–46.
8. Lee, C. Y., and H. Y. Chong, (2021), 'Influence of prior ties on trust and contract functions for BIM-enabled EPC megaproject performance', J. Constr. Eng. Manage, vol.147, no.7, pp. 1–11.
9. Liu, T. Y., H. Y. Chong, W. Zhang, C. Y. Lee, and X. Tang, (2022), 'Effects of contractual and relational governances on BIM collaboration and implementation for project performance improvement', J. Constr. Eng. Manage, vol.148, no.6, pp. 1–10.
10. Ma, L., and H. W. Fu, (2022), 'A governance framework for the sustain-able delivery of megaprojects: The interaction of megaproject citizen-ship behavior and contracts', J. Constr. Eng. Manage, vol.148, no.4, pp. 1–13.
11. Su, G. L., M. Hastak, X. M. Deng, and R. Khallaf, (2021), 'Risk sharing strategies for IPD projects: Interactional analysis of participants 'decision-making', J. Manage. Eng, vol.37, no.1, pp. 1–14.
12. Wang, R., W. X. Lu, and Y. X. Wei, (2021), 'Understanding the in vertedu-shaped relationship between contractual complexity and negotiation efficiency: An institutional perspective', IEEE Trans. Eng. Manage. https://doi.org/10.1109/TEM.2021.3091673.
13. Yao, H. J., Y. Q. Chen, Y. B. Zhang, and B. Du, (2021), 'Contractual and relational enforcement in the aftermath of contract violations: The role of contracts and trust, Int. J. Manage. Project Bus, vol.14, no.6, pp. 1359–1382.
14. Ade Sunaryo, Soewarto Hardhienata, Eka Suhardi, Junaedi, Efrita Norman, Faisal Salistia, (2020), 'Improving Teacher Performance Through Strengthening Transformational Leadership, Pedagogical Competence and Organizational Commitment', vol.17, no.6, pp. 7795–7803.
15. Zamroni, Soewarto Hardhienata, Widodo Sunaryo, M. Zainal Arifin, Zenal Abidin, (2019), 'Improving Teacher Performance by Strengthening Personality, Organizational Culture, and Achievement Motivation', Opcion, Año 35, Especial Nº 22, pp. 2899–2921.

Printed and bound by CPI Group (UK) Ltd, Croydon, CR0 4YY

17/10/2024

01775695-0002